Severe Storm Engineering for Structural Design

Severe Storm Engineering for Structural Design

Michele Melaragno

University of North Carolina
Charlotte

Gordon and Breach Publishers

Australia Austria China France Germany India Japan
Luxembourg Malaysia Netherlands Russia Singapore
Switzerland Thailand United Kingdom United States

3 Boulevard Royal
L-2449 Luxembourg

British Library Cataloguing in Publication Data

Melaragno, Michele
Severe Storm Engineering for Structural
Design
I. Title
624.176

ISBN 2-88449-150-3 (hardcover)
2-88449-126-0 (softcover)

To

Deborah

CONTENTS

PREFACE

To introduce this book from the right perspective, I should probably date its inception as beginning in the mid-1960s. At that time, I started my first investigations at Kansas State University exploring the effects of high-velocity winds on structures within the velocity spectrum occurring during tornadoes-in-action. A small research grant provided the basic minimum funds to support my work at the time. Eventually in 1968, the first publication, *Tornado Forces and Their Effects on Buildings*, presented the basics of my findings. Later, in a succession of other publications, I expanded the scope of building interactions with high-velocity winds to include a more general field of applications, concurrent to what was actually taking place worldwide, in a relatively new discipline referred to as *wind engineering*. This second phase produced my work *Wind in Architectural and Environmental Design* (Van Nostrand Reinhold, New York, 1982), which in practical terms went even beyond the scope of wind engineering, probing into a wider range of wind-induced effects. In so doing, the structural effects of wind forces were only a part of the overall subject, which included practical applications into the design of the built-up environment — both at the micro and macro scales. From interior ventilation to the wind circulation around buildings, from the local effects of air flow in urban spaces to the effects of shelter belts abating wind speeds over agricultural land, this work enlarged the scope of my studies to more than was anticipated; hence, I felt the need for refocusing. In my subsequent work, in fact, I gradually narrowed my interests again, going back to the point of origin that centered on the most violent of the storms — tornadoes.

However, in restructuring my work since 1982, I felt that in again converging toward the effects of high-velocity winds, I should not restrict attention only to tornadoes, but that I should include the effects of tropical cyclones. These constitute an area of interest which affects the built-up environment in wider geographical area to a much higher level. Statistical data on hurricanes, typhoons, and tropical cyclones, in general, indicate a major disproportionality between them and tornadoes because of the larger number of fatalities that the former produce. Hence, this current book includes the action of tornadoes, all types of tropical cyclones, and high-velocity winds in general into one comprehensive presentation. Yet, to justify the grouping together of these different weather phenomena, it should be pointed out that the basic interests of this work are not the meteorological aspects per se, but rather the responses of buildings subjected to the action of wind loads. These have much more in common with one another despite the large disparity between storm types.

While the progression toward the conception of this book followed the path previously described, the assembling of data to arrive at an organic manuscript was definitely enhanced by the impact that Hurricanes Hugo and later Andrew exerted on the public at large and more specifically on the design professions. Engineering and architectural aspects of design, including planning, were confronted with fundamental needs of restructuring design concepts for wind forces, and the need for reviewing construction practices and updating building codes. In this context, the book reflects the needs of coordinating existing information with new data that have recently been gathered by governmental agencies and private research groups in their examination of damage. Even more inspirational in the structuring of this book is the more dramatic effects that violent storms have in other parts of the world. For instance, figures counting the fatalities that have occurred so often in Bangladesh alone must leave any reader with a sense of disbelief concerning the risk that weather phenomena present to human lives. A violent tropical cyclone in Bangladesh killed 300,000 in 1970, while the most destructive storm that ever occurred in the United States produced only 6,000 deaths in Galveston, Texas, in 1900. Moreover, we notice that, still in Bangladesh, such an incredible number of casualties is not rare at all — as demonstrated by the storms of 1991 and 1985 that killed, respectively, 70,000 and 10,000 inhabitants. In the United States, the next two most destructive hurricanes — 1938 and 1989 (Hurricane Hugo) — killed 600 and 504, respectively.

Apart from the violence of the storms per se, the major causes for such a high number of casualties must be found within characteristics of structural strength reflected in construction practices and minimum standards required by building codes. Obviously, it is because of the lower standards of building construction in Bangladesh that we find such a disproportionately larger number of casualties in comparison to the fatalities that occur in the United States. The problem is therefore, most of all, a problem of technology that must stimulate the search for solutions within the disciplines that involve building technology, from design to construction. Considering the problem from a design point of view, it was extremely interesting to follow the philosophy of building codes enforced in Australia which encompasses a particularly inspiring coherent approach.

In this case the structuring of codes implies subdivision of the entire area of concern into three categories. The first includes those structures not particularly sensitive to wind forces and of low economic value which constitute two basic reasons for not requiring a sophisticated engineering analysis. Most predominant in this category are residential dwellings of one or two stories which do not induce high wind effects in comparison to tall structures. These buildings also are less sensitive to wind because of the stiffness of their structural assembly. Usually subdivided into small spaces, they include many vertical diaphragms constituted by exterior walls and partitions tied together by horizontal diaphragms formed by floor slabs and roofs. Dwelling structures usually built empirically without engineering analysis can continue to be constructed intuitively and still be adequately resistant to wind forces if regulated by appropriate building codes that take into account recommendations formulated through experience and basic structural concepts.

In the second category, codes include those structures that do indeed require specific analysis but are not particularly sensitive to wind forces because of their structural and aerodynamic characteristics. These structures with low structural slenderness have heights not larger than five times their width. For them codes offer specific guidelines that any structural engineer could follow with safety.

Finally, in the third category, codes include those structures that require a more sophisticated design procedure due to their response to wind-induced oscillations both in wind direction and perpendicularity. These structures with heights larger than five times their width require special design criteria that codes also provide for the structural engineer to follow in minute detail. Through these codes, a clear sense of direction through the intricacies of wind-engineering as a whole can be obtained. On one side are building codes that guide the design and construction of the largest number of buildings around the world; on the other side stands engineering methodology subdivided into two parts that fit efficiently into different types of buildings. In such a schematization, each category of building is well-defined, and equally well-defined are the methodologies to be used in design. This offers an invaluable prototype for formulation of codes in different countries that require special regulations to fit individual needs.

This work has been organized into three basic sections. The first chapter covers the fundamental principles of the action of high-velocity winds on structures; the second and third chapters contain the details of specific storms. In the second chapter, hurricane and typhoon effects on buildings are grouped together, while in the third chapter, tornadoes are considered separately. Such a division of the material follows a basic logic of addressing the subjects within a hierarchical order that proceeds from lower to higher velocity levels, culminating with tornadoes — at the summit of such an order. Proceeding in the same ascending order, the subject, as related to buildings, is structured in a pyramidal arrangement starting at the base, where we group low-rise buildings quite insensitive to wind forces. This constitutes the largest number of structures. Expanding into more sensitive buildings, the number decreases, eventually reaching the top of the pyramid, where we place those few extremely sensitive structures, such as nuclear power plants, which are designed to sustain the highest wind speeds found in tornadoes.

ACKNOWLEDGMENTS

In concluding this work, I found quite a list of people to whom I owe a debt of gratitude for their kindness and invaluable help through the course of working on this project. Aiming not to exclude anyone and hoping at least to include the majority, I would like to acknowledge the following: Charles Blazek, Department of Community Affairs, State of Florida, Tallahassee; Greg Reardon, James Cook University, North Queensland Townville, Australia; Dr. Yotochi Koga, Director of Nuclear Power Safety, Tokyo, Japan; Judy Jestes, DuPont, Wilmington, Delaware; Dennis Graber, Florida Concrete and Products Association; Charles F. Harper, architect, Texas; James G. Partlow, U.S. Nuclear Regulatory Commission, Washington, DC; Dr. Mary Meachum, Librarian, NOAA (National Severe Storm Laboratory), Norman, Oklahoma; Robert C. Sheets, Director of the National Hurricane Center, Coral Gables, Florida; Dr. Mark Powell, Hurricane Research Division, Miami, Florida; Barbara Brown and Jeremy Drier, Duke Power Company, Charlotte, North Carolina; Patrick Newman, American Institute of Steel Construction, Inc., Chicago; Marilyn LeMoine, the American Plywood Association, Tacoma, Washington; Dr. Ronald F. Zollo, Department of Civil and Architectural Engineering, the University of Miami, Florida; W.L. Burmeister, Georgia Power, Waynesboro; Joseph Scopelliti, Susquehanna Energy Information Center, Berwick, Pennsylvania; M.J. Gahan III, Baltimore Gas and Electric, Lusby, Maryland; Jerry A. Reiniger, American Electric Power, Columbus, Ohio; D. Standerwick, AEA, Cheshire, United Kingdom; Dr. Pieter Huyberg, Tu Delft University, Delft, The Netherlands; and Sam Houston, Hurricane Research Division, NOAA, Miami, Florida.

Special recognition is due to my research assistants who have worked with me throughout the work on this book, including Britt Quisenberry, Heather Kreider, Karen Ghera, and especially Rebecca Pinkerton. I owe special gratitude to my wife Deborah for her active participation in editing the work and in contributing to the numerous photographic illustrations that are an integral part of this work.

CHAPTER ONE

High-Velocity Winds

1. INTRODUCTION

In the history of the evaluation of the building arts, wind forces have never dominated the area of concerns imposed by nature. Gravitational forces, of course, characterized by the dominance of mass, have traditionally controlled the static equilibrium of structures. Masonry, in fact, which has embodied building technology in almost all civilizations, has always retained an inherent capacity to withstand lateral wind loads, exhibiting occasional weakness under suction forces when buildings have included timber roof structures rather than masonry vaults.

Vulnerability to lateral wind forces is a problem that has emerged within the building technology of this century for a variety of concomitant factors. New materials and the consequent new construction techniques have reduced the structural mass and thus eroded the intrinsic inertia that masonry exhibits against wind forces. Light framing systems, in steel or in reinforced concrete, have allowed creation of slender structures that emphasize the need for structural mechanisms to resist lateral wind forces. Hence, wind forces have finally appeared as a major force to be factored into the process of building design, and also as the most critical force in the case of tall slender structures.

In addition to reduction of mass and the increased slenderness of modern structures, a factor which further emphasizes the importance of wind forces is the growth of the built-up environment on geographical areas frequently affected by exceptional climatic conditions, areas where high-velocity winds and high-intensity storms occur with particular frequency. This last aspect of the problem is indeed a sociological phenomenon not new to the history of social development. It always has had an impact on the reality of the physical world.

Any demographic growth expands the human habitat outside the limits of ideal climatic conditions. Such population growth forces humanity to adapt to less and less desirable climates as the need for expansion increases. Consequently, the recent exceptional increase in population density has extended the presence of humans outside the region of the hypothetical Garden of Eden into an ever-expanding territory that is almost ubiquitous.

Seismic activity and floods are extraordinary natural forces that exert substantial effects on the habitat and against which many building technologies do not provide adequate protection. Yet, although their measured impacts are recognized as major factors present on a global scale, they are just acknowledged herein and purposely excluded to reduce the scope of investigation to the effects of high-velocity winds.

The development of knowledge on wind forces per se has followed an interesting path that began in practical terms with Eiffel's studies during the nineteenth century to calculate the wind pressure on his unique innovative structure, which could be critically affected by such a lateral load. This was followed by a similar problem connected to the monument that the people of France donated to the people of America. Thus, wind forces were studied in civil engineering prior to the development of aerodynamics prompted by aviation. In fact, though interest in wind engineering gained momentum after the Tacoma Bridge disaster in 1940, this did not mark the beginning of a new discipline, but rather its second phase, which used the substantial amount of data accumulated from aerodynamics, which had substantially progressed in the meantime.

Wind engineering has constantly progressed since the end of World War II, entering another phase during the 1960s: severe storm engineering. Major attention was then given to the extreme high-velocity forces developed by tornadoes with regard to design of a new special breed of buildings: nuclear power plants. These structures constituted a category of buildings never before encountered. Structural building failure here can affect (indeed, take) the lives of millions. The Chernobyl disaster is a dramatic example of the risk involved. Hence, formulation of wind engineering specifications for design of structures capable of protecting nuclear reactors from the action of tornado forces became a necessity that opened the way to a new level of knowledge, which we refer to as "severe storm engineering." Obviously, the level of safety required by nuclear reactors is not economically feasible for application to conventional buildings; however, substantial risk reduction can be achieved in conventional structures by following some general design criteria established for nuclear installations. Much more is now known about tornado forces than was a few decades ago, when tornado forces were considered outside the realm of any possible control measures.

2. WIND GENERALITIES

In the design of the built-up environment, architects, engineers, and planners are involved in applied meteorology, necessary when dealing with wind forces, and thus require only general information on that which the physical phenomenon concerns. Attention is concentrated on the tangible effect that the natural phenomenon of air motion has on structures. We thus include a short introduction on the generalities of wind here. We briefly touch on the fundamentals of air motion around the earth so as to introduce the subject to those practitioners studying high-velocity wind forces for the first time. For more detailed data on the meteorological aspects of wind, the reader is referred to more specific sources.

Air Motion

Wind has been defined as air in motion since the sixth century B.C. Wind energy is provided by the sun. The mechanics of air motion is based on temperature differences between various points in the atmosphere. Warmer air has less density and thus rises, whereas cold air falls with respect to adjacent warmer air. In addition to such vertical motion, there is horizontal

Figure 1.1. The Coriolis force.

motion due to the movement of cold dense air toward zones of lower pressure and higher temperature.

How can temperatures differ within the atmosphere from one point to another? There are two basic mechanisms. As the sun's radiation enters the atmosphere, part of it is absorbed to different degrees by the air and water vapor, and part of it is either reradiated or absorbed by the earth's surface. Different types of earth surfaces (e.g., bodies of water, land formations, types of soil, and vegetation species) reflect heat in different amounts, producing areas of unequal temperature in the air. For instance, we can imagine vertical air currents that have different speeds as the terrain changes from one type to another. Under such conditions, a light glider flying over these vertical currents would encounter varying lifting forces.

The Coriolis Force

The Coriolis force (named after Gaspard Gustav de Coriolis, who discovered it) is caused by rotation of the earth (see Fig. 1.1). It makes the path of moving bodies, and therefore of winds, deviate toward the right in the Northern Hemisphere and toward the left in the Southern Hemisphere. As wind speed increases, so does the Coriolis force; it is zero at the equator and maximum at the poles, and it increases proportionally to the sine of the latitude.

Geostrophic Winds

Geostrophic winds are controlled by the simultaneous action of pressure, gravity, and the Coriolis force. Wind speed wind is given by the following expression:

$$V = \frac{dp/dn}{2\omega\,\rho\,\sin\varphi},$$

where V is the velocity of the wind above the gradient height; dp/dn is the horizontal pressure gradient along horizontal line n; ρ is the average air density; ω is the angular velocity of the earth; and ϕ is the angle of latitude.

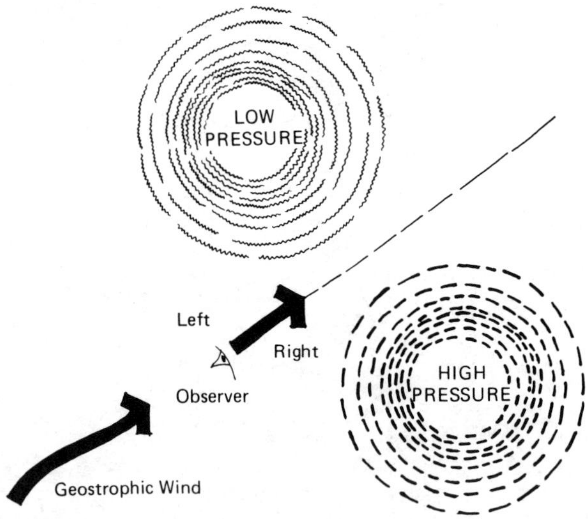

Figure 1.2. Buys Ballot's law.

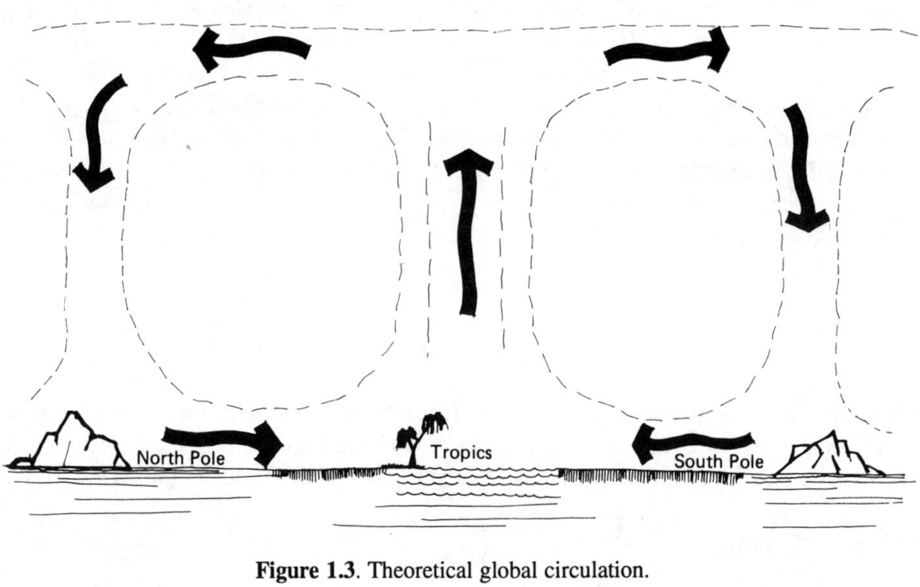

Figure 1.3. Theoretical global circulation.

Buys Ballot's Law

In a simplified visual format, Buys Ballot's law states that an observer in the Northern Hemisphere with his or her back to the geostrophic wind will have low pressure to the left and high pressure to the right. The reverse is true for the Southern Hemisphere (Fig. 1.2).

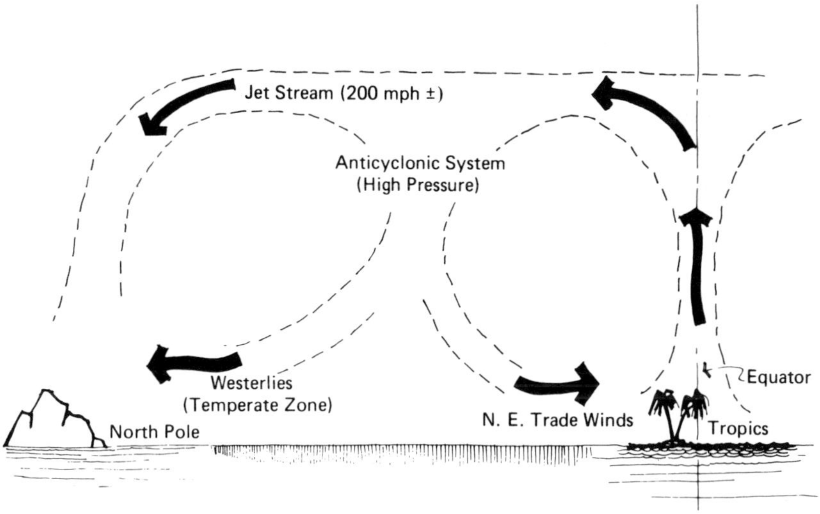

Figure 1.4. Circulation in the Northern Hemisphere.

Theoretical Global Circulation

In theory, overall circulation in the atmosphere can be visualized as follows (see Fig. 1.3): hot air over the tropical regions rises to the upper atmosphere and moves toward the North and South poles, deflected by the Coriolis force. At the poles, the air cools and descends, and then moves toward the tropics on a deflected path, again caused by the Coriolis force. Because of seasonal changes and the different amounts of heat which radiate from continents and ocean surfaces, such a simplistic circulation system does not exist.

Global Circulation of the Atmosphere

Hot air over the tropics rises to the upper atmosphere, forming jet streams that move toward the poles at speeds over 200 mph (Fig. 1.4). Some of these winds descend on subtropical areas, producing anticyclonic systems of high pressure. This air descends further down toward the earth's surface and partially proceeds to the north and south. In the Northern Hemisphere, the northbound air is deflected to the east and forms the middle-latitude westerlies, while southbound air is deflected to the west and forms the northeasterly trade winds.

Gradient Wind

Gradient wind height is the term used to indicate that portion of a moving air mass starting at elevation z_G (where the ground no longer has a friction effect), which has the following values: open country = 900 ft; suburban area = 1200 ft; metropolitan area = 1500 ft.

Below the gradient height, wind speed is affected by the earth's friction and is calculated as follows:

$$V_z = V_G(z/z_G)^{1/\alpha},$$

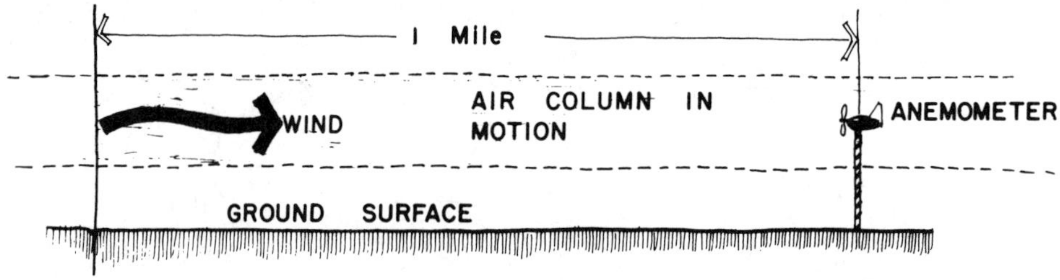

Figure 1.5. Fastest-mile wind speed.

where V_z is the wind speed, mph; z is the height above the ground, ft; z_G is the gradient height, ft; and $1/\alpha$ is an exponent dependent on the terrain: $1/\alpha = 1/7$ for open country (typical farmland with few trees); $1/\alpha = 1/4.5$ for suburban areas (with buildings from 30 to 50 ft high or wooded areas); $1/\alpha = 1/3$ for metropolitan areas (urban centers with tall buildings or terrain with pronounced orography).

Related to gradient height and different wind speed distributions below z_G, it is useful to consider the approach used by the Standards Association of Australia (SAA) (see section 6 in this chapter).

It is now logical to introduce the basic wind speeds recommended for design parameters for different geographical locations. Wind-speed maps are available for all parts of the earth. They are prepared In the United States based on observations recorded by many meteorological stations. Annual extreme fastest-mile speeds 30 ft above the ground (see Fig. 1.5) have been indicated on charts based on mean recurrence intervals of 25, 50, and 100 years (see Fig. 1.6).

The basic wind speed V used to determine design wind loads on buildings and other structures can be found in *Minimum Design Loads for Buildings and Other Structures* (American Society of Civil Engineers 1990, §6.5.2, p. 13).

3. HIGH-VELOCITY WINDS

The term "high-velocity wind" must imply some sort of specific definition, even if it may not be officially recognized in a purely meteorological context. Since the present work is not primarily dedicated to meteorology but to the effects that meteorological phenomena have on building design, for the sake of expediency, we assume that an arbitrary wind velocity value can be established as the threshold beyond which the term can be applied. If the value of 75 mph is selected, we should point out that this value is indicated in the Beaufort scale as the lower limit for wind of hurricane strength. This also corresponds to the definition used by the U.S. Weather Bureau, which refers to a wind velocity of 75 mph (i.e., 65 knots) with the following description of the sea: "The air is filled with foam and spray. Sea completely white with driving spray; visibility very seriously affected." Under another heading in the Beaufort scale, the probable mean height of waves is given as 45 feet or more, while the sea is described as "phenomenal."

The Saffir–Simpson scale for hurricane intensity also uses the same threshold velocity value of 75 mph. Currently in use for classification of hurricanes, this scale indicates

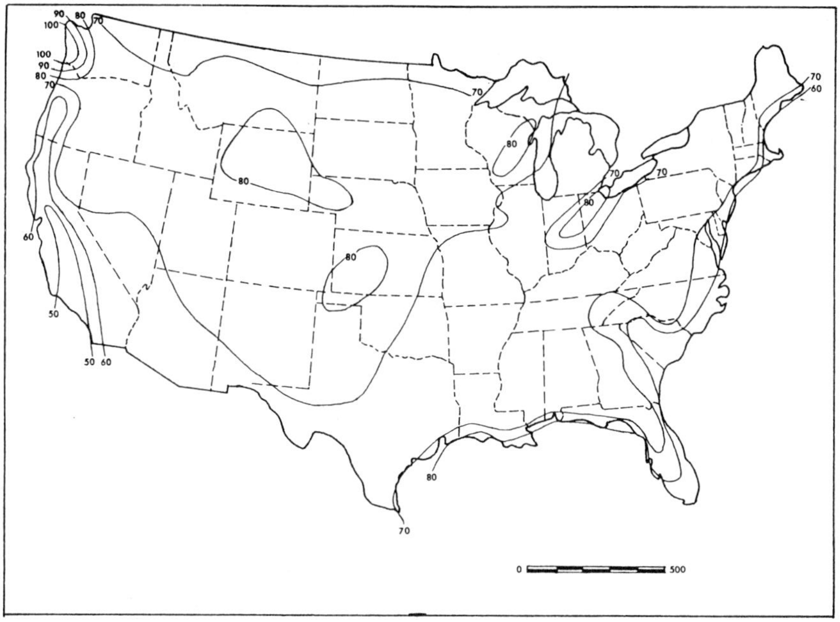

Figure 1.6a. Basic wind speed (in mph) 25-year mean recurrence interval (annual extreme fastest-mile speed 30 ft above ground).

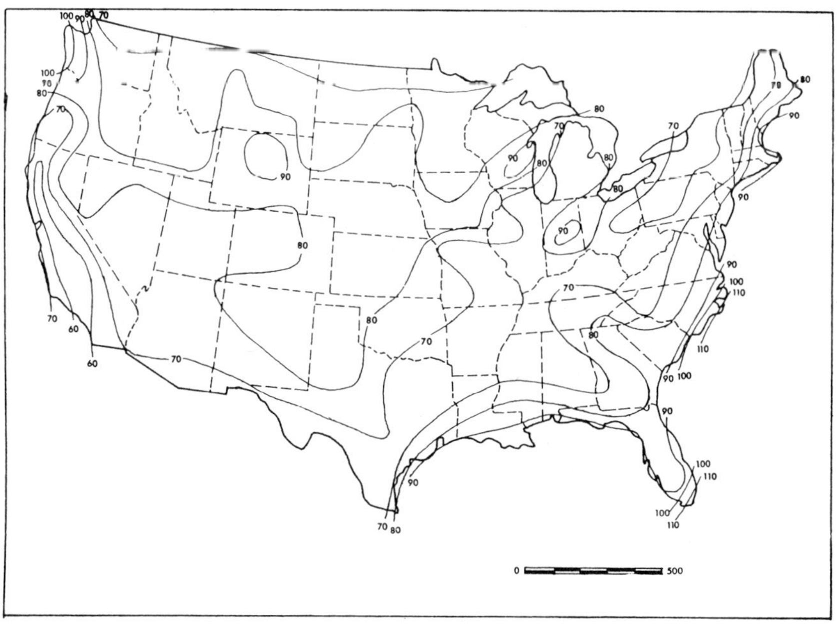

Figure 1.6b. Basic wind speed (in mph) 50-year mean recurrence interval (annual extreme fastest-mile speed 30 ft above ground).

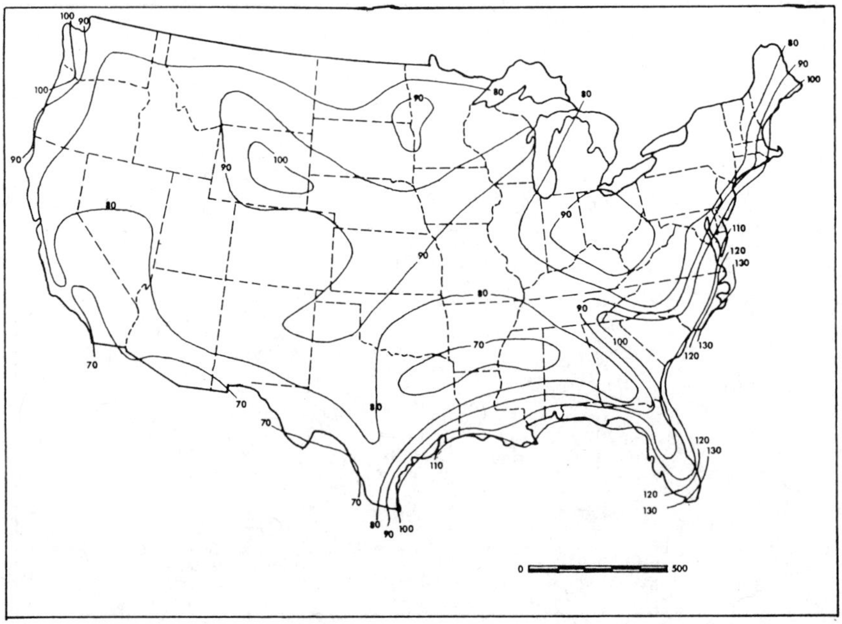

Figure 1.6c. Basic wind speed (in mph) 100-year mean recurrence interval (annual extreme fastest-mile speed 30 ft above ground).

Figure 1.6. Basic wind speed for the United States. (From American Society of Civil Engineers 1990, §6.5.2, p. 13.)

hurricanes categorized as "weak" can include winds with velocities ranging between 75 and 95 mph, which corresponds to an *F*-scale value between 1.0 and 1.4. In describing the damage typically incurred by this type of hurricane, the scale states: "Damage is primarily to shrubbery, trees, foliage, and unanchored mobile homes. No real damage occurs to building structures. Some damage is done to poorly constructed signs." Also in terms of storm surge for this category, this scale indicates that it is nominally 4 to 5 feet above normal. A description of the effects of such a category of hurricane includes: "Low-lying coastal roads are inundated; minor pier damage occurs; some small craft in exposed anchorages break moorings."

Having established the definition for high-velocity winds to be used in the present context, the arbitrary threshold value of 75 mph does not stand in total isolation from any other values but corresponds to two specific values contained in different scales and is thus more easily identifiable by referring it to the description included in each scale.

4. SEVERE STORM ENGINEERING

Loads applied to buildings for the purpose of performing stability analysis of the structure and consequential structural design of its components include a variety of physical forces, among which are the effects of high-velocity winds, the subject of the present text. When talking about wind forces, this term is too general to be understood intuitively through the entire variety of possible conditions for basic fundamental reasons that reside in part in the nature of the winds per se and in part in the nature of the structure being considered. Focusing first on the former case, wind forces must be considered within the context of the atmospheric disturbance that produces them. Of course, this implies entering into the complex mechanics of meteorology and atmospheric science, which, although unavoidable, will be kept to a minimum in order to maintain our focus mostly on engineering aspects rather than on the original causes of the loading in consideration. In particular, the wind forces considered herein are determined by high-velocity winds that are associated with specific atmospheric disturbances that we will refer to as "storms." As such, one can see that, from the variety of different kinds of storms which are clearly identified in meteorology, there are a variety of different wind conditions, each of which can be defined in accordance with the nature of the individual storm. We can refer in meteorological terms to "severe local storms," which are first taken into account in this context to establish some sort of terminology to be used in connection with the engineering aspects of wind loads due to high-velocity winds.

"Severe local storms" have been defined quite specifically in the pertinent literature. An event, or atmospheric disturbance identified with such terms, includes one of four specific cases (Greiner et al. 1990):

- *Tornadoes* — a rotating circulation touching the ground and associated with a convective cloud.
- *Hail* — equal or greater than 3/4 inch (1.9 cm) in diameter.
- *Thunderstorm wind gusts* — of at least 50 knots (93 km/hr).
- *Thunderstorm wind damage.*

It is obvious that the terminology used is significant and cannot be used loosely without creating major confusion, even if within the engineering context one could feel free to determine an independent series of terms. Yet, in order to establish a meaningful vocabulary that will not generate confusion, we use the term "severe storms" herein and exclude the term "local" to identify all storms that generate high-velocity winds. In this category we include thunderstorms, hurricanes, typhoons, tornadoes, and so on, which are different individual atmospheric phenomena all of which have in common high-velocity winds that could affect the stability of a structure, aside from other specific loading conditions (e.g., pressure drop, missile impacts). With such a premise, we assume that the reader will not be induced into drawing erroneous conclusions and will be able to differentiate between the terminology adopted in the present work and a specific term used in the context of atmospheric signs that will also be followed when this applies. Naturally, when the term *high-velocity wind* is used, one must make a distinction between wind loads in general and high-velocity wind loads by establishing a wind velocity threshold for each.

5. U.S. CODES OF STANDARD PRACTICE

The significance of information derived from codes has its foundation in understanding the spirit in which that information has been gathered and organized. Any design consultant realizes the major difference between legally enforceable building codes and codes of standard practice. The latter, to be used as an indispensable guide in practical application of professional design, have immense practical value, yet to constantly update them they demand continuous attention, so that professionals can keep abreast of the accumulated scientific knowledge and be able to evaluate previous designs inspired by code recommendations which were valid when the design was executed.

We will cite and illustrate several codes related to basic recommendations concerning application of high-velocity wind forces. We provide a short introduction to place them in their proper perspective and to make it possible for the reader to better understand the significance of each source.

ANSI/ASCE-7

A short chronological description of the development of the American National Standards Institute (ANSI) is necessary to give an overall view of the background that generated the most significant present-day code: "Minimum Design Loads for Buildings and Other Structures," Revision and Redesignation of ANSI A58.1-1982, Approved December 1988, American Society of Civil Engineers, Published July 1990 (ANSI/ASCE 7-88).

We first consider a report of the Department of Commerce Code Committee entitled "Minimum Live Loads Allowable for Use in Design of Buildings" published by the National Bureau of Standards in 1924. In 1945, the American Standards Association's Sectional Committee on Building Code Requirements for Minimum Design Loads in Buildings published a new report in 1945 (A58.1-1945) which covered a wider area. This was revised in 1955 under the name "American Standard Building Code Requirements for Minimum Design Loads in Buildings and Other Structures" (A58.1-1955). A 1972 revision of this last publication was entitled the "American National Standards Building Code Requirements for Minimum Design Loads in Buildings and Other Structures" (A58.1-1972).

Referring specifically to wind forces, focus of the present work, it should be noticed that the section on wind loads in the latter document was completely revised. Quoting from the abstract of the publication, "New distributions of extreme fastest mile winds in the United States for various mean recurrence intervals are based on the latest National Weather Service data. A major innovation is the concept of varying load requirements, both with the dynamic properties of the building and with the nature of the terrain."

The ASCE document "Minimum Design Loads for Buildings and Other Structures" (Revision and Redesignation of ANSI A58.1-1982, Approved December 1988, Published July 1990) specifically includes wind loads in addition to dead and live soil, snow, rain, and earthquake loads. In practical terms, this code is directed toward architects, structural engineers, and those engaged in preparation of local building codes.

Participation of the ASCE in wind engineering is evidenced by two major publications: "Wind Bracing in Steel Buildings" (Final Report, ASCE Subcommittee No. 31, Committee on Steel, *ASCE Transactions*, 105:1713, 1940, and "Wind Forces on Structures" (Final Report of the Task Committee on Wind Forces of the Committee on Loads and Stresses of the Structural Division), 1961. This last document is one of the most useful compendia of information for guiding structural design practitioners in matters of wind engineering.

With this introduction, the reader will find it easier to understand the interrelationship of the various sources in connection with topics derived from the codes.

The Uniform Building Code (UBC)

The Uniform Building Code was initiated in 1927 and sponsored by the International Conference of Building Officials at their sixth annual business meeting (October 18–21). After its first writing, the code has been periodically revised every three years. The last edition of the code was published on May 1, 1994.

An important characteristic of this code is its high adaptability to the constant changes that take place in the building sciences in terms of development of technology and design improvements. Its high adherence to the most up-to-date concepts in the practice and theory of design and construction is illustrated not only by the three-year frequency of code revisions but also by the yearly revisions published in supplement form. The International Conference of Building Officials (headquartered in Whittier, California) is a highly recommended international resource for professionals looking for guidance on wind engineering for the UBC and for other literature that the organization publishes on the various disciplines of the building sciences.

The UBC deals specifically with wind engineering in the 1992 editions in Part II: Wind Design (section 2311, Generalities; 2312, Definitions; 2313, Symbols and Notations; 2314, Basic Wind Speed; 2315, Exposure; 2316, Design Wind Pressures; 2317, Primary Frames and Systems; 2318, Elements and Components of Structures; 2319, Open-Frame Towers; 2320, Miscellaneous Structures; 2321, Occupancy Categories).

We make special note to section 2321 on "Occupancy Categories," which classifies buildings with respect to their type of occupancy in terms of importance factors, following similar criteria used in seismic design. We reproduce two self-explanatory tables here from the UBC (Tables 1.1 and 1.2) as an example of practical application of the code to basic concepts in wind engineering, especially for professionals from different parts of the world where practical codes are not readily available.

ASCE-7, 1995 Amendments

The newest update of ASCE-7 was published in May 1995, but the final approved revisions are not yet officially available. The preliminary data indicate a series of changes which would be partially discussed in this context in order to keep the present work up to date with the latest available information. Although not official at the time of the present writing, a second draft of the document dated February 1994 has been circulated among interested parties so that it is possible to speculate on a few points without a confirmed final version of the document.

Main Wind Force Resisting System Coefficients (MWFRSCs) will be included following the same criteria used by the Canadian Code derived from research conducted at the University of Western Ontario. However, this coefficient will not be affected by the reduction factor of 0.8 as in the Canadian Code.

Another major change is in the wind speed database used to arrive at numerical values. To avoid confusion, we list the different systems currently used: peak gust (3- to 5-second average); fastest-mile (typically used in the previous codes); and sustained wind (1-minute averaging period). The new code will use the 3-second database values. With such a change, the wind speed values will generally increase with respect to those previously used. In using such values, the engineering units will coincide with the values used by the National Weather Service, which are also those reflected by the news media in general, thus eliminating the discrepancies that have occurred when fastest-mile wind speed values were used. Notice,

Table 1.1. Occupancy Categories

Occupancy categories	Occupancy type or functions of structure
I. Essential facilities	Hospitals and other medical facilities having surgery and emergency treatment areas. Fire and police stations. Tanks or other structures containing housing or supporting water or other fire-suppression materials or equipment required for protection of essential or hazardous facilities, or special occupancy structures. Emergency vehicle shelters and garages. Structures and equipment in emergency-preparedness centers. Standby power-generating equipment for essential facilities. Structures and equipment in government communication centers and other facilities required for emergency response.
II. Hazardous facilities	Structures housing, supporting, or containing sufficient quantities of toxic or explosive substances to be dangerous to the safety of the general public if released.
III. Special occupancy structure	Covered structures whose primary occupancy is public assembly; capacity > 300 persons. Buildings for schools through secondary or day-care centers; capacity > 250 students. Buildings for colleges or adult education schools; capacity > 500 students. Medical facilities with 50 or more resident incapacitated patients, but not included above. Jails and detention facilities. All structures with occupancy > 5000 persons. Structures and equipment in power-generating stations and other public utility facilities not included above, and required for continued operation.
IV. Standard occupancy structure	All structures having occupancies or functions not listed above.

Essential facilities are those structures which are necessary for emergency operations subsequent to a natural disaster.

Source: International Conference of Building Officials 1991, p. 185.

however, that while the numerical values for wind speed will increase with respect to the previous ones, the resulting pressure values will not reflect any significant change. It is further interesting to notice that the maximum designed wind speed for the inland regions of the United States will be mostly within the 90 mph range, which reflects adjustments in the evaluation criteria.

Table 1.2. Occupancy Requirements

Occupancy category	Importance factor *I* Earthquake	Wind
I. Essential facilities	1.25	1.15
II. Hazardous facilities	1.25	1.15
III. Special occupancy structures	1.00	1.00
IV. Standard occupancy structures	1.00	1.00

Source: International Conference of Building Officials 1991, p. 186.

6. AUSTRALIAN CODE

Generalities

The national standards in Australia for the design of structures under wind loads were first initiated in 1952. The first document on this topic was published in 1952 by the Standards Association entitled "Interim 350."

"Australian Standard CA34 Part II 1971" was an elaboration of the 1952 document and included a map of Australia with graphical indications of the recommended wind speed to be used for design shown similarly to contour lines. Four different terrain divisions were recognized, and an elaborate table of pressure coefficients for different building configurations was also included. "Australian Standard 1170: Part 2, the 1973 Edition" introduced metric units and a discussion on wind forces on tall slender buildings.

The 1975, 1978, 1981, and 1983 codes were all affected by the violent storm which struck Australia in 1974, prompting a reevaluation of wind forces. This storm was named Tracy, and is usually referred to as a cyclone in the technical literature. It produced a reevaluation of Australian wind maps, including establishment of a cyclone-prone area along a strip of the northern coast. In addition, the negative pressure coefficients for the sidewalls of tall buildings were further increased. New negative coefficient pressures were also introduced for flat roofs and those with pitches of less than ten degrees.

The latest code, published in 1989, was entitled "Australian Standard: Minimum Design Loads on Structures, Part 2, Wind Loads" (AS 1170 Part 2). This document constituted a major revision in the sense that it was introduced within the context of the Australian structural design codes. Some of its major goals include more precise application of wind loads to structures not requiring complex engineering analysis (e.g., low-cost structures like housing) and buildings typically insensitive to wind forces (e.g., reinforced concrete buildings up to three stories in height). It also furnishes a simplified analysis of wind loads that takes account of the dynamic properties of slender buildings sensitive to wind forces.

It includes three sections: "Simplified Procedure," "Detailed Procedure: Static Analysis," and "Detailed Procedure: Dynamic Analysis."

a. **Simplified Procedure**. Here the code groups elements derived from the "Detailed Procedure: Static Analysis" section using a more empirical and simplified approach to determining wind loads which can only be applied to insensitive structures no higher than 15 m. Wind forces are determined by using specified basic pressures multiplied by specified factors. Such factors are determined by the following conditions: wind speeds specified by local geographical conditions; conditions of the terrain lying under the incoming wind; building height; sheltering effect caused by other structures; and general topographical conditions of the site.

b. **Detailed Procedure: Static Analysis**. Here wind speed is used to represent the peak gust value modified by such local conditions as topography and height (quasi-steady assumption). Dynamic pressure is then obtained from the wind speed value. Wind loads are then obtained by multiplying the dynamic pressure by the pressure of force coefficients (determined by geometrical characteristics). The applied forces are considered to act statically; however, application of this methodology is limited to wind-insensitive structures, which includes buildings with a height-to-width (aspect) ratio not greater than 5. Furthermore, the procedure can only be applied if the fundamental frequency of natural vibration (*f*) is larger than 1 Hertz, i.e., has a fundamental period (*T*) less than 1 second. In practical terms, such limitations include buildings up to 50 m in height; however, buildings up to 150 m in height can be included if their structural slenderness is greatly minimized and their mass is particularly light.

c. **Detailed Procedure: Dynamic Analysis.** This section applies to design of tall slender structures with a low fundamental frequency of natural vibrations (< 1 Hz), i.e., high fundamental periods of natural vibrations in excess of 1 second, since these structures can have a dynamic response in resonance with the frequencies of wind-induced vibrations. This procedure also applies to structures with a height-to-width (aspect) ratio larger than 5.

Wind-Sensitive Buildings

As stated above, the design of wind-sensitive buildings in the Standards Association of Australia (SAA) Code is discussed separately under the title of "Detailed Procedure: Dynamic Analysis." Some general discussion on this subject is in order, starting by emphasizing the random variations of wind velocity that can create wind forces which vary randomly. Such variable forces can act as buffeting loads which excite buildings to vibrate, especially buildings that because of their structural slenderness are more prone to oscillations. The response of a building to dynamic excitation obviously depends on its ability to vibrate in a natural free state. Critical conditions arise when resonance occurs between the frequency of the exciting forces and the natural frequency of free vibration of the building. The critical frequencies involved are those corresponding to the first mode of vibration, because it is with the first mode that the building undergoes oscillations of the largest magnitude. The energy involved in the process is furnished by the kinetic energy of the wind itself. An observation from the SAA commentary is clear: "In most cases, the major part of energy of excitation is at frequencies much lower than the fundamental natural frequency of the structure and decreases with increasing frequency." Therefore, only a response in the first mode is needed in practical terms for the design of buildings under critical conditions. The average response of a structure may be taken within a period of observation varying between 10 minutes and 1 hour and be assumed to be a steady response. On the other hand, the gustiness and turbulence of the wind generate an instantaneous dynamic response which can produce resonance, while the kinetic energy of the wind is absorbed and eventually dissipated by the building in motion. In this respect, it should be noticed that high-velocity winds tend to accentuate the difference between the steady and dynamic response in a building. Thus, the values of forces, accelerations, deflections, etc., in a building are described either as the average measured within periods varying between 10 minutes and 1 hour or as peak response when dynamic conditions are changed for convenience into equivalent static ones. Such peak values can be calculated from a probability distribution of the random variables in question. The peak value of a variable can be obtained as

$$\hat{x} = \bar{x} + g\sigma_x,$$

where g is the peak factor, $\hat{x}$ is the peak deviation, $\bar{x}$ is the mean deviation, and σ_x is the standard deviation, assuming that all values are measured within a period of 1 hr.

The response of a building is better understood if it is subdivided into two components, one of which is along the direction of the wind (the "along-wind" component) and the other is perpendicular to it (the "cross-wind" component). Of course, combining the two will generate an elliptical configuration of response. Such double perpendicular responses are generated by two different mechanisms independent of each another.

The fundamental frequency of natural free vibration which occur in the first mode can be estimated from different sources. However, this yields approximate results since the value of the factors has to be estimated initially. The results are particularly less dependable when applied to tall slender buildings. One such formula (Jeary and Ellis 1983) for buildings with a floor plan close to square in shape is

$$n_1 = 46/h,$$

where n_1 is the first mode frequency of the structure (in Hz), and h is the building height (in m).

Jeary and Ellis also recommend a formula for buildings considered as vertically cantilevered structures with a uniform distribution:

$$n_1 = \sqrt{[(EI/n)]},$$

where E is the effective elasticity modulus, I is the second moment of area (m^4), and m is the mass per unit length (kg/m).

Their formula for a substanding lattice tower is

$$n_1 = 1500(w_a/h^2),$$

where h is the height of the structure (in m), and w_a is the average width in meters.

Effects of Terrain and Heights

The effects of turbulence produced by friction with the ground within a boundary layer are treated in a different manner by the procedures issued by the Standards Association of Australia than the approach followed by the codes of standard practice in the United States. In subsection 2.5.2 of the "Simplified Procedure" (Standards Association of Australia 1990), the SAA code introduces a multiplication factor (B_2) to account for terrain conditions and

Table 1.3. Terrain and Height Multiplying Factors (B_2)

Building height (h_t) m	Factors (B_2)			
	Suburban		Open rural terrain	
	Exposed	Sheltered	Regions A and B	Regions C and D
<4	0.85	0.55	1.15	1.10
7	0.90	0.60	1.30	1.20
10	1.00	0.65	1.45	1.30
15	1.15	0.75	1.60	1.45

For intermediate values of h_t, linear interpolation is permitted.
Source: Standards Association of Australia 1990, p. 15.

for the height of buildings upstream. Such a coefficient multiplied by p' (the net basic wind pressure, in kilopascals) gives p_d (the ultimate limit state design wind pressure, in kilopascals). Such values of B_2 are given in Table 1.3.

The terms used in the table are briefly explained as follows: (1) open rural terrain applies to self-standing structures in open fields or next to large bodies of water; (2) suburban terrain applies to structures less than 10 m in height within urban areas and located at a distance not less than 0.5 km from open rural terrain, as well as to structures 10 m or more high within and located at a distance not less than 2/5 km from open rural terrain.

Buildings in this type of terrain are subdivided into sheltered and exposed types. Sheltered buildings are those surrounded by two rows of buildings of equal or greater height, and, furthermore, the density of buildings of the site should be not less than 10 buildings per hectare. Exposed buildings are considered those surrounded by other buildings of equal or greater height spread out over the surrounding area with a density of less than 2.5 buildings per hectare.

The code then clearly specifies that the coefficient can be appropriately interpolated for buildings located in areas that fall in between the two types of terrain. Indeed, the above-mentioned detailed classifications in this code are extremely specific and can be used to generate other codes of standard practice which are less precise on this subject.

The next section the code introduces B_3, the topographic multiplication factor, which applies to structures sited on hilltops, on mountainous terrain features, or on the edge of steep slopes. Values of this factor are listed in the Table 1.4.

In the procedure outlined in the Australian code, variations of terrain are subdivided into four terrain categories: **Category 1** applies to open terrain with no obstructions and to large water surfaces. **Category 2** applies to open terrain with sparse obstacles not higher than 10 m. **Category 3** applies to terrain occupied by obstructions spaced close together as individual self-standing dwellings, typical of suburban zones. **Category 4** applies to terrain occupied by tall closely spaced buildings, typical of major urban centers.

Having selected the type of terrain, the code furnishes an approximate expression for the value of the wind velocity (V) at a height z above grade:

$$V_z \approx u^*/0.4 \log_e (z/z_0),$$

Table 1.4. Topographic Multiplying Factors (B_3)

Topographic features	Maximum incline	Location	Factor (B_3)
Escarpment	<1 in 7.5	B	1.0
Escarpment	<1 in 15	A	
Hill or ridge	<1 in 7.5	B	
Hill or ridge	<1 in 10	A	
Escarpment	≥1 in 4	B	1.3
Escarpment	1 in 7.5	A	
Hill or ridge	1 in 4	B	
Hill or ridge	1 in 7.5	A	
Escarpment	≥1 in 4	A	1.6
Hill or ridge	1 in 3	B	
Hill or ridge	1 in 5	A	
Hill or ridge	>1 in 1.5	B	2.0
Hill or ridge	1 in 2	A	
Hill or ridge	≥1 in 1.5	A	2.4

Source: Standards Association of Australia 1990, p. 15.

Table 1.5. Roughness Length (z_0)

Terrain category	Roughness length (z_0), m
1	0.002
2	0.02
3	0.2
4	2.0

Source: Standards Association of Australia 1990, p. 20.

where V_z is the mean hourly wind speed at height z (in m/s); u^* is the friction velocity; z is the distance or height above the ground (in m); z_0 is the characteristic terrain roughness length (in m). (See Table 1.5 for standard roughness lengths.) For lengths that fall between terrain categories one uses the following formulas:

$$z_0 = 2 \times 10^{(\text{terrain category} - 4)} \quad \text{and} \quad z_g = \text{the gradient height.}$$

7. OTHER CODES

Aside from the major emphasis placed on the design procedures for wind loading in the United States and on the procedures mandated by building codes in Australia, we now briefly consider the code specifications of two other countries: Canada and Japan. We assume that the reader can easily obtain access to normative data for specific countries with which he or she is involved.

The Canadian Code

The latest publication on design procedures for wind loads on structures controlled by building codes in Canada includes the "Supplement to the National Building Code of Canada 1990" and is published by the Associate Committee on the National Building Code (1990). In Chapter 4 ("Commentaries on Part 4 of the National Building Code of Canada 1990"), Commentary B focuses on wind loads and generally follows the conceptual path of the equivalent U.S. and Australian codes. The commentary addresses a topical sequence which includes: reference wind speeds and basic velocity pressure, exposure factors and dynamic response, speed-up over hills and escarpments, vortex shedding effects, pressure coefficients for external and internal conditions, pressure on roof membrane layers, pressure on rounded structures, wind and ice effects on cables, loads on frames and shielding, torsion on buildings, building motions, and pressure differences across interior walls and partitions. Covering these subjects in only 32 pages, this work is concise yet effective in its guidance to designers in terms of basic criteria. The code subdivides the procedure for wind design into three approaches:

a. The simple procedure is the most general application that also includes low- and medium-rise buildings in terms of the structure itself as well as the cladding. In this case, it is assumed that the building under load is stiff and rigid, insensitive to any dynamic response, so that the wind load is considered to act in a static condition. To this effect, appropriate pressure coefficients for most common building shapes are within the commentary itself.

b. The experimental procedure involves the use of wind tunnel testing or other similar experimental analysis. This approach is to be adopted when more specific information is needed concerning the dynamic response of the building. This approach is also used in the event that the pressure coefficients furnished by the commentary do not apply to the specific geometry of the building in question.

c. The detailed procedure is needed when the overall dynamic response of the building under load including resonance magnification is required instead of specifics on the cladding and structural components. Part of the process includes determination of the intensity of wind turbulence for the site in question, reflecting the roughness of the terrain and height above ground level, as well as determination of the dynamic response of the building based on its geometry, type of structure, fundamental period of vibration, and damping characteristics.

Final output of such analyses is used to determine the static design pressure that would produce the same effect as the one generated by the dynamic response of the building in the condition of resonance. To obtain such a pressure, the dynamic pressure, q, must be multiplied by such factors obtained through this process as the gust effect factor and the exposure factor.

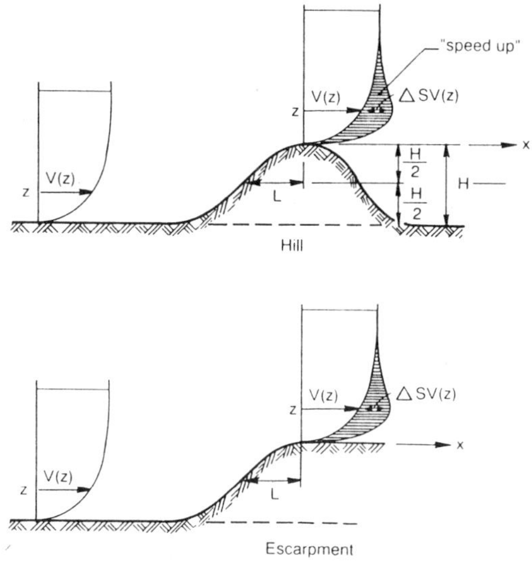

Figure 1.7. Wind speed-up over hills and escarpments.

Speed-up over hills and escarpments

As indicated above, an interesting issue covered by the supplement to the National Building Code (NBC) of Canada involves an increase in wind speed that takes place on inclined ground when the slope is more than 10°. Thus, when buildings are sited on the crest of a hill facing a slope greater than this critical value, the wind pressure on the structure is considerably increased. Introducing the "speed-up factor" that applies in such conditions, the supplement states: "The exposure factor at height z above the local ground elevation is then equal to that over flat terrain multiplied by a factor $(1+\Delta S(z))^2$, where $\Delta S(z)$ is the 'speed-up factor' for the mean wind speed." Here, ΔS is the change in wind speed, and z is the height above flat ground level. The code specifies that, if the building is near the crest, within a distance $x < kL$, the exposure is modified as follows:

$$C_e(z) = C_e(z)\ \{1 + \Delta S_{max}\ [1 - x/kL]\ e^{(-\alpha_z/L)}\}^2,$$

where $C_e(z)$ is the exposure factor over flat terrain, ΔS_{max} is the relative speed-up factor at the crest near the surface, and α is a decay coefficient for a decrease in speed-up with height. The values for α and ΔS_{max} depend on the shape and steepness of the hill (see Fig. 1.7). The exposure factor C_e for different heights z varies for three different types of exposures: A, open or standard exposure, open level terrain with only scattered buildings, trees, or other obstacles, open water or shorelines; B, suburban and urban areas, wooded terrain, or centers of large towns; and C, centers of large cities with heavy concentrations of tall buildings where at least 50% of the buildings exceed four stories.

Pressures on roof membrane layers

An interesting section of the code supplement which particularly applies to high-velocity wind conditions is dedicated to roof membrane layers and considers the case of rigid roof insulation which is not effectively bonded to the waterproofing membrane constituted by the structural deck. The code indicates that the insulation is not subject to the same wind pressure as that applied to the structural roof beneath it because of leakage of air between the upper layer and the roof assembly, and due to pressure equalization between the upper and lower parts of the insulation layer. This condition means that the external wind pressure or suction will not affect the insulation but will affect the roof assembly below because it is this barrier which separates the enclosed interior space from the exterior, thus becoming the resisting structural component subjected to a pressure differential. The supplement indicates that additional data on this subject can de found in works by Quigley and Vogan (1970) and Hardy (1957) and their bibliographies.

Design Procedures in Japan

The official design procedure for determination of wind forces on buildings is codified in Japan in the Japan Building Standard Law. Article 87 of the code indicates that wind pressure should be calculated as follows:

$$\text{Wind force} = \text{Velocity pressure } (q) \times \text{Wind force coefficient} \times \text{Surface area.}$$

The velocity pressure is calculated as follows:

a. For those parts of buildings that do not exceed a height of 16 m above grade:

$$q = 60(h)^{1/2},$$

where q is the pressure (in kg/m^2), and h is the height above grade (in m).

b. For those parts of buildings that exceed a height of 16 m above grade:

$$q = 120(h)^{1/4}$$

(the symbols were defined above).

The values of the wind force coefficient contained in the Building Standard Code of Japan indicate that there are seven main groups of structures which are further subdivided into different categories. The main groups are the following:

a. The shape of flat surfaces as found in billboards carrying advertisements in open areas, on top of buildings, etc. To be noticed is that the board is considered independent of its thickness; consequently, the coefficient includes both the effects of positive pressure on the windward side and simultaneous suction pressure on the leeward side.
b. Buildings consisting of an airtight envelope enclosing an interior space.
c. Buildings opened either on the windward or leeward side.
d. Structures consisting of a roof over a totally open space.
e. Lattice structures permeable to wind flow.
f. Net structures such as wire netting.
g. Cylindrical structures, such as slender towers and chimneys.

In considering application of wind forces for the entire country, the government regulations take into account that there may be areas where expected wind velocities are lower, so that reductions can be applied in certain areas. Such reductions cannot be greater than 40% except in area where heavy snow is expected, in which case the maximum reduction is 60%. Also, when a building is sheltered on the windward side by another building or by a tree windbreak, the maximum wind pressure reduction is 50%.

Gradient wind velocity

Japanese codes also recognize the gradual increase of surface winds with height above ground due to the frictional force exerted by ground roughness or the presence of obstacles, including buildings. This increase in velocity with height is expressed by the following formula (proposed by Douglas Archibald):

$$V = V_0(h/h_0)^{1/n}.$$

This general equation can be further reduced by entering specific data. For instance, with $h_0 = 15$ m (the typical height where anemometers are placed in weather bureaus at airports) and $V_0 = 63$ m/s (velocity measured during the Muroto typhoon of 1934), considering that $1/n$ varies between 1/2 and 1/5, and taking a rounded average of 1/4, the previous equation becomes

$$V = 63(h/15)^{1/4}.$$

Substituting this value into the equation for velocity pressure, we obtain

$$q = 60(h)^{1/2},$$

where q is in kg/m^2 and h is in m.

8. WIND VELOCITY AS A FUNCTION OF TIME

The average or mean wind speed of course depends on the length of the time period during which wind speeds are measured. As this period shortens, the average wind speed increases. Simiu and Scanlan (1986) offer the following expression:

$$U_t(z) = U_{3600}(z) + c(t)\overline{u'^2}^{\,1/2},$$

where U_t is the wind speed average over t seconds, U_{3600} is the hourly speed, and $c(t)$ is a coefficient that depends on t that is determined by statistical wind speed data. Its value corresponding to open terrain conditions is applicable when z_0 is equal to or less than 1.0 m (Durst 1960) and $z = 10$ m. Other values of t are as follows:

t (sec)	1	10	20	30	50	100	200	300	600	1000	3600
$c(t)$	3.00	2.32	2.00	1.73	1.35	1.02	0.70	0.54	0.36	0.16	0.00

u' is the longitudinal turbulent fluctuation. Through basic substitution, we derive a new expression:

$$U_t(z) = U_{3600}(z)[1 + (0.98c(t)/\ln(z/z_0)],$$

where z_0 is the height of obstacles on the terrain (in m), and z is the elevation (in m).

In practical terms, $U_t(z)$ is expressed in the United States in hourly mean speeds, but the official records available for various locations are expressed as the fastest-mile wind speed (U_f). To convert from one to the other, consider that the time period to be considered is

$$t\ (\text{sec}) = 3600/U_f,$$

where U_f is expressed in mph.

9. AVERAGE VELOCITIES

Since wind velocities vary constantly in direction and magnitude, any measurement is in practical terms an average of the various readings taken within a certain time period. It is usual to consider three different observation periods, each of which is assigned a different term:

Peak Gust Velocity is averaged within 2- to 5-second periods.

Sustained Wind Velocity is averaged within 1- to 2-minute periods.

Fastest-Mile Wind Velocity is averaged within the period of time necessary for a column of air 1 mile long to pass over the anemometer used for the measurement (standard engineering reference conditions).

To be able to compare different types of velocity measurements, it is necessary to have formulas to convert one to the other. Two of the most widely used methods for such conversions are named after the researchers who have developed them: the Durst (1960) method and the Krayer–Marshall (1992) method.

The Durst Method

This method was developed by C.S. Durst ("Wind Speeds Over Short Periods of time," *Met. Mag.* 89:181, 1960). The relationship between different velocity types is derived from the following expression:

$$-V_f/V_t = (1 + [0.185c(t_f)])/(1 + [0.185c(t_t)]),$$

where V_f is the fastest-mile winds speed (mph); V_t is the wind speed as an average speed over t seconds (mph); t_f is the fastest-mile wind speed duration, $(3600/V_f)$ (sec); t_t is the duration used for averaging wind speed (sec); $c(t)$ is a coefficient obtained as follows:

t (sec)	1	10	20	30	50	100	200	300	600	1000	3600
$c(t)$	3.0	2.32	2.0	1.73	1.35	1.02	0.70	0.54	0.36	0.16	0.0

The Krayer–Marshall Method

This method was developed by William R. Krayer and Richard D. Marshall in 1992 ("Gust Factors Applied to Hurricane Winds," *Bulletin of the American Meteorological Society* 73/5, May 1992).

10. DYNAMIC PRESSURE (q)

The dynamic pressure (q) can be found easily by applying Bernoully's equation between a point P in free air and a point P_1 on a structure that lies in the path of the wind. We consider p, v and p_1, v_1 are, respectively, the air pressure and velocity of the wind at points P and P_1; r is the air density. We write

$$p + 1/2(r\,v^2) = p_1 + (1/2)(r\,v_{-361}{}^2)$$

and

$$(p_1 - p) = (1/2)r\,v^2 - (1/2)r\,v_{-361}{}^2.$$

However, v_1 is zero, because it is assumed that the air particles lose their velocity as they strike a structure placed in the path of the wind. Also, the quantity $(p_1 - p)$ is the overpressure above the atmospheric pressure generated by the wind on the structure at point P_1, which is referred to as velocity pressure q. Substituting, we obtain

$$q = 1/2(r\,v^2).$$

Substituting the values for air density and introducing the necessary conversion factor, we obtain

$$q = 0.00256v^2,$$

where v is expressed in mph and q in psf. Using SI system units and assuming an air density equal to 50×10^{-6}, the above expression becomes

$$q = 50 \times 10^{-6} \times V^2,$$

where q is in kPa and V is in k/hr. Of course, the average air density varies for each geographical region. The value chosen above is an example chosen by the National Building Code of Canada.

This expression is theoretical since it is based on an assumption that air particles lose all their kinetic energy as they come into contact with a structure, which in reality does not occur.

Velocity Pressure (q_z)

The velocity pressure (q_z) is a more realistic quantity than the dynamic pressure (q). In fact, while q does not include any factor related to the object exposed to wind pressure, q_z does. q_z is defined as the velocity pressure at height z above ground. It is written as

$$q_z = 0.00256K_z(IV)^2,$$

where I is the importance factor, furnished by the following:

Category	I at 100 miles from hurricane oceanline and in other areas	I at hurricane oceanline
I	1.00	1.05
II	1.07	1.11
III	1.07	1.11
IV	0.95	1.00

and K_z is the exposure coefficient, given in Table 1.6. Basic wind speed (V) for Hawaii and Puerto Rico is given in Table 1.7.

Table 1.6. Velocity Exposure Coefficient K_z

Height above ground level z, ft	K_z			
	Exposure A	Exposure B	Exposure C	Exposure D
0–15	0.12	0.37	0.80	1.20
20	0.15	0.42	0.37	1.27
25	0.17	0.46	0.93	1.32
30	0.19	0.50	0.98	1.37
40	0.23	0.57	1.06	1.46
50	0.27	0.63	1.13	1.52
60	0.30	0.68	1.19	1.58
70	0.33	0.73	1.24	1.63
80	0.37	0.77	1.29	1.67
90	0.40	0.82	1.34	1.71
100	0.42	0.86	1.38	1.75
120	0.48	0.93	1.45	1.81
140	0.53	0.99	1.52	1.87
160	0.58	1.05	1.58	1.92
180	0.63	1.11	1.63	1.97
200	0.67	1.16	1.68	2.01
250	0.78	1.28	1.79	2.10
300	0.88	1.39	1.88	2.18
350	0.98	1.49	1.97	2.25
400	1.07	1.58	2.05	2.31
450	1.16	1.67	2.12	2.36
500	1.24	1.75	2.18	2.41

Linear interpolation for intermediate values of height z is acceptable. Tables 1.6–1.8 and Figures 1.8–1.10 are from *Minimum Design Loads for Buildings and Other Structures*, American Society of Civil Engineers, November 27, 1990. They are reproduced by permission of the publisher (ASCE).

Gust Response Factor

The gust response factor (G_z) takes into account the additional loading effects caused by turbulence and those caused by the amplification effect due to structure flexibility. It is important to note that this factor does not take into account cross-wind deflections generated by formation of eddies, by galloping, or by flutter.

Table 1.7. Basic Wind Speed, V, for Hawaii and Puerto Rico

Location	V (mph)
Hawaii	80
Puerto Rico	95

The unique topographical features common to the islands of Hawaii and Puerto Rico suggest that it may be advisable to adjust the values to account for locally higher winds for structures sited near mountainous terrain, gorges, and ocean promontories.

Table 1.8. Gust Response Factors G_h and G_z

Height above ground level, z, ft	G_h and G_z			
	Exposure A	Exposure B	Exposure C	Exposure D
0–15	2.36	1.65	1.32	1.15
20	2.20	1.59	1.29	1.14
25	2.09	1.54	1.27	1.13
30	2.01	1.51	1.26	1.12
40	1.88	1.46	1.23	1.11
50	1.79	1.42	1.21	1.10
60	1.73	1.39	1.20	1.09
70	1.67	1.36	1.19	1.08
80	1.63	1.34	1.18	1.08
90	1.59	1.32	1.17	1.07
100	1.56	1.31	1.16	1.07
120	1.50	1.28	1.15	1.06
140	1.46	1.26	1.14	1.05
160	1.43	1.24	1.13	1.05
180	1.40	1.23	1.12	1.04
200	1.37	1.21	1.11	1.04
250	1.32	1.19	1.10	1.03
300	1.28	1.16	1.09	1.02
350	1.25	1.15	1.08	1.02
400	1.22	1.13	1.07	1.01
450	1.20	1.12	1.06	1.01
500	1.18	1.11	1.06	1.00

For main wind-force resisting systems, use building or structure height $h = z$. Linear interpolation is acceptable for intermediate values of z. Value of gust response factor should not be less than 1.O.

Gust response factor G_z is derived as follows: $G_z = 0.65 + 3.65T_z$, where $T_z = [2.35(D_0)^{1/2}]/[(z \times 130)^{1\alpha}]$; D_0 is the surface drag coefficient that varies according to type of terrain (urban: $1/\alpha = 1/3$, $D_0 = 0.025$; suburban: $1/\alpha = 1/4.5$, $D_0 = 0.01$; open country: $1/\alpha = 1/7$, $D_0 = 0.005$).

The gust response factor, G_z, at elevation z above the ground can be found directly from Table 1.8 as a function of exposures A, B, C, and D. Exposure coefficients are used in connection with building cladding, not for main structural features. Notice that z is the height (in ft) above ground level. A gust response factor denoted G_h is used for structural members constituting the mainframe which resists wind forces (h denotes the height of the structure). Therefore, there is only one value of G_h calculated for a given structure. The values of G_h

given in Table 1.8 can be safely used for buildings that have a fundamental period of vibration ≥1 Hz because dynamic amplification can be neglected at high frequencies (since in these conditions gust energy is minimal).

Notice that the above-mentioned gust response factors only apply to rigid structures. The gust response factor for flexible structures (G) has a different character and is obtained following a different procedure. G for flexible structures which are subjected to dynamic amplification is found by means of two different formulas (one for buildings and another for an open framework). For buildings,

$$G = 0.65 + [P/\beta + ((3.32T_1)^2 S)/(1 + 0.002c)]^{1/2}.$$

For an open framework,

$$G = 0.65 + [1.25P/\beta + ((3.32T_z)^2 S)/(1 + 0.001c)]^{1/2},$$

where

$$P = f J Y,$$

and

$$f = (10.5 f h)/(s^V),$$

f is the fundamental period of vibration of the structure (in Hz); h is the height of the structure; V is the design wind speed (mph); and s is found as follows:

Exposure	s	λ
A (urban)	1.46	$8.20/h$
B (suburban)	1.33	$3.28/h$
C (open country)	1.00	$0.23/h$

Other values of f can be found in Standards Association of Australia (1990) for three different types of structures according to Jeary and Ellis (1983), where the symbol λ is used in place of f (see below).

Example. The following sample problem is presented to illustrate calculation of mean gust response factor $\overline{G}$. Given are basic design wind speed $V = 90$ mph, type of exposure = B, building height h = 600 ft, building width c = 100 ft, building fundamental natural frequency $f = 0.15$ Hz, and building damping coefficient = 0.02.

The calculations are as follows:

$$\bar{f} = \frac{10.5 \times 0.15 \times 600}{1.33' \times 90} = 7.89,$$

$$\frac{c}{h} = 0.166,$$

$$\gamma = \frac{3.28}{600} = 0.00547,$$

$$J = 0.0105,$$

$$P = 0.00795,$$

$$T_1 = 0.13,$$

$$S = 0.78,$$

$$\overline{G} = 0.65 + \left(\frac{0.00795}{0.02} + (3.32 \times 0.13)^2 \times \frac{0.78}{1 + 0.002 \times 100}\right)^{1/2},$$
$$= 1.37,$$

where J is the pressure profile factor, which is a function of λ and is found from Figure 1.8; resonance factor Y is found from Figure 1.9; c is the width of the structure; S is the structure size factor, which is found from Figure 1.10 as a function of h and exposure; B is the damping coefficient of the structure (B = 0.01 for steel and 0.02 for concrete); and $T_z = 2.35(D_0)^{1/2}/(z/30)^{1/\alpha}$. Other values of $\bar{f}$ can be found in the SAA code for three different types of structures (according to Jeary and Ellis 1983), where the symbol n is used in place of f (see section 6, above).

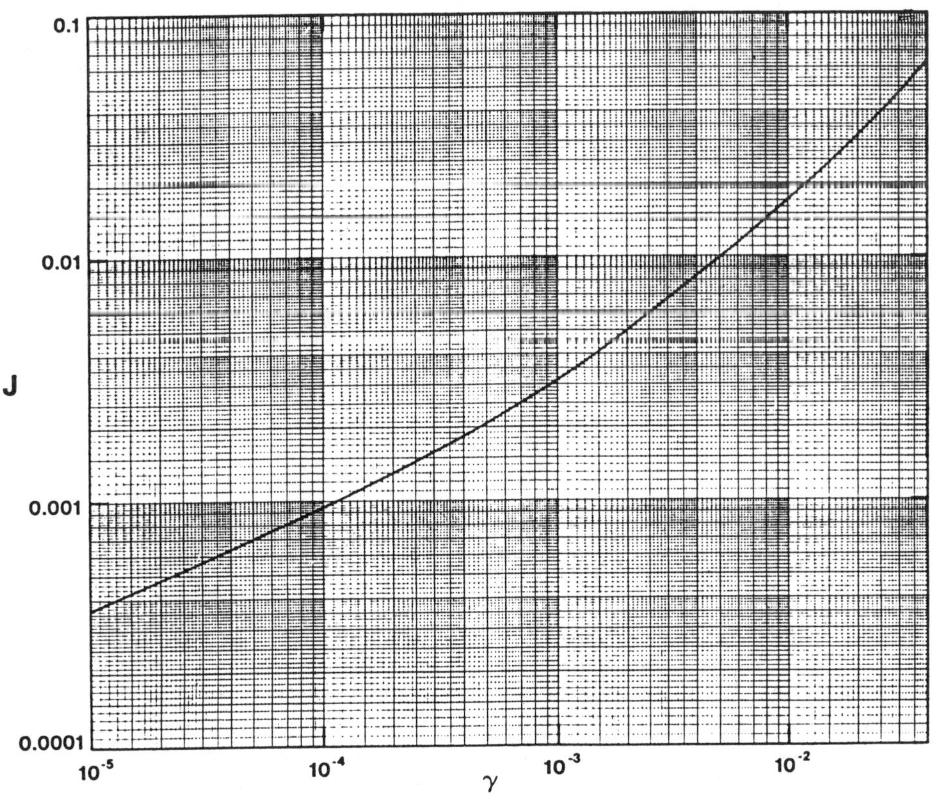

Figure 1.8. Pressure profile factor J as a function of γ.

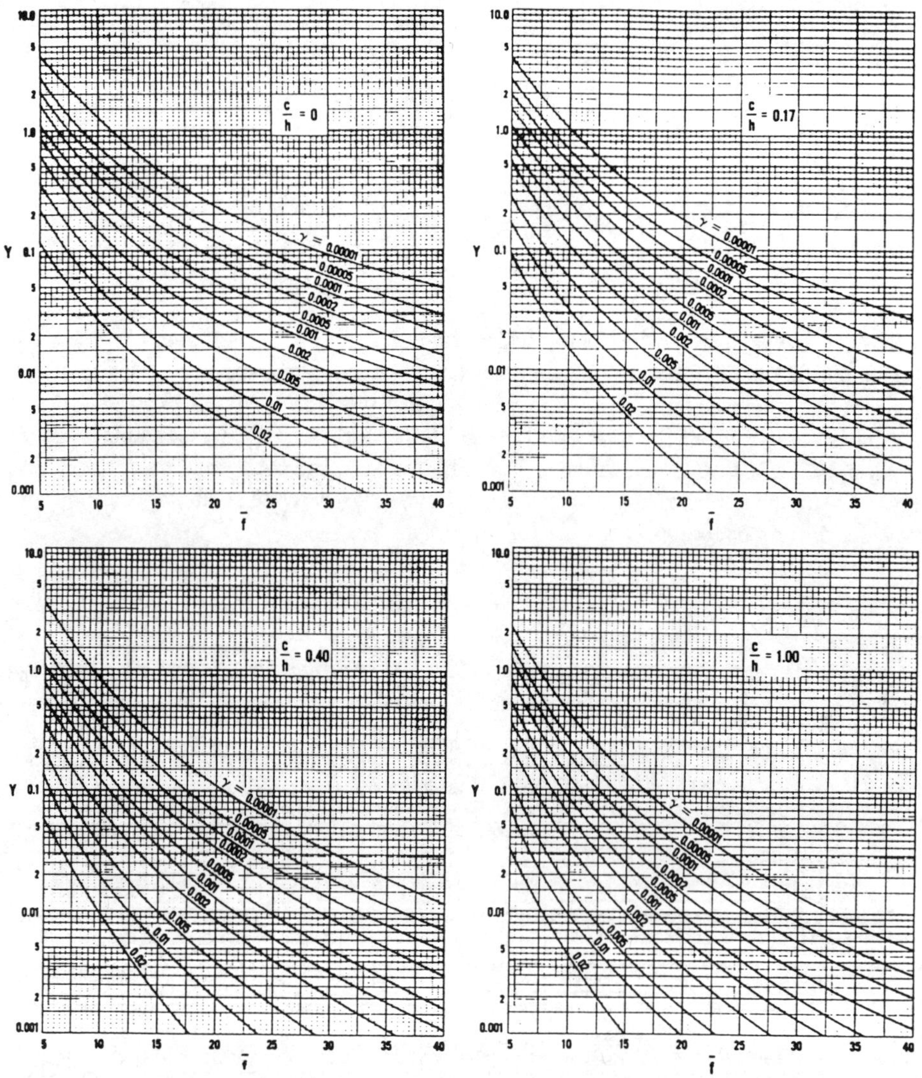

Figure 1.9. Resonance factor *Y* as a function of γ and ratio *c*/*h*.

11. BOUNDARY LAYER WIND TUNNELS

Wind tunnel apparatuses typically used for aircraft aerodynamics reached a high level of development by the time of World War II and were occasionally used for application to civil engineering structures, as in the case of testing conducted in 1933 on a model of the Empire State Building. The requirements for buildings undergoing tests for wind analysis do not coincide with those for aircraft. The basic differences between the two involve the condition of the airstream, which for buildings involves a substantial amount of turbulence and for aircraft a minimum amount of turbulence. The turbulence required in building tests is a reflection of the existing turbulent flow that occurs in the field in upwind regions due to

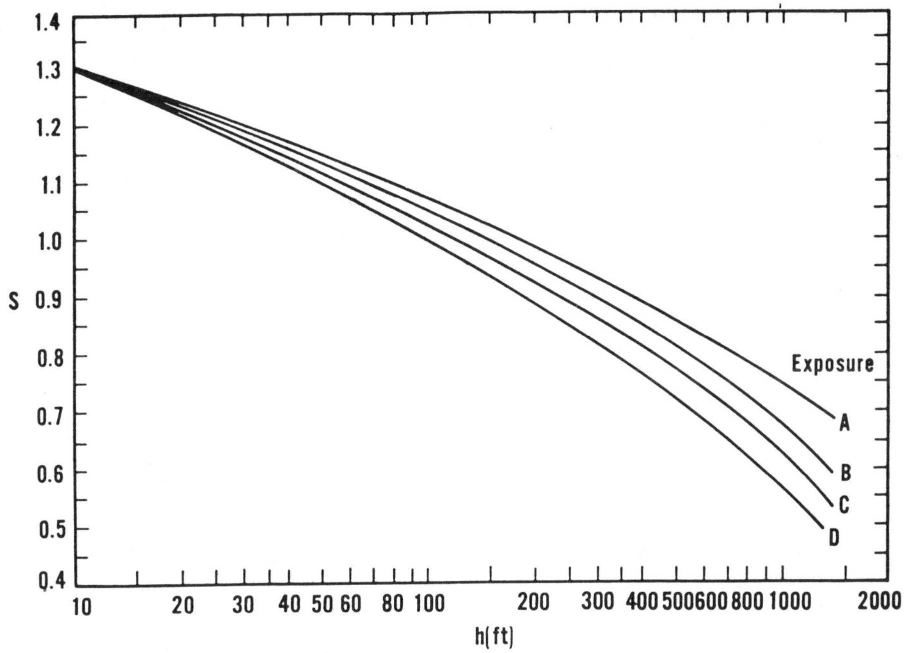

Figure 1.10. Structure size factor *S* as a function of *h* and exposure.

friction between the roughness of the ground and the free flow of the wind. In such regions, not only the roughness of the terrain but the presence of structures and buildings of different heights can produce particularly intense turbulent flows, which need to be duplicated in the testing apparatus. When testing buildings, it is necessary to reproduce an inappropriate scale (the so-called boundary layer), which represents the portion of the atmosphere where the wind flow is affected by the presence of the ground, the presence of the building in question, and the presence of other buildings not yet built. Called boundary layer wind tunnels (BLWTs) because of such conditions, these apparatuses are much larger than those used for aeronautical aerodynamics so that they can generate the necessary conditions to test the strength of a building.

BLWTs require working sections with a length varying between 50 and 100 ft, a width varying between 6.5 and 13 ft, and a height varying between 6.5 and 10 ft. The optimal length/height ratio is between 5 and 15; the optimal width/height ratio varies between 1 and 3. For short tunnels with lengths not larger than five times the height, it is still possible to simulate a boundary layer by adding screens or grids that will produce equivalent turbulence. Working chambers with large widths are desirable so that other structures lateral to the one in question can be included to create the most accurate replica of the existing on-site conditions. BLWT apparatuses are characterized by low-velocity airflows (compared to those used for aircraft testing), and they are built as either closed circuit and open circuit types, of which the latter is the most popular. The wind speed for BLWT testing is selected with the assumption that the Reynolds number (Re) is equal to or greater than 500. This condition is usually attained when the mean wind speed (U_g) is not less than 33 ft/s (or about 22.5 mph).

There are two different kinds of modeling techniques for wind tunnel testing in BLWT apparatuses. One uses a rigid building which will not deflect under loading. These can be subdivided into static rigid, rotating rigid, and spring-mounted rigid models. The other uses aeroelastic building models, which do deflect under loading, and can be subdivided into static aeroelastic and dynamic aeroelastic types.

Rigid models employ non-deformable materials. Various points of special interest are identified on the surface, and the pressure at such points is gauged by means of pressure taps. Aeroelastic models use buildings which deflect under the action of a steady flow and oscillate under the action of typical aerodynamic mechanisms. Peak and mean values of fluctuating moments, as well as values of deflections and accelerations of the models, are measured by the use of strain gauges and accelerometers mounted at specific points on the model surface.

While determination of positive or negative wind pressures with a rigid building model does not require further attention, testing of aeroelastic building models does require additional discussion. Aeroelastic models can be schematically subdivided into two groups: a) those that reproduce the swaying response in the direction of the flow and in the transverse direction, and b) those that reproduce the swaying response in the direction of the flow and in torsional mode.

Models are mounted on adjustable springs which control elastic deformation. These springs can be adjusted so that the ratio between the fundamental frequency of vibration of the model and the fundamental period of vibration of the prototype are the same for both longitudinal and transverse oscillations. Several tests are conducted. In each case, the response is kept within 0.5 to 5% of critical damping. The peak and mean values of moments at the base and sway deflections are measured by strain gauges mounted on the springs.

Peak accelerations are measured by accelerometers mounted on top of a model. Vertical torsional motions are reproduced in a model by means of vertically mounted springs. When tall slender buildings do not have constant slenderness (structural or geometrical) throughout their height, a model can be built to represent such changes by dividing the model vertically into sections of constant slenderness and by connecting these parts by appropriate springs.

In all cases, the scale to be selected for construction of models to be tested in BLWT apparatuses first depends on the size of the working chamber and other specific parameters: a) the ratio of the boundary layer height and the height of the building must be equal for both the prototype and the models; b) in considering cross-sectioning of the model and that of the working chamber, the area of the former should not exceed 10% of the area of the latter. Since the gradient height, the height of the building, and the height of the working chamber are given, the height of the model cannot be too flexible. On the other hand, the restrictions mentioned under parameter b can present problems in terms of the width of the working chamber and the width of the area surrounding the building needed for rendering a true representation of the field.

Other restrictions that relate the characteristics of a model to that of a prototype are expressed as follows:

$$m_1 D_{\text{model}} = D_{\text{prototype}},$$

$$m_{-361}{}^{2} \Delta_{\text{model}} = \Delta_{\text{prototype}},$$

$$m_2 E_{\text{model}} = E_{\text{prototype}},$$

$$e_{\text{model}} = e_{\text{prototype}},$$

$$\sqrt{m_2} \cdot V_{\text{model}} = V_{\text{prototype}},$$

$$\frac{\sqrt{m_2}}{m_1} n_{\text{model}} = n_{\text{prototype}},$$

$$\delta_{\text{model}} = \delta_{\text{prototype}},$$

Δ is area, D is the typical major length, V is the air velocity, m_1 and m_2 are proportionality factors, n is the natural frequency of vibration, e is damping, and δ is density.

Stresses and moments can be derived from the following relationship:

$$m_2 \sigma_{\text{model}} = \sigma_{\text{prototype}}$$
$$m_1 d_{\text{model}} = d_{\text{prototype}}.$$

Wind tunnel tests on models can be related to the actual condition of structures exposed to the wind if the following relationship between the model and prototype of the structure is respected:

$$\text{Re}_{\text{model}} = \text{Re}_{\text{prototype}}.$$

Testing Technology

Pressure taps are placed at specific points in testing pressure values on the surface of rigid models. These consist of 1 mm in diameter holes connected by metal tubes inserted from inside the surface, which in turn are connected to flexible plastic tubes attached to a rotary pressure switch. The tubing from tap to pressure switch must be of equal length to provide for an equal frequency response. Starting from the pressure switch, the tubes reach one side of a differential pressure transducer, which registers the pulsating pressures from each of them. The other side of the transducer is connected to a tap placed away from the model in an area of the working chamber where the airflow is steady and the pressure is constant. This is used as the reference pressure. The transducer compares each measurement of pulsating pressures coming from individual tubes with the value of the reference pressure. Each comparative measurement is translated into an electrical voltage measurement, which is recorded by a data-processing system. Several taps within a given zone of a model are sometimes grouped together in order to obtain an average pressure, called zonal "pneumatic" averaging.

12. BUILDING ENVELOPES

The structural components of a building have generally been identified by the essential parts necessary to maintain its stability as a whole. All buildings include vertical and horizontal components, whether of skeletal frame type or masonry weight-bearing wall construction. In the first case, whether the frame is steel, reinforced concrete, or even timber, the frame includes vertical columns and horizontal beams, girders, trusses, etc., constituting the primary structure. Other members constitute the secondary structures (e.g., joists, purlins).

The building envelope of framed structures has not been considered part of the primary frame, since portions of it can be removed without endangering overall stability.

In the second case, the bearing walls are not just part of the building envelope but constitute part of the primary structure and are thus treated as such. With such a typical "de facto" attitude, many wall envelopes that enclose framed buildings are not considered structural entities in structural engineering design work, which presents a paradox in the design sequence. Omitting a detailed structural design for building facades has left the design and detail of this important building component to designers mainly concerned with construction details which are nonstructural in nature.

Such a practice can be easily understood. For centuries, the dominance of masonry structures has meant enclosures consisting of bearing masonry walls with inherent structural strength not only for vertical gravity loads but also for sheer resistance to horizontal wind forces. With the advent of present-day skeletal building frames, the exterior envelope can include lighter and lighter paneling systems with less inherent strength which means a more fragility when glazing surfaces dominate the cladding envelope.

With respect to glazed surfaces, both glass panels and mullions have demonstrated vulnerability to wind forces, as could have easily been predicted. Wind gusts as well as sustained wind pressure in many cases have quite often overcome the ultimate strength of the glass panels themselves or of the connecting devices holding the glass, as demonstrated by hurricane damage. The missile impact of flying debris during severe storms has also been a major culprit in damage to glazed surfaces.

The problem related to the structural characteristics of building enclosures thus comes down not to transferral of loads but to sustaining them without violating structural integrity. Comparative analysis of damage done to building envelopes by severe storms has allowed for accumulation of a substantial amount of data for different kinds of cladding systems.

To this effect, we now consider the individual components typically employed in present-day building construction.

Structural Glazing

The inherent vulnerability of glass under static, and even more so under dynamic, loading is a major concern in the design of building structures which include large glass surfaces. The problem of the structural strength of glass has increased in relevance with the increased use of glazed surfaces in twentieth-century architecture. The basic mechanical properties of typical glass material, although well known, indicate some fundamental peculiarities of this ancient material which help to interpret the behavior of structural glazing in its fenestration applications. To begin with, one notices the substantial difference in strength that glass exhibits in compression and tension — ultimate compression stresses range between 90,000 and 180,000 psi, while ultimate tensile stresses vary much more widely: between 4,000 and 1,500,000 psi — which creates somewhat of a confused notion as to the significance of these values. When conducting tensile tests, ultimate stresses increase considerably as specimen cross-section decreases. What this means in reality is that, as the cross-section gets smaller, the probability of occurrence of flaws in the material is drastically reduced, and that its strength increases proportionally because it is through such flaws that fractures occur.

Additional structural factors which describe the mechanical characteristics of glass include the elasticity modulus (equal to 6,500,000 psi) and Poisson's ratio (ranging from 0.14 to 0.271, with a typical value of 0.22 for ordinary glass). A major increase in strength can be obtained for glass through thermal treatment, which consists of sudden cooling of the material through a precisely controlled process, rather than the conventional annealing procedure. This results in a product that has a strength on the order of eight times greater than that of untreated glass. The improvement of the structural strength of glass through present-day manufacturing technology has caused new products to emerge for practical application in buildings exposed to high-velocity wind forces and missile impacts.

Laminated glass panels have substantially greater structural strength than conventional glass. Laminated glass consists of resin sandwiched between layers of glass subjected to heat and pressure until it constitutes a monolithic structure. This material is the best choice for glazing in fenestration of building enclosures exposed to high-velocity wind effects.

Recent tests on glass products conducted at the University of Missouri at Rolla drew the following significant conclusions about glazing products:

a. The behavior of heat-strengthened laminated and fully tempered laminated glass demonstrate adequate resistance against the impact of relatively light missiles, so that, although fracture of impacted lamina may occur, the backing layer will remain intact.

b. The performance of laminated glass products with a silicone anchor bead sandwiched in between indicates that such material provides greater strength than mechanically fastened dry laminated glass plys.

c. Heat-strengthened laminated glass with a polyvinyl butyryl polymer/polyethylene terephthalate (PBV/PET) interlayer and a silicone anchor bead interlayer exhibit the highest strength under the most critical loading conditions (involving 190 high-velocity missile impacts and more than 1000 pressure/suction cycles).

These tests obviously confirmed the superior strength of heat-strengthened and fully tempered glass over typical annealed glass under impact loads and demonstrated that laminated glass has greater strength than commonly assumed under applied wind pressure. In the sandwiched panels, the outside layer of glass (the one that will potentially fracture), which receives the impact, is the one that absorbs all the kinetic energy of the missile, while the inner panels behind it usually survive undamaged.

In examining the overall behavior of laminated glass, we can observe the individual contributions of the two layers. On one hand, the structural interaction obtained by their bondage through the PVB interlayer causes monolithic structural behavior; on the other hand, the outer layer receiving the missile impact can sometimes spare the interlayer by itself absorbing the total energy without transferring it to the layer behind due to the separation that the interlayer affords.

The kinetic energy of an impact is sometimes greater than what the outer layer can absorb, and part of it is transferred to the inner layer, causing it to break. Yet even in this case when both layers break, they usually remain attached to the PVB lamina interposed between them. The PVB lamina bonding outer and interlayer together keep the fractured layers in place with great tenacity, thus preserving the integrity of the glazing. Tests have shown that, when fracture on both layers occurs and the two remain bonded to the flexible interlayer, the whole panel can deflect in and out by several inches and retain its overall integrity due to the resiliency of the PVB interlayer. Such behavior is of paramount importance, since it will prevent wind and rain from penetrating into the interior spaces, thus avoiding the severe consequences described above. The importance of the PVB interlayer is further emphasized by the fact that, the thicker it is, the more effective it is in protecting the inner glass panel layer: when a 0.032-inch inner layer was used in a sandwich panel, the chance of breakage of the inner panel was 52%; when the thickness of the interlayer was increased to 0.060, the chance of breakage was reduced to 7%.

Laboratory tests were conducted using 58-in × 98-in panels that were 3/8 of an inch thick and consisted of heat-strengthened, fully tempered, and film-coated annealed monolithic glass. There were two types of interlayers: a 0.060-inch layer of polyvinyl butyryl (PVB) polymer, and a combination of two 0.0015-inch layers of PVB sandwiching a 0.003-inch layer of fibrous polymer polyethylene terephthalate (PET).

Glass panels were mounted on frames during tests so as to simulate typical mounting of glass on windows. To this effect, the American Society of Testing Materials (ASTM) specifications were adapted, and panels were mounted in three ways:

a. Mounting was done employing a dry process using a gasket around the glass panel to secure it to the window frame.

b. Mounting was executed by interposing a 3/4-inch wide bead of silicone between the glass and the window frame, then adding and placing parametrial batten over the edges on the inside face, holding the glass securely in place even if fracture occurred.

c. Mounting was executed similar to as in case b, except that the width of the silicone bead was reduced from 3/4 to 3/8 inch.

The main intent of this test was to duplicate typical conditions occurring during severe wind storms. The effect of gravel carried aloft by wind and impacting against glazed window panels was simulated by blowing particles weighing 2.03 g (usually found on built-up roof systems) against the glass. The same panels were then subjected to 1096 alternate repetitive cycles of positive and negative pressure, simulating the typical process which occurs during a 4-hr storm and reflecting a storm intensity which is likely to reoccur every 100 years. The wind pressure applied to the panels reflected speeds up to 80 mph, while the alternate pressure and suction values were 45, 60, and 75 psf.

Protective Shutters

The full exposure of glass surfaces in the fenestration of building facades in the North American architecture is a phenomenon which generates a sense of wonder to visitors unused to such a cityscape. The absolute absence of traditional building components associated with the concept of fenestration in the European tradition does not seem to have an explicable rationale, whether applied to commercial or even residential buildings. However, while certain contemporary aspects of commercial architecture may indicate a clear rejection of such building components, some decorative expression in the residential architecture of United States and Canada at times includes imitation of shutters on both sides of windows, with the intent of creating an ornamental motif, often without even attempting to simulate the ability to function or operate as one would expect. In fact, often displaying a size incompatible with the window opening, shutters are nailed to the siding without any obvious intent except to appear as simple decoration. Thus, the puzzle remains as to why shutters have not been adopted for their basic use, in spite of the knowledge of their usefulness and of the demonstrated consideration explicitly displayed.

Yet, despite the existing status of the fenestration of buildings in residential and commercial applications in present-day architecture, a reevaluation of the utilitarian aspects of shutters calls for immediate attention in terms of the significant benefits that this building component could provide. Focal interest in the present context merges on some specific effects that shuttering devices could project in building technology with respect to protection of glazed areas on building facades against damage of wind loads. Whether for direct wind pressure or airborne missile impacts, most shuttering devices could offer brilliant solutions not only for these problems per se, but also in association with many other design factors related to window protection. The most significant factors include the following:

a. Protection against the vandalization and breaking-and-entering which has become a critical issue in the present-day social climate of many urban and suburban cultures.

b. An additional sound barrier to muffle penetration of sound, especially in noisy congested areas.

c. A thermal barrier against winter heat losses by providing thermal insulation, and an additional thermal barrier formed by the air space generated between closed shutters and window glazing.

d. A thermal barrier against heat gains by providing a sunlight screen against direct radiation.

e. Additional protective action for window frames and glazing against wind and rain throughout the life cycle of the window itself, minimizing maintenance (including painting and glass washing).

Structural Shutters

Through its long history, shutters have received more attention in the building technology of this century than at any other time. The diversification in construction of the new industrialized building methodologies can be accounted for by the innovations in the new generation of shutters currently available on the market. Shutters that could conceivably be incorporated in design could include almost any type of old or new prototype, because any traditional model from the past can easily be fabricated, in addition to the shutters currently in production. Traditional wood shutters of various configurations are low-technology building products that can be fabricated or handcrafted at reasonable cost. Thus, there is no practical limit to implementation of shutters not only for new construction but also for retrofitting traditionally designed buildings that require special configurations to match or be compatible with the architecture of the structure.

Traditional shutters are made of wood and consist of two panels hinged on either side of the window unit. Rotating around a vertical axis, the panels operate between closed and opened positions and remain fixed in the desired state by means of special hardware. When closed, the two panels remain securely fastened by means of a latching mechanism operable from the inside, so that full protection can be attained against intruders attempting to open them from the outside.

Even the hardware used to keep the panels open in a flat position along the facade walls on each side of the window merits some special attention as an architectural detail. The wood panel can display various configurations of fitting the window by acting as solid panels impermeable to the sun's radiation, air, or rain penetration without any possibility of adjustment, or by acting as adjustable panels that can regulate the passage of light and air to different degrees. In the latter case, a typical panel configuration involves using louvers usually mounted in a fixed position. Some adjustment can be achieved by keeping the two panels totally shut or partially opened in an intermediate position, but still loosely attached. Some louvered shutter panels can be operated by means of sections usually placed in the lower part of panels which are hinged horizontally on top, allowing the section to swing outward to different degrees, allowing more air to circulate and permitting an occupant to look downward and out while the overall body of the shutters is closed.

An option to the vertical hinged type are horizontally sliding shutters that can be opened by retracting them into a slot in the wall on both sides of the windows. The configuration of the individual panels with this type of operating shutters can be solid or louvered, and can include any other mechanism for partial control. The importance of this type is that it represents an alternative opening and closing operation to that used by the vertically hinged type. However, we should note that horizontally sliding retractable shutters are not easily retrofitted.

The shutter types discussed above represent traditional methodologies which have centuries of practical application that differentiate themselves from the vertically rolling types currently available with new technology, and which can be coincident from a different point of view in the context of architectural design. These shutters can be considered an historical evolution of the past or as a functional construction device for protection of windows that also reflects the personal choice of an architect. In either case, the character of the artifact can vary substantially not only in terms of mechanical specifications for construction and operation but also for all other design factors dealing with form, color, finish, etc. From the crafted and colorful wood shutters typical of the vernacular architecture of the Alpine regions to those mounted on the facades of Italian *palazzi*, the gamut of possible variations leaves quite a wide degree of freedom of expression for the creative mind.

When vertically rolling type shutters eventually appeared around the turn of the century, a new generation of shutters eventually emerged which offered a more versatile system and became the focus of attention for the new industrialized modern technology. This latest type operates by sliding the panel vertically through U-shaped metal guiding profiles on each side of the window. The panel, consisting of individual horizontal slots hinged to one another, opens by sliding vertically upward, rolling over a horizontal axis mounted at the top of the window. In the open position, the articulated panel remains rolled in an enclosed overhead compartment, accessible only from the inside. Opening and closing can take place by rotating the horizontal axis by means of a manually operated side-mounted strap. The strap mounted on the side of the window lies flat against the wall and retracts automatically by means of a self-winding mechanism. More recent applications include electrical mechanisms (replacing mechanical ones), available for commercial use for wide shutters as well as for any other conventional shutter type if so required by the specifications, even for small residential types.

The original old models of rolling type shutters were fabricated from wood slats and employed different species of wood, especially those most resistant to weathering (e.g., chestnut) because of constant exposure. Less expensive species have been used for the sake of economy, including pine. The most modern types have expanded the spectrum of materials used to include plastic (PVCs) and aluminum. These materials allow for construction of tubular cross-sectional shapes fabricated by extrusion. A large variety of configurations is obtainable using such a process. In obtaining different results, a tubular cross-section can be designed to increase its moment of inertia to obtain higher bending resistance to pressure loads. However, enlarging the cavity area through proper design of the cross-section can solve other problems as well (e.g., increased thermal resistance within the shutter body). Individual tubular slats can create a thermal barrier by enclosing a dead air space in the cavity or by filling the cavity with insulation (e.g., styrofoam or similar products). The most prominent advantage of the latest models of rolling shutters is that these two materials allow one to obtain almost any desired shape with ease, so that both structural strength and thermal insulation can be substantially increased at limited costs, considering the major advantages they offer. Furthermore, plastics and aluminum are relatively much less maintenance-intensive, and their life span is definitely much longer than that of the corresponding wood counterparts. Even in terms of unit weight per square foot of shutter, the new models are much lighter and much easier to operate, especially with manual operation.

The old as well as the new rolling type shutters guarantee flexible operation by adjusting the shutter position either in a tightly closed position or in an intermediate one. In the former, the slots are engaged by one another in by trough and grove interlocking, which occurs when the roller is allowed to reach the bottom. The weight of the slots keeps them engaged downward into the groove of the slot below. In such a closed and interlocked position, the roller becomes a monolithic panel which prevents penetration of light or air and is also self-protective against the action of intruders (who may try to compromise shutter integrity even with intrusive tools). In such a position security as well as such functions as sun radiation reflection, light blockage, sound barrier, and thermal barrier are guaranteed, On the other hand, when the shutter is suspended in an intermediate position, slots are separated from one another at a certain spacing established during the manufacturing process. With approximately a 1/4-inch spacing, a certain degree of light penetration is allowed, filtering external daylight by a desired percentage, which creates an adjustment of interior daylight illumination not obtainable with other devices. What can be said about light control also applies to of outdoor air for natural ventilation of the interior space, adding quite a number of additional environmental control advantages to the main one: protection against wind forces.

On the basis of the present state of affairs, in which window shutters are an old and firmly entrenched tradition in the architecture of Europe and of other parts of the world but completely rejected in North American traditions, it is obvious that the problem of implementing shutters in the building technology of countries like the United States and Canada is just a matter of cultural adjustment. Therefore, defending implementation of shutters on the technological level is definitely an incorrect approach which would not lead to any meaningful achievement. The validity of shutters for protection against the effects of high-velocity winds is self-evident without and does not require further proof. Similarly, the major advantages of control of sound, control of light penetration, and provision of thermal comfort provided by shutters are also evident in European architecture and well known in North American design structures. The problem is not so much that of communication of information, but rather one of opening the eyes of responsible designers to the obvious. In fact, the obstinate position of depending on glass alone for protection against high-velocity wind effects on the fenestration of building facades is hard to sustain when one of the most instinctive protective actions usually taken by laymen in preparation of an incoming hurricane is that of boarding up glass with plywood at the last minute. An adequate design of new fenestration which includes shutters in an integrated fashion together with other fundamental modern requirements of windows is definitely what logic demands.

The New Window

The use of particular building components is strongly related to individual architectural styles — a trend which can be seen throughout history. Each historical period displays a typical character of the local architecture, which reflects an expression of the specific culture of that part of the world. Among other components, building facades display quite distinct morphological delineations of particular details which identify the exact point in time when the building was constructed with astonishing precision. Window units, for instance, stand out as major indices of the vocabulary of the mannerism of the times. They mechanically complete the fenestration carved out from the continuity of the heavy masonry walls of articulated facades. In their mechanical structure, the operable window components are primarily designed to hold the glass panes, considered the most valuable part of a window throughout the centuries prior to present-day glazing inexpensive production.

It is interesting to note that the essence of the window has not really changed that much over the long course of architectural development. In fact, while structural fenestration has undergone a large number of mutations in terms of details and basic configuration, the design of window units has remained relatively constant. They have remained nothing more than flat operable panels in wooden frames, operating either by rotation along the vertical axis or sliding vertically into grooves, retaining essentially the same construction concept up to the present. The present state of development, still rooted in old traditions, is not really difficult to understand if one considers it within the context of a building technology that has always been conservative and resistant to new technologies. Only the strong input of industrialization seems to have been an effective force for change, mostly due to economic factors, while pure rationalization seems to have had little effect.

In this context, a new interpretation of the window is needed to respond logically to the needs of the present and at the same time utilize the benefits of new technologies. The new window demanded by the present cannot be restricted by the limits of old concepts. The new window should emerge from optimization of the several tasks that the building demands, which basically include the following: afford transparency that allows a view of the exterior and light penetration; provide surface integrity against air and rain penetration; provide a surface (shutters) over the glazing which protects against high-velocity wind pressure and missile impacts; act as thermal insulation to reduce heat losses; reflect sun radiation during summer to reduce heat gains; screen for partial ventilation; provide a sound barrier to control penetration of exterior noises; and form a dependable barrier against vandalism and potential intruders. As we have seen, some of the devices associated with windows perform relatively well, but not all simultaneously. Thus, the most important point in this context is need to design windows which respond comprehensively and efficiently to the above-mentioned demands.

The windows of today somehow respond to the many demands through individual action without a fully integrated design by spontaneous adjustment to environmental demands. Shutters are still much used in many countries to perform specific tasks. Partial remedies are employed in some countries where building technologies have abandoned their use. The protective action of shutters against intruders coming through windows can be provided by an effective system of law enforcement (which in the past was somewhat able to guarantee safety). Protection from wind pressure and missile impacts against glass are usually obtained by boarding up windows using temporary plywood shields prior to an expected incoming severe storm (hurricane, tornado). Shading against sun radiation is sometimes obtained through exterior awnings, while light penetration is often controlled by adjustable interior blinds. The amount of light can be controlled with expensive draperies which, at a substantial cost, can also provide attenuation of sound as well as enrichment of interior decor. When we combine all these components, it is obvious that the individual components are not acting in unison or reinforcing each other. Most important, the position of the architect in this context is unclear, since no logical response seems to emerge.

While no specific solutions are offered in this context (since they would be logically out of place), the only recommendation is to emphasize the need for a drastic departure from conventional stereotypes. The window can no longer be considered as just a flat structure which glazes a gap cut out of the facade of a building envelope. The combined functions that windows are expected to perform demand not a surface but a volumetric structure that contains several components capable of operating and regulating performance to various degrees. The window of the future could even add other complementary tasks, such as providing incremental daylight penetration through reflective components, further defusing or refracting light, acting as a greenhouse, alternating window configuration between day and night hours, etc., incorporating all into one integrated composition. To accomplish this, however, we should not necessarily concentrate on high-technology solutions. Emphasis should be placed on the opposite approach of maximizing the use of passive systems implemented through relatively simple technologies. Nor should the concept in question be interpreted as a forced conglomeration of tasks just for the sake of compacting more and more functions into a single package. The new windows advocated in this context should be regarded as a general problem to be solved within the rules of design that have always formed the basis of any meaningful architecture.

13. MAPPING THE WIND

When dealing with airflow through the inner spaces between buildings, the most common problem is predicting the direction and magnitude of the flow before the project is realized. The designer of the outer spaces needs to know the final effect of the siting of buildings in a project from its inception. Wind modeling by various processes is the approach conventionally followed to this effect. De facto verification after construction is a practice aimed at assessing actual wind flows and devising retrofitting strategies for modifications and improvements. The best term to describe this operation is "wind mapping." Of course, wind mapping of existing sites has such a wide range of applications that in reality it does not take second place to wind analysis through modeling. In fact, the existing built-up environment around the world, including the agglomerations of buildings within urban centers and isolated rural zones, constitutes a vast field of operation for wind mapping.

Wind mapping problems in architectural education arise not only in environmental control courses within the area of building sciences and technologies but also within the site analysis studio context. Formulation of basic problems within this scope includes some fundamental points which are briefly discussed here.

An existing site which accommodates several clusters of building masses that would be readily available for a classroom assignment is the university campus where the class is conducted. Complex inner spaces between buildings, through which the campus circulation develops via an articulated pattern of walks and landscape areas, often constitute ideal conditions for a classroom project involving several groups of students. The class can be divided into groups of two. Each group is assigned wind mapping of a certain area. Wind mapping of the entire campus can be accomplished by combining the various assignments. At the same time, the compatibility between the work of different groups will verify the correctness of the individual maps.

Furthermore, it would be particularly productive to combine work on an individual basis (two-student teams) with work done on the communal basis where the entire class participates.

A site map of the campus can be subdivided among the groups, which eventually enlarge individual maps to a scale compatible with field operations. Each group is equipped with a small hand anemometer, a small weather vane, and a stopwatch. The small anemometer (with one or more dials) reads the length (in meters or feet) of a column of air passing through the blades of the anemometer. The switch allows the blades to rotate freely, and the same switch blocks them. The stopwatch measures the time elapsed between turning the switch on and off. Dividing the length measured by the time measured on the stopwatch, the wind speed is calculated in terms of an average over a preestablished time period. Prior to measuring the speed, the team must identify the exact location where measurements are taken on the map. The direction at a specific point on the map is recorded with a small arrow using any kind of wind direction indicator (e.g., holding a light piece of string). At the end of the field operation, a substantial number of map points are marked which indicate wind speed with an approximate tolerance and wind direction indicated by an arrow.

Elaborating field data on a drawing board, additional wind speed values at other points can be identified by interpolation. The direction of the wind flow obtained from the arrows will identify the streamlines. When these converge, becoming closely spaced, a velocity increase is to be expected; vice versa, as they become more widely spaced, wind velocity is decreased. Zones of turbulence can also be mapped by observing the velocity and randomness of wind patterns. At the beginning and end of the surveying operation, the direction of free wind above building heights should be measured so that the survey and the free wind direction can be correlated.

Repeating the survey for the same location under various conditions allows for comparative analysis. By adding specific wind barriers at strategic locations to obtain increased or decreased wind conditions, it is possible to evaluate their effects by comparing the different surveys. The study of effective windbreakers to mitigate wind speed in winter and the effect of wind-driven rain is a typical problem with wide practical application. Similarly, experimenting with baffles to increase the speed of summer breezes in breezeways can be effectively measured only with wind surveys of this sort. The instructor, however, should not have any difficulty in creating specific conditions to interact with the design features of existing buildings to improve the human comfort provided by any outdoor space.

The significance of such exercises in an undergraduate curriculum needs to be considered as a first step toward incorporating wind effects in the overall design of buildings and outdoor spaces in a de facto situation, where wind effects have usually been omitted. Students in undergraduate civil engineering programs have been exposed mainly to wind forces in terms of static loads, usually given as uniform pressures applied to buildings within the context of structural design and lateral loads. Undergraduates in architecture and landscape architecture programs are barely familiar with wind effects as structural or environmental design factors, so that introduction of basic wind design concepts need to be addressed to keep abreast with present-day design. Starting with the average wind speed values commonly found in conventional climatology and expanding sequentially to high-velocity values, wind effects need to be presented in design education.

CHAPTER TWO

Tropical Cyclones, Hurricanes, and Typhoons

1. INTRODUCTION

This chapter focuses on tropical cyclones that form over the oceans. We select those which reach high intensity, typically referred to as hurricanes in the North Atlantic and North Pacific and as typhoons (or simply cyclones) in the West Pacific and Indian oceans. In terms of damage and fatalities, these storms are at the top of hierarchy of atmospheric perturbations (even higher than tornadoes that develop higher-velocity winds). The reason for this is the extraordinary size of the storms, which involves a much larger geographical area than that in the case of tornadoes. Their catastrophic impact on human habitats is expressed by the extraordinary cost of the damage they produce and, more important, the high number of fatalities they cause.

In terms of the casualties they cause, there is an incredible difference between the numbers caused in those which occur in North America and Asia. The difference in number of casualties appears inconceivable if one limits attention to storm intensity itself; however, other factors must be considered, such as the different building technologies used in the two continents. Total annual casualties due to North American hurricanes typically do not exceed 100, with the notable exception of the August–September 1900 hurricane in Galveston, Texas (which killed over 6000 deaths). In contrast, 300,000 died in the November 13, 1970 cyclone in Bangladesh, and 70,000 perished there in the April 30, 1991 cyclone. Astonishingly, a third cyclone in the same area took 10,000 lives on March 25, 1985. Such numbers clearly characterize the unfortunate relationship between wind force strength and construction methods which traditionally focus on gravitational forces alone while considering resistance to wind forces a secondary attribute. Surely, these numbers of fatalities also underscore the relevance of building regulation anywhere in the world.

The threat that these storms pose to the worldwide human habitat can be better understood by studying official statistics (see Table 2.1). The data for 1888–1991 include records of major storms for different geographical areas, with the number deaths caused by each storm. These numbers require some interpretation which reflects other circumstances concurrent

Table 2.1. Strong Hurricanes, Typhoons, and Cyclones, 1888–1991

Date	Location	Deaths	Date	Location	Deaths
1888 Mar 11–14	Blizzard, Eastern US	400	1967 Jul 9	T. *Billie*, SW Japan	347
1900 Aug–Sep	H. Galveston, TX	6,000	1967 Sep 5–23	H. *Beulah*, Carib., Mex., TX	54
1906 Sep 21	H., LA, MS	350	1967 Dec 12–20	Blizzard, SW US	51
1906 Sep 18	T., Hong Kong	10,000	1968 Nov 18–28	T. *Nina*, Philippines	63
1926 Sep 11–22	H. FL, AL	243	1969 Aug 17–18	H. *Camille*, MS, LA	256
1926 Oct 20	H. Cuba	600	1970 Jul 30–	H. *Celia*, Cuba, FL, TX	31
1928 Sep 6–20	H. S FL	1,836	Aug 5		
1930 Sep 3	H., Dom. Rep.	2,000	1970 Aug 20–21	H. *Dorothy*, Martinique	42
1938 Sep 21	H. Long Island, NY, New Eng.	600	1970 Sep 15	T. *Georgia*, Philippines	300
1940 Nov 11–12	Blizzard, NE and Midwest	144	1970 Oct 14	T. *Sening*, Philippines	583
1942 Oct 15–16	H., Bengal, India	40,000	1970 Oct 15	T. *Titang*, Philippines	526
1944 Sep 9–16	H., NC to New Eng.	46	1970 Nov 13	Cyclone, Bangladesh	300,000
1952 Oct 22	T., Philippines	10,300	1971 Aug 1	T. *Rose*, Hong Kong	130
1954 Aug 30	H. *Carol*, NE US	68	1972 Jun 19–29	H. *Agnes*, FL to NY	118
1954 Oct 5–18	H. *Hazel*, E US, Haiti	347	1972 Dec 3	T. *Theresa*, Philippines	169
1955 Aug 7–21	H. *Diane*, E US	400	1973 Jun–Aug	Monsoon rains in India	1,217
1955 Aug 12–13	H. *Connie*, Carolinas, VA, MD	43	1974 Jun 11	Storm *Dinah*, Luzon, Philippines	71
1955 Sep 19	H. *Hilda*, Mexico	200	1974 Jul 11	T. *Gilda*, Japan, S. Korea	108
1955 Sep 22–28	H. *Janet*, Caribbean	500	1974 Sep 19–20	H. *Fifi*, Honduras	2,000
1956 Feb 1–29	Blizzard, Western Europe	1,000	1974 Dec 25	Cyclone leveled Darwin, Aus.	50
1957 Jun 25–30	H. *Audrey*, TX to AL	390	1975 Sep 13–27	H. *Eloise*, Caribbean, NE US	71
1958 Feb 5–16	Blizzard, NE US	171	1976 May 20	T. *Olga*, floods, Philippines	215
1959 Sep 7–19	T. *Sarah*, Japan, S. Korea	2,000	1977 Jul 25, 31	T. *Thelma*, T. *Vera*, Taiwan	39
1959 Sep 26–27	T. *Vera*, Honshu, Japan	4,466	1978 Oct 27	T. *Rita*, Philippines	c. 400
1960 Sep 4–12	H. *Donna*, Caribbean, E US	148	1979 Aug 30–	H. *David*, Caribbean, E US	1,100
1961 Sep 11–14	H. *Carla*, TX	46	Sep 7		
1961 Oct 31	H. *Hattie*, Belize	400	1980 Aug 4–11	H. *Allen*, Caribbean, TX	272
1963 May 28–29	Windstorm, Bangladesh	22,000	1981 Nov 25	T. *Irma*, Luzon, Philippines	176
1963 Oct 4–8	H. *Flora*, Caribbean	6,000	1983 Jun	Monsoon rains in India	900
1964 Oct 4–7	H. *Hilda*, LA, MS, GA	38	1983 Aug 18	H. *Alicia*, S TX	17
1964 Jun 30	T. *Winnie*, N. Philippines	107	1984 Sep 2	T. *Ike*, S Philippines	1,363
1964 Sep 5	T. *Ruby*, Hong Kong and China	735	1985 May 25	Cyclone, Bangladesh	10,000
1965 May 11–12	Windstorm, Bangladesh	17,000	1985 Oct 26–	H. *Juan*, SE US	97
1965 Jun 1-2	Windstorm, Bangladesh	30,000	Nov 6		
1965 Sep 7–12	H. *Betsy*, FL, MS, LA	74	1987 Nov 25	T. *Nina*, Philippines	650
1965 Dec 15	Windstorm, Bangladesh	10,000	1989 Sep 16–22	H. *Hugo*, Caribbean, SE US	504
1966 Jun 4	H. *Alma*, Honduras, SE US	51	1990 May 6–11	Cyclones, SE India	450
1966 Sep 24–30	H. *Inez*, Carib., FL, Mex.	293	1991 Apr 30	Cyclone, Bangladesh	70,000

Names of hurricanes and typhoons in italics. H = hurricane, T = typhoon.

Source: Funk and Wagnall 1993.

with the physical damage caused by high-velocity winds. For instance, in the case of the November 13, 1970 storm in Bangladesh, the number of victims escalated above a million due to the famine that followed the cyclone. Official statistics by themselves can never paint the full picture of the disaster or afford a comprehensive and mature evaluation of the facts. All official statistics should be regard from this point of view, and one should always be open to realistic reevaluation.

2. CYCLONES AND HURRICANES

Tropical Cyclones: Hurricanes and Typhoons

A cyclone is a system of winds which rotate counterclockwise in the Northern Hemisphere with a diameter varying from 50 to 1000 mi and a barometric pressure that diminishes toward the center. Other features of cyclones include vertical ascent of winds around the center, formation of a cloud above the center, and condensation with precipitation. They can be so mild that they pass over inhabited areas without being detected, though registered by meteorological stations. However, they can reach extremely high intensity, as in the case of hurricanes, which are the most violent tropical cyclones. Tornadoes are similar to hurricanes, but more violent. They are at times produced by hurricanes themselves, yet are not part of the cyclone family. Tornadoes have a diameter much smaller than 50 mi, which is the minimum for cyclones.

Hurricanes, typhoons, and cyclones, as they are usually referred to in the North Atlantic, North Pacific, and West Pacific/Indian oceans, respectively, are tropical cyclones of particularly high intensity. As cyclones, they consist of rotating masses of air that spirally converge toward a center where the atmospheric pressure is low. The wind speed gradually increases toward the center and the barometric pressure drops. Winds at the periphery of the storm are relatively horizontal, but rise gradually as they approach the center (which they never reach). At a certain distance from the periphery, their vertical velocity component is maximum. As they rise, the winds form a cloud which produces heavy torrential rains. Maximum wind speed occurs at the edge of the central zone called the eye, where there is sunshine and relative calm.

Most hurricanes are formed in the southwest part of the North Atlantic Ocean, which includes the Caribbean Sea and the Gulf of Mexico. With few exceptions, most hurricanes occur from June through October. The average life span of a hurricane is about 9 days, though hurricanes forming in August have a 12-day average life. The annual frequency of tropical cyclones has been slowly increasing since the 1930s (presently this number ranges between 8 and 10). This increase can be partly attributed to better detection systems and to gradual warming of the atmosphere.

Most tropical cyclones attain hurricane strength: approximately 60%. Other storms similar to North Atlantic hurricanes are:

- Pacific typhoons,
- tropical cyclones in the Bay of Bengal,
- tropical cyclones in the Arabian Sea,
- Australian willy-willies.

For an average hurricane, the diameter of the area where winds reach gale force (40 mph) is 350–400 mi, while the diameter where winds reach hurricane force (75 mph) is 100 mi. The diameter of some hurricanes has been estimated as large as 600 mi; of some typhoons, 1000 mi.

Hurricanes vs. Tropical Cyclones

Tropical cyclones within the Atlantic zone can produce sustained wind speeds above 74 mi/hr (119 km/hr), which is the minimum velocity for a storm to be called a hurricane. Thus, not all tropical cyclones that develop in this area are officially designated as hurricanes. To further clarify some of the confusion in layman usage of meteorological terminology, it should be noted that similar kinds of storms which occur in the West Pacific are referred to as typhoons. Hurricanes are tropical cyclones with a maximum wind velocity between 74 mi/hr (119 km/hr) and 200 mi/hr (320 km/hr) that occur in the North Atlantic and in the eastern North Pacific. The South Atlantic and eastern South Pacific are excluded because their local wind conditions prevent formation of such storms.

The Structure of a Hurricane

A typical hurricane includes a central circular eye which is relatively calm and has a cloudless sky, which is surrounded by an annular zone in the form of a vertical cylindrical tower of air that includes dense cumulonimbus clouds. A dense layer of clouds in the upper level of this cylinder constitutes a cap and has a diameter that varies between 15 and 50 miles. The diameter at the lower layer near the ground ranges between 200 and 400 mi, and the cylinder here is configured of a low vortex. Throughout the height of the cylinder, the hurricane is structured as an inward spiraling band of cumulus clouds.

The structure of wind circulation varies with altitude. The structure in the upper portion (between 2 and 9 miles, the troposphere) differs from that in the lower part. The winds spiral with outward rotational motion and with a vertical upward component, so that the wind rises and spreads outward. Rotation is counterclockwise or cyclonic in this zone in the Northern Hemisphere.

The structure of the hurricane which interests engineers the most is that of the lower atmosphere (below an altitude of 2000 ft), since it is this is where the hurricane interacts with the human habitat. Here the winds rotate in a spiral fashion and accelerate from the periphery to the center of the storm. Tangential velocity increases from the outside edge, reaching a maximum at the edge of the eye. At the periphery the storm diameter varies between 200 and 400 miles; around the eye, where the rotating wind peak velocity, the diameter varies between 10 and 20 miles. Indeed, this is the zone where maximum destruction occurs. In addition, a peculiar dynamic equilibrium occurs in this zone surrounding the eye, where wind velocity and destruction peak, which forces the wind to veer vertically upward. In this annular zone there is a balance between the centripetal forces generated from the low pressure in the eye and the centrifugal (Coriolis) forces caused by the rotational winds. The two opposite forces neutralize each other in the horizontal plane, leaving no alternative for the wind but to change its course in vertically upward. The rising wind creates a vertical circulation that supports the deck of clouds at the top of the storm. Note that the horizontal wind velocity near the ground in this zone is almost equal to the vertical velocity in the upper level of the cylinder due to conservation of the momentum of the air.

Thus, the most active part of the storm is within the cylindrical wall that surrounds the eye: the horizontal tangential velocity is maximal and the vertical upward velocity is equally high. Low-pressure air from the eye is sucked in at the base of the cylindrical wall, activating a downdraft of dry warm air within the eye starting from the upper level downward.

General Characteristics of a Hurricane

The energy required to form and sustain a hurricane is derived in part from the atmosphere and in part from the ocean over which it forms and propagates. When hurricanes move above land, this sustaining mechanism is incomplete, and storm duration is greatly reduced since one of its two basic energy sources is no longer available. The dynamic mass of a hurricane includes particles of air and water vapor in motion and can accumulate about 2.2 million tons (2.0 million metric tons) of material per second. This mechanism can be considered a thermal engine whose energy is provided by latent heat released in the atmosphere by formation of cumulus clouds, which involves condensation of water vapor, as well as by condensation that produces torrential rains. The energy contributed to this engine by the ocean is much smaller in quantity and involves heat flux from the water surface. Obviously, warm tropical oceans are necessary substrates for hurricane formation.

As stated above, not all tropical cyclones develop into hurricanes, but an average of 70 tropical cyclones annually reach either hurricane or gale strength wind speeds. Such storms originate is the monsoonal trough starting approximately 5 degrees of latitude north of the equator. The original disturbances from which hurricanes develop are referred to as "seedlings," which usually form over the ocean (there are rare occasions when they form over land). However, not all seedlings generate hurricanes: in the North Atlantic 100 seedlings form on average form each year, only 9 of which develop into hurricanes.

Propagation of hurricanes basically follows a general path that proceeds westerly while tending toward polar regions. Their average translational speed usually varies between 3 and 15 mph (5–7 km/hr). Hurricanes with larger diameters usually have a wider range velocity translation: 0–10 mph (0–16 km/hr). When storms move within the band of the trade winds, they assume a westerly velocity of 10–20 mph (15–30 km/hr). After leaving this zone, they continue to move north toward the pole, their speed increasing to 40 mph (65 km/hr).

The Eye of a Hurricane

The central part of a hurricane, the eye, is an area of relative (though absolute) calm. Wind speed there is not higher than 15 mph. There is no rain, the cloud deck is very thin, and occasional sunshine. Birds are sometimes carried into an eye by the fury of peripheral winds. The eye of hurricanes have often been penetrated by observers, including ships. This is unlike the eye of a tornado, where such penetration would be impossible. Many observers describe the eye as oppressive, as the low pressure there can make one feel as if they are suffocating. Their average diameter is about 14 mi, and the smallest diameter on record is about 4 mi. The size of an eye tends to increase as a hurricane passes over land. The shape of the eye is usually circular, but it tends to become elliptical when the hurricane passes over land. The long axis of the ellipse is in the direction of translation (forward motion) of the storm. Elliptically shaped eyes sometimes reach 90 mi in diameter. As the storm progresses into its life cycle, the size of the eye varies. The temperature in the eye is much higher than in outer areas, especially at higher altitudes.

The highest wind speed ever recorded instrumentally is 188 mph, with gusts of 229 mph, measured at Mount Washington Observatory in New Hampshire on April 12, 1934, and was not related to a hurricane. This speed was partially caused by the configuration of the local terrain. The following are measurements of high hurricane wind speeds:

Speed	Place	Date
186 mph	Mount Washington, New Hampshire	January 11, 1878
186 mph	Mount Washington, New Hampshire	September 21, 1938
175 mph	Chetumal, Mexico	September 27–28, 1955
163 mph	Havana, Cuba	October 18, 1944
155 mph	Hillsboro Lighthouse, Florida	September 17, 1947
150 mph	San Juan, Puerto Rico	September 13, 1928
132 mph	Jupiter Lighthouse, Florida	August 26, 1949
128 mph	Miami Beach, Florida	September 18, 1926
121 mph	Milton, Massachusetts	September 21, 1938

All of these measurements were recorded over land. Wind speeds at sea are appreciably higher due to the smaller surface friction.

Hurricane Precipitation

It is impossible to measure the exact amount of rainfall in a hurricane since the wind does not allow the rain to be collected by a measurement gauge. The biggest rainfalls are usually associated with hurricanes and tropical cyclones.

The number of inches of rainfall at a given point depends on several factors. The slower the forward motion of the storm, the longer the duration and amount of precipitation at that location. The rate of rainfall increases over land. The following is a list of precipitation associated with tropical cyclones and hurricanes:

Inches	Days	Place	Date
96.5	4	Silver Hill, Jamaica	November 1909
88.0	3	Baguio City, Philippines	July 1911
81.5	3	Funkike, Formosa	July 18–20, 1913
63.0	3	Mount Malloy, Queensland, Australia	Date unknown
47.0	7	Reunion, Mauritius	February 15–21, 1896
46.0	–	Baguio, Philippine Islands	July 1911
41.0	7	La Marie, Tamarind Falls, and L'Étoile, Mauritius	February 15–21, 1896
33.7	5	Crowley, Louisiana	August 6–10, 1940
31.66	5	Abbeville, Louisiana	August 6–10, 1940
29.65	5	Lafayette, Louisiana	August 6–10, 1940
29.60	2	Adjuntas, Puerto Rico	September 13–14, 1928
26.0	7	Pamplemousse, Mauritius	February 15–21, 1896
23.11	2	Taylor, Texas	September 9–10, 1921
22.22	2	Altapass, North Carolina	July 14–15, 1916
21.4	2	Alexandria, Louisiana	June 15–16, 1886

Hurricane Surges

A *hurricane surge* is a rapid rise of water level caused by winds and low barometric pressure. The water rises approximately 1 ft for each inch of pressure drop. This number is usually 3–4 ft during a hurricane. Wave action increases surge height above these heights, especially near the shoreline, where water is shallow, or in channels, sounds, bays, etc.

Hurricane Forerunners

A forerunner is a characteristic rise in water level near the coast that precedes arrival of the center of a hurricane. This is explained by the fact that water is being transported and accumulated toward the shore. Forerunners can extend for long distances over shores, whereas a surge will cover only a small length of shoreline. A surge is usually located to the right of the storm center, where the wind and forward motion are parallel and additive, creating the most severe wind action. The action of a surge is combined with that of forerunners to raise the water level.

If the path of a hurricane approaching a coastline forms a 90° angle, the surge will produce the highest water rise. The water rise decreases as the angle diminishes. Waters too shallow can reduce amplification and surge effects, but both increase when the sea bottom slopes gradually, thus extending the area of shallow water far into the ocean. Naturally, if the surge coincides with meteorological high tides, the water level will increase at high tide.

Hurricane Waves

A *hurricane wave*, which is a rare phenomenon, is a particularly rapid and high surge which approaches land. It is usually described as a "wall of water." Its effects are always disastrous. Such a wave was responsible for 100,000 deaths in the Bengal region of India in 1878 during the Backergunger tropical cyclone. Another hurricane wave killed almost 2000 on the South Carolina coast on August 27, 1893, and another killed 1800 near New Orleans on October 1, 1893.

Wave Action During Hurricanes

The dynamic action of the waves produced by hurricanes can cause severe damage. Breaking waves can reach forward speeds of 60 mph as they impact on land. Buildings and sand dunes can be easily destroyed by the impact of water as well as by repetitive pounding of heavy debris carried by waves (e.g., rocks, trees, and building components like bricks and plaster).

Hurricane Rating

A hurricane rating system similar to the Beaufort scale was proposed by the National Weather Service. Its scale includes five categories, defined by wind speed and expected type of damage. As indicated in Table 2.2, category 1 is limited to 74–95 mph, while category 5 includes wind speeds above 155 mph.

Hurricane Effects

Because of their size and duration, hurricanes generate extensive damage and severe loss of human life. They are thus of major interest in engineering terms for mitigation of their effect on the human habitat. Since the onset of global weather satellite observation, hurricane reporting is much more precise and comprehensive than it used to be. To better understand the impact of hurricanes on a global scale, we include here a brief mention of those storms that have produced most damage and largest loss of human life in this century.

The most violent hurricanes include: August–September 1900, Galveston, TX (6000 deaths); September, 21, 1906, Louisiana and Mississippi (350 deaths); September 11–22, 1926, Florida and Alabama (243 deaths); October 20, 1926, Cuba (600 deaths); September 6–20, 1926, south Florida (1836 deaths); September 3, 1930, Dominican Republic (2000 deaths); September 21, 1936, Long Island, NY and New England (600 deaths); October 15–16, Bengal, India (40,000 deaths); September 9–16, 1944, North Carolina to New England (46 deaths); August 30, 1954 (Carol), NE United States (68 deaths); October 5–18, 1954 (Hazel), eastern U.S. and Haiti (347 deaths); August 12–13, 1955 (Connie), The Carolinas, Virginia, and Maryland (43 deaths); August 7–21, 1955 (Diane), eastern U.S.

Table 2.2. Hurricane Disaster-Potential Scale (National Weather Service)

Category	Wind speed (mph)	Probable property damage
5	155	Shrubs and trees blown down; considerable damage to roofs of buildings; all signs down. Very severe and extensive damage to windows and doors. Complete failure of roofs on many residences and industrial buildings. Extensive shattering of glass in windows and doors. Some complete building failures. Complete destruction of mobile homes. Small buildings overturned or blown away.
4	131–155	Shrubs and trees blown down; all signs down. Extensive damage to roofing materials, windows, and doors. Complete failure of roofs on many small residences. Complete destruction of mobile homes.
3	111–130	Foliage torn from trees; large trees blown down. Practically all poorly constructed signs blown down. Some damage to roofing materials of buildings; some window and door damage. Some structural damage to small buildings. Mobile homes destroyed.
2	96–110	Considerable damage to shrubbery and tree foliage; some trees blown down. Major damage to exposed mobile homes. Extensive damage to poorly constructed signs. Some damage to roofing materials of buildings; some window and door damage. No major damage to buildings.
1	74–95	Damage primarily to shrubbery, trees, foliage, and unanchored mobile homes. No real damage to other structures. Some damage to poorly constructed signs.

(400 deaths); September 19, 1955 (Hilda), Mexico (200 deaths); September 22–28, 1955 (Janet), Caribbean (500 deaths); June 25–30, 1957 (Audrey), Texas to Alabama (390 deaths); September 4–12, 1960 (Donna), Caribbean, eastern U.S. (148 deaths); September 11–14, 1961 (Carla), Texas (46 deaths); October 31, 1961 (Hattie), Belize (400 deaths); October 4–8, 1963 (Flora), Caribbean (6000 deaths); October 4–7, 1964 (Hilda), Louisiana, Mississippi, and Georgia (38 deaths); September 7–12, 1965 (Betsy) Florida, Mississippi, and Louisiana (74 deaths); June 4–10, 1966 (Alma), Honduras and SE U.S. (51 deaths); September 24–30, 1966 (Inez), Caribbean, Florida, and Mexico (293 deaths); September 5–23, 1967 (Beulah), Caribbean, Mississippi, and Texas (54 deaths); August 17–18, 1969 (Camille), Mississippi and Louisiana (256 deaths); July 30–August 5, 1970 (Celia), Cuba, Florida, and Texas (31 deaths); August 20–21, 1970 (Dorothy), Martinique (42 deaths); June 19–29, 1972 (Agnes), Florida to New York (118 deaths); September 19–20, 1974 (Fiji), Honduras (2000 deaths); September 13–27, 1975 (Eloise), Caribbean and NE U.S. (71 deaths); August 30–September 7, 1979 (David), Caribbean and eastern U.S. (1100 deaths); August 4–11, 1980 (Allen), Caribbean and Texas (272 deaths); August 18, 1983 (Alicia), southern Texas (17 deaths); October 26–November 6, 1985 (Juan), SE U.S. (97 deaths); September 16–22, 1989 (Hugo), Caribbean and SE U.S. (504 deaths); August 24–26, 1992 (Andrew), southern Florida and Louisiana (14 deaths); September 11, 1992 (Iniki), Kauai, Hawaii (3 deaths).

We can see that the astonishing number of fatalities which hurricanes produce cannot fail to have a profound impact on the design profession. The most obvious reaction to such catastrophic effects is to facilitate construction of a physical habitat that would more efficiently protect life from the fury of the elements. Understanding the forces involved is one of the first steps of interest in any engineering study concerned with the effects of these violent storms. Structural and construction techniques should then be developed within the discipline of "Severe Storm Engineering."

3. TYPHOONS

Generalities

We now study the effects of typhoons while minimizing the meteorological component and focusing on the engineering aspects. "Typhoon" is the official term used in the English language to identify a certain type of severe storm. The word itself is of Cantonese origin and was derived from two words, which, combined together, mean "great wind." Typhoons are tropical cyclones that occur in the Indian and West Pacific oceans. More generally, typhoons are storms which are the same type as hurricanes but which are usually considered somewhat less violent than hurricanes. They are observed on both sides of the equator within a band which extends between 6 and 35 degrees of latitude north and south of the equator. Their rotational motion is clockwise in the Northern Hemisphere and counterclockwise in the Southern Hemisphere, having a tangential velocity as high as 150 mph. Areas particularly prone to typhoons include the vicinity of the Philippines, the South China Sea, the Ryukyu Islands, and the Taiwan Strait. Typhoons usually occur between July and November in the Northern Hemisphere when the doldrums are farthest away from the equator. They form over open water and can travel to a coast. When moving over land their intensity usually dissipates gradually.

As they storms travel in the band north of the equator, typhoons first move in a northwestern direction as they travel within the zone of the trade winds. The path then veers approximately 90 degrees, forming a curvilinear trajectory which proceeds and accelerates in a northeasterly direction in the zone of the westerly winds. While the translational velocity in the northwestern direction is no higher than 5 mph, after the change in direction the speed of the storm increases to over 20 mph.

The meteorological disturbances associated with typhoons involve large areas: up to 2000 mile in diameter. Low dark clouds dominate the scene, which includes exceptionally intense precipitation and a surge of the sea surface. The general climatology preceding the arrival of a typhoon includes rapid pulsating variation of atmospheric pressure, creating an overall unpleasant sensation for humans that any change in barometric pressure usually produces in the population at large. The climatological zone of action of a moving typhoon usually has a toroidal-shaped configuration, with an annular projection on the earth's surface that has a minimum diameter of 50 miles which can range up to several hundred miles. This shape implies a cloudless central zone, referred to as the eye, usually 5–25 in diameter. The average annual number of typhoons is not greater than 30.

We list here some of the most violent typhoons of this century: September 18, 1906, Hong Kong (10,000 deaths); October 22, 1952, Philippines (300 deaths); September 17, 1959 (Sarah), Japan and South Korea (2000 deaths); September 26, 1959 (Vera), Honshu, Japan (4466 deaths); June 30, 1964 (Winnie), northern Philippines (107 deaths); September 5, 1964 (Ruby), Hong Kong and China (735 deaths); July 9, 1967 (Billie) SW Japan (347 deaths); November 18, 1968 (Nina), Philippines (63 deaths); September 15, 1970 (Georgi), Philippines (300 deaths); October 14, 1970 (Senine), Philippines (583 deaths); October 15, 1970 (Titang), Philippines (526 deaths); August 1, 1971 (Rose), Hong Kong (130 deaths); December 3, 1971 (Theresa), Philippines (169 deaths); July 11, 1974 (Gilda), Japan and South Korea (108 deaths); May 20, 1976 (Olga), Philippine floods (215 deaths); July 25 and 31, Thelma (T. Vera), Taiwan (39 deaths); October 27, 1978 (Rita), Philippines (400 deaths); November 25, 1981 (Irma), Luzon, Philippines (176 deaths); September 2, 1984 (Ike), southern Philippines (1363 deaths); November 25, 1987 (Nina), Philippines (650 deaths).

A Case Study

Since the early 1950s, in-depth investigation of tropical cyclones in the Atlantic (hurricanes) has been conducted through instrumented aircraft and reported in many studies by a number of researchers (e.g., Jorgensen 1982; Hawkins and Imbembo 1976; Shea and Gray 1973; and Willoughby 1981). Investigations of typhoons had not been conducted, since all studies focused on hurricanes. New investigations on typhoons were begun using U.S. Air Force C-130 aircraft equipped with new recording systems. The major goals of these new investigations included comparing the structure of mature typhoons in the West Pacific with that of mature hurricanes in the Atlantic, and comparing the quality of data gathered through aircraft in the West Pacific and the Atlantic. A better understanding of these storms has been derived based on this new course of investigation of typhoons. Jorgensen (1982) reported data gathered by an instrumented aircraft on the development of typhoon Irma in November 1974. These data covered the period of investigation which constituted the meaningful life span of the storm and focused on eight meteorological parameters that illustrate the storm cross-section and profile: relative tangential wind, actual and relative wind, adjusted temperature, relative humidity, mixing ratio, adjusted D value, and adjusted equivalent potential temperature.

The last to occur in the 1974 season, Irma came to life on November 21 in the West Pacific south of Guam and dissipated by December 2 after passing over Manila. It was originally considered a weak storm, but it gained strength to reach the following wind speeds: 50 mph on November 23, 70 mph on November 24, 85 mph on November 25, 105 mph on November 26, 115 mph on November 27, 90 mph on November 28, 60 mph on November 29, 70 mph on November 30, 65 mph on December 1, and 40 mph on December 2. Irma first translated in a northwesterly direction until November 26, and then veered, moving westerly until December 1, when it changed course again moving in a northerly direction. These data describe the full cycle of the storm, but meteorological data were gathered via aircraft flights only on November 24, 25, 27, 28, and 29. These reconnaissance flight dates represent the different phases of typhoon development: weak, moderate, intense, and return to weak. Data collected on each flight included: horizontal analyses of streamlines, isotachs, adjusted temperature, wind profile plots, and thermodynamic quantities from radial penetrations.

Comparing the data gathered on Irma in this detailed study with that of many other Atlantic hurricanes demonstrated the extreme similarity of the structure of typhoons and hurricanes. This strong similarity holds true despite the fact that the instrumentation used on board Atlantic aircraft was different from that used on Pacific flights, and despite the fact that the diameter of the storm and the diameter of the eye of Irma were much larger than those measured in most other hurricanes studied.

4. SAFFIR–SIMPSON SCALE

A damage scale was desperately needed because of the need to compare the effects of high-velocity winds found in storms such as hurricanes on the built-up environment around the world, even if the hurricane intensity scale had to be based on subjective observations. No such scale existed in 1970 for wind storms or hurricanes, though similar scales had been used to measure earthquake damage (e.g., the Mercalli and Richter scales).

The Saffir–Simpson Scale, the first damage scale for hurricanes, appeared in 1972 (it has been modified since), and was named for its inventors, Herb Saffir, a structural engineer who spent many years investigating hurricane wind effects on structures, and Dr. Robert H. Simpson, a former director of the National Hurricane Center. This scale is best described by Sheets (1992), director of the National Hurricane Center in Coral Gables, Florida.

Saffir designed the wind portion of the scale with reference to different levels of damage. There has been some debate as to whether the original definition of the wind scale referred to gusts or sustained winds. Actual usage, including the storm surge aspect, is clearly related to sustained winds (defined as a 1-minute average at the standard 10-m level).

The storm surge scale was added to the wind scale by Simpson. It refers to a range of storm surge heights for a given storm category. However, this range only gives a relative feel for the potential storm surge for a given basin. A category 5 storm will generally produce a surge at least twice that of a category 3 storm. Absolute values of the potential storm surge for a given Saffir–Simpson category storm is basin-dependent, and there are large variations which depend on wind direction, speed of approach, and wind field distribution. According to Robert C. Sheets:

> In actual practice, we have developed a catalogue of storm surge values for a given family of storm conditions for each coastal region. Emergency Managers then can refer to the catalogue when an actual hurricane threatens, selecting the correct family of storms for their basin to determine the expected storm surge value for this particular category of storm in their area. That is, a category 3 hurricane approaching from the southwest at 10 mph will give a different storm surge value than an identical storm approaching from the southeast, etc.

The scale in its final form is presented in Table 2.3.

5. GOVERNMENT RESEARCH ON HURRICANES IN THE UNITED STATES

The efforts of the U.S. Government on the specific topic of hurricanes are primarily executed by the Hurricane Research Division (HRD) of the National Oceanic and Atmospheric Administration (NOOA). Its structure and objectives are concisely described as follows (NOOA 1993):

> The Hurricane Research Division is NOAA's primary focus for studies of tropical storms. The research program is directed at improved prediction through improved physical understanding of the structure and dynamics of these storms. A large portion of HRD's resources is devoted to the conduct of an annual hurricane field program that uses the NOAA Aircraft Operations Center (AOC) WP-3D aircraft to acquire data sets for the research program. HRD interacts with the National Hurricane Center (NHC) in all phases of its research, with the National Meteorological Center (NMC) and the Geophysical Fluid Dynamics Laboratory (GFDL) in research concerned with numerical modeling of hurricanes, with the National Severe Storms Laboratory (NSSL) in the study of landfalling hurricanes, and with other NOAA groups, Federal agencies, and universities.

Table 2.3. The Saffir–Simpson Hurricane Scale[1]

Category One Hurricane — Weak

Winds[2]: 75–95 mph (65–82 kt) at standard anemometer elevations. F-scale is 1.0–1.4. Damage is primarily to shrubbery, trees, foliage, and unanchored mobile homes. No real damage occurs to building structures. Some damage is done to poorly constructed signs.[3]

Storm surge: Nominally is 4–5 ft (1.2–1.5 m) above normal. Low-lying coastal roads are inundated, minor pier damage occurs, some small craft in exposed anchorages break moorings.

Category Two Hurricane — Moderate

Winds: 96–110 mph (83–95 kt) at standard anemometer elevations. F-scale is 1.5–1.9. Considerable damage is done to shrubbery and tree foliage, some trees are blown down. Major structural damage occurs to exposed mobile homes. Extensive damage occurs to poorly constructed signs. Some damage is done to roofing material, windows, and doors; no major damage occurs to building structures.

Storm surge: Nominally is 6–8 ft (1.8–2.4 m) above normal. Coastal roads and low-lying escape routes inland are cut by rising water 2–4 hr before arrival of center. Considerable pier damage occurs, marinas are flooded. Small craft in unprotected anchorages break moorings. Evacuation of some shoreline residences and low-lying island areas is required.

Category Three Hurricane — Strong

Winds: 111–130 mph (96–113 kt) at standard anemometer elevations. F-scale is 2.0–2.4. Damage occurs to shrubbery and trees: foliage is blown off trees, large trees are blown down. Practically all poorly constructed signs are blown down, some roofing material damage occurs, some window and door damage occurs, and some structural damage occurs to small residences and utility buildings. Mobile homes are destroyed. There is a minor amount of curtainwall failure.

Storm surge: Nominally is 9–12 ft (2.7–3.7 m) above normal. Serious flooding occurs at the coast, with many smaller structures near the coast destroyed. Larger structures are damaged by battering of floating debris. Low-lying escape routes inland are cut by rising water 3–5 hr before the center arrives. Terrain continuously lower than 5 ft (1.5 m) above sea level may be flooded inland 8 mi (12.9 km) or more. Evacuation of low-lying residences within several blocks of the shoreline may be required.

Category Four Hurricane — Very Strong

Winds: 131–155 mph (114–135 kt) at standard anemometer elevations. F-scale is 2.5–2.9. Shrubs and trees are blown down, all signs are down. Extensive roofing material damage occurs, extensive window and door damage occurs, complete failure of roof structures occurs on many small residences, and complete destruction of mobile homes occurs. Some curtainwalls experience failure.

Storm surge: Nominally is 13–18 ft (3.9–5.5 m) above normal. Terrain continuously lower than 10 ft (3 m) above sea level may be flooded inland as far as 6 mi (9.7 km). Major damage occurs to lower floors of structures near the shore due to flooding and battering action. Low-lying escape routes inland may be cut by rising water 3–5 hr before the storm center arrives. Major erosion of beach area occurs. Massive evacuation of all residences within 500 yds (457 m) of the shoreline may be required and of single-story residences on low ground within 2 mi (3.2 km) of the shoreline.

cont'd

Category Five Hurricane — Devastating

Winds: Greater than 155 mph (135 kt) at standard anemometer elevation. F-scale is 3.0 or greater. Shrubs and trees are down, roofing damage is considerable, all signs are down. Very severe and extensive window and door damage occurs. Complete failure of roof structures occurs on many residences and industrial buildings. Extensive glass failures occur, some complete buildings fail, small buildings are overturned and blown over or away, and complete destruction of mobile homes occurs.

Storm surge: Height is nominally greater than 18 ft (5.5 m) above normal. Major damage occurs to lower floors of all structures located less than 15 ft (4.6 m) above sea level and within 500 yd (457 m) of the shoreline. Low-lying escape routes inland are cut by rising water 3-5 hr before the storm center arrives. Massive evacuations of residential areas situated on low ground within 5-10 mi (8-16 km) of the shoreline may be required.

---

[1]The Saffir-Simpson Hurricane Scale (SSHS) does not apply to Pacific Islands. [2]A sustained wind is one that persists for the minimum time period to establish optimal dynamic forces on a nominal building structure. [3]Fujita 1971.

6. THE NATIONAL HURRICANE CENTER

The National Hurricane Center (NHC), located in Miami, is one of three national centers operated by the National Weather Service (NWS). The other two are the National Severe Storm Forecast Center (NSSFC) in Kansas City, Missouri, and the National Meteorological Center (NMC), located near Washington, DC. The National Hurricane Center also interacts with the Hurricane Research Division, which is also located in Miami.

The NHC's responsibilities include tropical cyclone (hurricane) forecasts and warnings; tropical analysis; and national and international forecasts (public, marine, and aviation). The agency is best known for its hurricane forecast and warning program. The agency operates in the tropical and subtropical regions of the North Atlantic and eastern North Pacific oceans, the Caribbean Sea, the Gulf of Mexico, and adjacent land areas.

Its areas of concern have international repercussions of interest to professionals involved in the engineering aspects of high-velocity winds. It is certainly a major source of information on hurricane-related design factors.

7. SEEDING OF HURRICANES

The concept of hurricane modification through seeding was first formulated in the early 1960s in Project Stormfury conducted by the U.S. Weather Bureau. Its first application involved Hurricane Debbie in 1969, which was seeded with silver iodide crystals. Basic changes in intensity and structure attributed to the seeding operation were observed. Of course, these results generated great hopes for future control of these storms. As a result, the National Hurricane Research Laboratory (NHRL) was transferred from the Atlantic Oceanographic and Meteorological Laboratories (AOML) to the Environmental Research Laboratories (ERL) Weather Modification Program Office (WMPO) in Boulder, Colorado.

The focus of this program between 1970 and 1972 was to collect data on the effects of storm seeding to confirm the results obtained during Hurricane Debbie and prove its efficacy for application to other storms. It was decided to move the theater of operation from the Atlantic to the Pacific, where suitable storms occur more frequently, so as to accelerate data communication. A moratorium was imposed on Project Stormfury from 1973 to 1975 to await procurement of more suitable aircraft.

By 1975, the Experimental Meteorology Laboratory (EML) was involved in enhancement of rainfall using silver iodide seeding of cumulus clouds, and Project Stormfury was still strongly supported. However, movement operations from the Atlantic to the Pacific never occurred due to international complications, and the Atlantic remained the area of experimentation. As fate would have it, storms capable of being seeded formed over the Atlantic during the late 1970s, which left the activity of Project Stormfury stagnant. The coup de grace occurred when observations conducted after 1977 proved that many hurricanes spontaneously presented the same modifications observed in Debbie without being seeded. Project Stormfury was eventually put to rest in 1982.

Project Stormfury was a remarkable event in the history of applied research. The possibility of penetrating the structure of such a powerful mechanism would have been a dramatic turning point in the history of meteorological science difficult to compare to other scientific achievements, so that the merit this work was amply justified within the framework of scientific progress.

8. URBANISM AND HURRICANES

Hurricane-prone areas in the United States constantly undergo population increases, not only in terms of permanent year-round habitation, but also the vacation and resort population, which dramatically increase the potential damage and loss of life that hurricanes can inflict there. The population of such areas as Ocean City, Maryland, Gulf Shores, Alabama, and Padre Island, Texas, has been reported to increase by up to 1000% during the vacation season. The total population of the hurricane-prone coastal counties of Virginia, Maryland, Delaware, and New Jersey increased from 2.3 million in 1900 to 8.3 million in 1985, as documented by Robert C. Sheets (1990).

The safety of the inhabitants of barrier islands and coastal regions, not to mention the amount of property damage, is dependent upon ample warning and rapid evacuation, which pose a major challenge to the planning profession in terms of formulation and implementation of viable strategies and suitable plans. Population increases augment the potential number of casualties and affect the percentage of potential victims due to increased congestion of evacuation routes. And, in fact, the increase in traffic capacity along inland escape routes has not increased in proportion to U.S. population growth.

It is useful to focus on some of the details that a planner should envision in formulation of preparedness planning when considering a possible scenario of activities prompted by the approach of an oncoming overland hurricane. First, in terms of the geographical scale involved, the average frontal coastline exposed to an incoming hurricane with a perpendicular approach is 300 miles. Sheets (1990) estimates the cost of preparedness for such an area at $50 million, regardless of whether the storm actually hits the region. Preparation activities include boarding up building glass surfaces, terminating commercial and industrial activities, and many other minute details commonly taken care of by the population at large. The cost incurred also includes untransacted business, such as losses in sales, hotel/motel rentals, travel tickets, and other tourist income. Of course, these costs increase as the warning area expands, so that low accuracy in prediction of the landfall site can increase the size of the warning area and thus the overall cost of preparedness. A lack of accuracy in prediction of hurricane strength can also generate greater preparedness costs. In addition, in case of doubt, cautious overestimation of storm strength means more precautions and a larger number of evacuees, both of which result in higher costs.

Sheets points to other dramatic consequences that can result from inaccurate hurricane prediction. The tendency to overestimate the size and intensity of an incoming storm because of a lack of precise information can produce a loss of credibility with the public in regard to the validity of warnings, so that many people may not take future warnings seriously and will be trapped in the high-risk situation of a severe storm. As more credibility is lost, fewer and fewer people will be convinced to take safety precautions or evacuate, especially when clear skies erroneously dispel the fear of a not yet visible incoming storm.

Evacuation of critical coastal areas and barrier islands in the United States at present takes between 24 and 36 hours, which coincides with the warning time permitted by currently available forecasting technology. Evacuation routes which lead to safe havens should thus be based on a realistic correlation between evacuation and warning time requirements, which of course must be determined according to the individual area.

Of course, evacuation plans for coastal regions and barrier islands in the United States cannot be applied to other geographical areas like Caribbean islands. Without an inland refuge zone away from the path of a hurricane, the inhabitants of these islands (excluding some tourists and vacationers) would probably not seek mass exodus by sea or air transport. In such a case, the focus of a preparedness plan would be securing vulnerable parts of buildings that would suffer the most damage. Also of major importance would be evacuation of areas likely to be inundated by rising sea surges and establishment of a safe refuge on high ground, where special facilities could be made available.

Vulnerability to flooding in coastal areas is determined using a model called "Sea Land Over Lake Surge from Hurricanes" (SLOSH), developed by Jelesnianski (Jarvinen and Lawrence 1985). This model involves several steps;

a. Developing a model of an individual basin with the Techniques Development Laboratory of the National Weather Service using topographical maps and bathymetry data.

b. Testing the model based on historical hurricane information (including data on path, wind, pressure field, and water levels). Values of rise and fall of water level computed through the model are then compared to historical data.

c. Surveying the area to verify data represented in topographical maps. This involves the Techniques Development Laboratory, the National Hurricane Center, the National Weather Service, and local emergency management.

d. Adjusting the model to any geographical feature that might have been excluded from topographical maps.

e. Final testing of the model, so that it is ready for use by the National Hurricane Center.

f. Simulating different scenarios for each water basin, with between 250 and 500 storm simulations.

g. Compiling maps of areas susceptible to maximum flooding generated using possible water surge values for various storm scenarios. These maps are composed for storms with equal strength and motion characteristics; different flood area maps are created when single storms are used for modeling.

h. Formulating evacuation plans for flood areas which involve transportation and shelter, and which take human behavior into account. This is accomplished with the participation of the Army Corps of Engineers and local government.

9. HURRICANE HUGO: A CASE STUDY

The effects of Hurricane Hugo over the southern United States is included herein as a significant case study to illustrate those aspects of wind engineering associated with hurricanes. Hugo reached the coast of South Carolina near Charleston at approximately midnight on September 21–22, 1989, and moved through South Carolina and then to North Carolina. It first landed near Charleston and proceeded north along the coast, reaching Myrtle Beach in less than six hours, then veering inland toward Sumter, Columbia, and Camden, South Carolina, and Charlotte, North Carolina.

Though Hugo caused more damage than any other recorded storm in the Carolinas, its strength was within the limits of expected magnitude. It has also been established that much less damage would have been done to buildings if local building codes had included more complete and stringent criteria based on wind engineering science. Furthermore, by observing the track followed by Hugo, we can see that much damage was avoided by fortuitous chance. If the storm landfall had been approximately 20 miles further south, Hugo would have caused the kind of destruction in Charleston that was inflicted on Bull Bay. In addition, if it had landed 40 miles further north, the Myrtle Beach area would have suffered much stronger damage.

Hugo's average translational velocity was measured at 27 mph (43 km/hr). Penetrating inland much further than expected, its maximum wind speeds gradually decreased due to friction and were estimated at under the expected design wind speeds for the regions involved. In terms of amount of damage caused, this hurricane was considered the most violent and destructive storm to ever batter the United States. Estimated total value of destruction reached $6 billion. According to surveys of damage covered by insurance policies, the calculated losses due to wind effects totaled $4 billion, while damage due to floods reached $500 million. It should be noted that these numbers do not include uninsured or underinsured losses. Military installations at Charleston, Sumter, and Myrtle Beach were also battered, suffering losses of approximately $300 million. Losses due to the felling of a great number of trees has been estimated at another $1 billion.

These hurricane-prone coastal areas have been hit with a certain regularity according to the historical record: 76 storms have been recorded in the Carolinas between 1686 to 1980. These hurricane-prone coastal areas have been hit with certain regularity according to historical records: 76 storms have been recorded in the Carolinas between 1686 to 1980, which yields an annual frequency of 0.25 storms. Great hurricanes have occurred at a frequency of 1 every 30 years, major hurricanes have occurred at an 8.5-year frequency, while hurricanes in general have been observed at a 4-year average frequency.

In judging the effects of Hugo on this area, we consider it in comparison to those of other preceding hurricanes. Precise comparisons can only be made for the period after 1870, when wind speed measurements by anemometers began. However, personal accounts seem to confirm that the hurricane of September 15, 1752 involved wind velocities higher than those recorded for Hugo. Other hurricanes with comparable strength include the storms of August 27–28, 1813, August 27–28, 1893, and August 27–28, 1911.

Wind Velocities

The maximum sustained surface wind velocity at landfall was limited to a small area within the northeastern portion of the eye of the hurricane. This velocity value was established at

121 mph in coastal Bull Bay, South Carolina, with gusts up to 145 mph. After passing over Puerto Rico, the storm had gradually diminished in intensity. However it soon gathered strength, attaining maximum strength as it approached the U.S. coast. As the storm proceeded inland 1 to 5 km, wind speeds gradually decreased due to ground friction. Maximum flight level wind speeds measured by aircraft were recorded at 0.5 to 2 km, and there were very strong overland wind shears.

Tides

The high tides in the surge associated with Hurricane Hugo was documented by measurements taken at different coastal points, including those at the Customs House in Charleston and at Winyah Bay near Georgetown. Among the various high tide level readings above maximum surge level (MSL) are the following: 11–12 ft at Folly Beach, 10.5 ft in Charleston Harbor, 13 ft on Sullivan's Island, 13–15 ft on the Isle of Palms, 18–20 ft in Bull Bay, 14–16 ft in McLellanville, and 10–13 ft at Myrtle Beach and surrounding beach areas. The surge at the center of the eye typical of all hurricanes did not reach a peak during passage of the eye because offshore winds surges back. However, the water level did rise sharply as these winds subsided. The highest surge occurred at the northeast side of the eye, and the surge was minimal at the southwest side, where offshore winds decreased it. Hugo did produce the highest officially recorded tide in the Atlantic of this century. Equally high values were recorded at Savannah Beach during a storm that hit in 1893, and even higher flood levels probably occurred during the exceptionally violent storm of 1752 in Charleston. The following are the frequencies of occurrence of exceptionally high tide levels of comparable magnitude caused by hurricanes for various coastal locations: 500 years, Bull Bay; 300 years, McLellanville; 200 years, Isle of Palms; 100 years, Sullivan's Island; 100 years, the stretch of beaches near Myrtle Beach; and 80 years, Folly Beach.

Structural Performance of Buildings

A survey of the extensive damage to buildings caused by Hugo can shed light on the ability of structures to withstand the severity of wind loads (e.g., see Fig. 2.1). Although the wind velocity of the storm did not exceed that anticipated by the building code, substantial damage was even done to structures that complied with code specifications. Obviously, the codes had failed to recognize all the implications of wind loads. To this effect, we will briefly examine various building typologies and evaluate their ability to resist wind forces.

The most common type of building failure, regardless of specific construction technology, derived from local building damage effects and developed within wall claddings or roof decks. Such failures caused dual action; a) wind penetration into the interior of structures with a consequent increase in internal pressure (determined by outward forces on walls and uplifting of roofs); b) penetration of rain causing direct damage and/or indirect consequent failure. In addition to such local effects, total collapses were also observed not only in zones fully exposed to maximum storm severity, but also within relatively sheltered areas.

Within the overview of building performance, it is useful to differentiate structural damage from nonstructural damage caused by penetration of rain and wind into structural interiors after local cladding and/or roof failure. While structural failures are more visually obvious, nonstructural damage was so widespread that it constituted more economic losses, thus demanding more attention. In fact, roof and cladding damage began to occur at 75 mph

Figure 2.1. Collapse of church steeple. Hurricane Hugo, Charlotte, NC. (Photo courtesy of the Duke Power Company.)

wind speeds. Other types of considerable damage were caused by the surge itself, which acted either independently or in concert with the winds. In fact, it was often difficult to differentiate between damage caused by the surge and that caused by the wind, and the effect of the surge included not only water damage due to flooding but also structural damage generated by wave impacts or the impact of floating debris against buildings.

Specifically in terms of building type, experts observed overall satisfactory performance of light-framed residential buildings, including single- or two-family self-standing dwelling units. The kind of damage generally visible in these buildings mainly included roof, shear, and foundation failures. On the other hand, masonry buildings free of metal reinforcement demonstrated an intrinsic weakness not only when exposed to high wind forces but also within relatively sheltered zones. Metal buildings demonstrated markedly different behavior, though older structures performed more poorly than newer ones. As expected, engineered frame structures, whether of structural steel or reinforced concrete, withstood all structural loads, and, when damage occurred, it was limited to the cladding.

We now examine the individual types of structures more closely and present the specifics which explain particular behavior under hurricane loads.

First, as noted above, three major components of typical residential dwelling structures suffered the most damage: roofs, walls, and foundations. Roof vulnerability caused the most common type of damage, which often caused additional damage, as in the case when wind and rain would penetrate a building through a damaged roof. Inadequate anchorage between the roof and supporting structural walls was also frequently observed. When windows and doors failed first and the wind could penetrate building interiors, internal pressure would increase, causing uplift forces to act upon the roof that would combine with suction forces produced by wind on the exterior. Walls were also particularly vulnerable in this type of construction, especially when the roof became structurally disconnected. Walls in floor plans with long unbrazed spans were unable to provide adequate shear resistance, especially compared to buildings with more numerous cross-bracing partition walls. Shear failure in walls was also observed when stud walls did not have adequate bracing. The practice of bracing stud walls with only one sheet of plywood at each end often proved inadequate to provide full bracing action. In considering foundation failures in this type of building, we must note that most of the building damage caused by Hugo was along the coast, where wind speeds were highest. Many buildings there had raised foundations to comply with flood insurance regulations (i.e., the floor level had to be raised above flood level). Such conditions required foundations which included masonry piers up to 8 ft in height, which is a system not usually found elsewhere. In judging this kind of foundation, it is therefore important to emphasize their peculiarity. Unreinforced masonry piers were obviously most vulnerable. Other types of raised foundations built empirically without a structural engineering design were also vulnerable. While these comments apply to lightwood-framed residential structures, heavy timber structures proved to be much more resistant, since buildings of this type located in areas exposed to high-velocity winds remained undamaged during Hugo's passage.

Multifamily residential structures are usually better engineered than single-family ones. This type of construction is not much different from that of isolated single-family structures as they both include lightwood framing, but they are usually more professionally engineered and are thus more structurally efficient than buildings that solely depend on empirical practices. Most of these buildings suffered mainly from door and window failure, since they include a higher percentage of fenestration than single-family structures. As mentioned above, increased internal pressure caused by wind penetration is eventually added to the uplift force applied to the roof, causing structural failure that in turn generates extensive water damage from rain penetration.

The general performance of masonry structures in the area battered by Hugo was indeed poor. Total failures were observed not only in buildings near the coast but also in areas as far inland as Sumter, South Carolina. A typical observation was weakness in the anchorage between masonry walls and timber roofs which had become structurally disconnected. Of course, masonry buildings lack the heavy inertial mass of traditional massive masonry structures, which depend on the interaction of many shear walls to create stable structural integrity capable of withstanding large horizontal forces. The masonry structures in question were usually light. We should also note that these structures did not include steel reinforcement, as is now required by most building codes. Furthermore, the ratio of wall height to wall thickness allowed by local building codes was definitely high, thus generating structures which were excessively slender.

Preengineered metal buildings constitute a separate building category because of their special construction characteristics. They are usually characterized by large surface areas exposed to normal wind forces. Their walls and decks consist of light-gauge metal sheets and light structural framing components. The structural lateral resistance of the whole building is often provided by vertical and horizontal diaphragms that include light-gauge sheeting in walls and roof decks. These buildings exclusively depend on engineering expediency and rationality of design for their stability and are thus at the opposite end of the spectrum of building technology from buildings that depend on empirical design. Light-gauge metal buildings have minimal mass, i.e., they use a minimum amount of materials and do not include structural redundancies as traditional heavy masonry structures usually do. When failure occurs at one point, the lack of redundant members means that failure progresses to final stage. Since they exclusively design, it is essential that the design criteria for such buildings be sound, starting with the minimum requirements imposed by local codes. The preengineered metal buildings examined in the area affected by Hugo can be divided into two categories: a) older buildings erected under more lenient building codes that allowed weaker buildings and permitted design loads a third to half as small as than those prescribed by new codes; and b) newer buildings erected under more stringent building codes that provide for buildings with much better structural performance. As expected, the latter survived the hurricane undamaged, while the former were severely damaged or destroyed. This type of building can probably offer the most valuable lessons for evaluation of building performance under high-velocity wind forces because of the clarity with which they act structurally. Wind engineering studies should thus concentrate on them.

Frame buildings constitute another category and a valuable source for evaluation of the structural behavior of engineered buildings under high-velocity wind forces. Accurate analysis of structural frames usually guarantees integrity against vertical and lateral forces with a large margin for safety. Such buildings were expected to perform extremely dependably, and indeed they did. The cladding, however, presented an unforseen weakness in several cases, leading to embarrassing developments. In fact, it is amazing how often such obvious problems occur in everyday practice without being detected. The reason for the oversight was that nearly all attention was paid to the structural strength of the building frame and very little to the cladding itself. Such components as frames, floor and roof slabs, and foundations were carefully designed for all possible situations, but exterior cladding was treated separately, with little regard for structural integrity. This was strongly reflected by the damage suffered by buildings under the slashing force of Hugo.

A very explicit example of this state of affairs is furnished by failure that occurred in several buildings where stucco-finished exterior walls consisted of metal studs with gypsum board and rigid foam insulation on the interior. A variation on this theme was another material in place of the stucco, without anything to compensate for the changed overall structural strength of the wall panel. These cladding systems, independent of their structural strength, in many cases did not resist positive or negative wind pressure. Many types of buildings suffered severe damage due to rainwater and wind penetration due to the weakness of this kind of cladding, including office buildings, hospitals, schools, and hotels. In this respect, it most important to note that the cost of damage caused by Hugo to all types of buildings due to rain penetration was 20 to 30 times higher than that of actual wind damage.

As expected, mobile homes performed very poorly under the force of Hurricane Hugo; however, the lives of many of their occupants were saved due to how easy it is to evacuate them. Structures raised above safe flood water level on slender foundations were generally more vulnerable to wind forces and gained no benefit from this unwise design decision.

The presence of trees on developed sites can have a positive or negative influence on the building safety during the wind action of hurricanes. While trees can offer a shelter-belt that ameliorates wind speed and protects buildings to a certain extent, they can also collapse and inflict heavy damage. Considering the problem in general terms, knowing the effect of Hugo and other hurricanes on the role of trees should help to clarify this issue. The presence of trees during Hugo was generally observed to have the positive effect of sheltering buildings. A similar conclusion can be drawn from the effect of Hurricane Elena on Dauphin Island in

1985. The extensive forests of the Carolinas consist mainly of pines, which typically succumb at a threshold wind velocity of 70 mph, thus providing a dependable defense. This was confirmed by a survey of the damage on Bull Bay, where wind speeds during Hugo's landfall were maximal and definitely exceeded the 70 mph limit. In fact, very few trees remained standing there. While the role of trees in hurricane zones can be considered to act beneficially as a shelter-belt, it is definitely possible that wind speeds during severe storms will exceed that which they can sustain, so that a safe distance between trees and structures should be considered in design to prevent impact from falling trees.

Utility Poles

The vulnerability of utility poles under hurricane wind forces can be understood by analyzing the damage left behind by a storm. Typical collapse mode indicates that the critical section in the majority of cases is not near the ground, where one would expect it to be, but above midspan high above the ground. Photographs of felled utility poles support this typical behavior, which is somewhat paradoxical considering that a utility pole is a vertical cantilever subjected to horizontal forces distributed along the pole or concentrated where the electric cables are connected to the pole. Photographs taken by the Duke Power Company after Hugo clearly depict this typical behavior (see Figs. 2.2 and 2.3).

Hugo Over Puerto Rico

Prior to reaching the United States, Hugo battered Puerto Rico on September 17, 1989. We briefly discuss the damage it produced there for its relevance to evaluating the behavior of building techniques different from those typical in the United States in terms of efficiency in resisting wind forces.

The storm landed at Ceiba, Fajardo, Luquillo, and Rio Grande in the northeast section of the island. These areas were the most hard hit; the rest of the island was by and large spared. A highest sustained wind velocity of 104 mph was recorded at Roosevelt Roads in Ceiba, including gusts up to 120 mph. Lower values were recorded at San Juan Airport (84 mph with gusts up to 92 mph). Damage was substantially reduced by the fact that there was an absence of heavy rains.

Reports on damage suffered by different types of buildings indicate that the main factor was a combination of poor design and poor construction. Residential structures with wood construction performed quite well. Reinforced concrete structures, in both housing and major buildings, also performed quite well. The major culprits in building construction involved a large spectrum of applications shared by U.S. counterparts. The facades of sturdy frame buildings were vulnerable to wind pressure and suction, while the main frames remained intact. Metal buildings also suffered due to the vulnerability of light-gauge metal sheets in wall cladding and roof decks. Flat built-up roofing and exposed mechanical systems on rooftops incurred substantial damage, and wood buildings without metal connectors were particularly weak. General comments found in reports on Hurricane Hugo related to typical buildings subjected to hurricane forces approve the content of existing local building codes and point out that the basic reason for most failures was lack of compliance with code regulations.

The technology used to construct wooden single-family residential structures which constitute self-standing units is different from that used in the United States, so that

Figure 2.2. Failure of successive utility poles, showing similar collapse mechanism, with critical section quite high above the ground. Hurricane Hugo, Charlotte, NC. (Photo courtesy of the Duke Power Company.)

wood-framed Puerto Rican houses do not follow a similar tradition as that of their American counterparts. Basic standards are not followed, and each structure is erected using individual empirical methods which defy generalization. It is interesting to compare three similar structures from three different locations struck by Hugo. Two of the houses, built completely

Figure 2.3. Collapse of utility pole with critical section at approximately half the cantilevered span, i.e., lower than typical poles, yet still quite high above ground. Hurricane Hugo. (Photo courtesy of the Duke Power Company.)

by a single contractor, behaved much better under wind loads than a third, which had been partially built by a contractor and partially built by the homeowner. Though exposed to lower-velocity winds than the other two, the latter house suffered more than the others due to poor workmanship. The structure of all these houses involved eight vertical posts embedded in the ground which supported double wood beams at the second floor connected by bolts at roof level. TECO metal connectors and galvanized two-inch nails connected plywood cladding and decks to supporting frames. The third house (which suffered major damage) was missing metal connectors and nails and had improperly installed windows and doors that failed, allowing wind penetration and increased internal pressure, which caused most of the damage. Another typical example of wood structure in residential use which performed poorly during Hugo included a locally acquired prefabricated package. Such structures were rationally engineered as prefabricated building components and were structurally dependable; nevertheless, they failed because they were usually assembled by unskilled persons (e.g., owners), who were generally not able to finish the job right. Wooden buildings in Puerto Rico thus behaved poorly in the majority of cases due to a lack of the basic engineering requirements necessary to resist wind forces. However, the absence of an established tradition in wood carpentry on this island cannot be considered a fundamental factor for two reasons: 1) many wood buildings behaved satisfactorily even without established traditional techniques; and 2) even in the United States, where building techniques for wooden structures are well established and traditional, poor resistance to wind forces has been widely observed. Wind resistance in wood structures is generally not an inherent byproduct of a traditional technology, so that it must be specially engineered into structures. In fact, builders have traditionally only dealt with vertical gravitational forces, especially with heavy masonry structures, and lateral resistance was taken care of by the heavy mass of the structure. Lateral resistance is not automatically obtained in light-framed wood structures, regardless of how well the structure responds to gravity loads.

Engineered metal buildings behaved quite well due to the rationality of the structure. These buildings were expected to respond satisfactorily to the forces for which they were designed, and most survived the hurricane. When damage did occur, it usually did not affect the main or secondary structure, and was limited to local effects within the metal sheeting in those areas where peak pressures occurred. Such failures could have easily been prevented if more attention had been paid to the cladding. This is one of the major findings of these reports. As mentioned above, local failures of this type can produce very costly additional damage due to rain penetration and increased internal pressure.

Multistory reinforced concrete buildings behaved with total satisfaction from a structural point of view. Frames purposely designed for horizontal forces successfully responded according to engineering expectations. However, many of these buildings did incur extensive damage due to weakness of the cladding, especially door and window failures. Many with extensive glass facades suffered extreme damage due to a high percentage of failure of individual windows. Basic modes of failure included breakage of glass surfaces due to missile impacts or pulling of unbroken glass panels out of frames via external suction. Then, similar to other cases of local failure, subsequent damage was caused by devastation of interior spaces by rain and wind. Furniture and other objects in rooms exposed to wind suction through collapsed windows were seen to fall from high-rise buildings into the streets below. The details of these types of failure were reported for many structures, including the Minillas Government Center (which consists of two 17-story structures), a 15-story building in the Hato Rey area, and the 14-story Environmental Quality Board Building. At the Government Center, 54 windows were damaged in the north tower and 74 in the southern tower. Failures also occurred due to the impact of flying debris. The glass doors in the lobby of the south tower were damaged when the wind ripped them out from the frame, which had been poorly installed. On the first floor, 25 show windows with half-inch thick glass were pulled out of their frames due to weakness of the plastic clips that held them in place. At the 15-story building in the Hato Rey area, show windows on the first and second floors were pulled out of their frames due to clip weakness. These windows were also made of half-inch glass and consisted of $4\frac{1}{2} \times 9$ ft and $5 \times 15\frac{1}{2}$ ft panels. The Environmental Quality Board Building included a facade which consisted of vertical glass and aluminum panels connected to concrete beams on each floor. The windows were again made of half-inch glass and consisted of 4×5 ft and 4×6 ft panels. Wind suction force pulled 184 panels out of their frames on the north, west, and south facades. The cause of this failure was determined to be the connections between aluminum frames and concrete beams. Total disruption of rooms occurred when windows failed and furniture was pulled outside due to extreme suction force.

Multistory apartment buildings suffered the same type of damage to exterior cladding. One example of such failure is a 14-story apartment building in Guaynabo with a 45×160

ft rectangular floor plan. All four facades suffered damage, with the west side suffering the most. The major cause of failure was determined to be weakness in the vertical and horizontal mullions that transferred wind pressure from glass panels to the main reinforced concrete frame, which happened because of a lack of continuity in the mullions. Another apartment building reported similar exterior damage. This was a 26-story structure with a 45×75 ft floor plan on Ashford Avenue in El Condado, San Juan. The east and west facades bent under the wind pressure, rotating 12°, and apartments on the southeast and southwest corners from the seventh floor and up were severely damaged. Damage to the east and west facades was caused by failure in the mullions. On the north side of the building, where positive and negative wind pressures were maximum, windows did not suffer as badly as the rest of the building because the mullions were properly designed and behaved as expected. The southern facade, comprised of glass panels, failed completely under suction due to joint weakness. In sum, most exterior damage to reinforced multistory concrete buildings can be attributed to improper joining of mullions.

Additional structural damage caused by Hugo in urban Puerto Rico included utility poles and lampposts. The damaged utility poles were made of southern yellow pine and tapered from 15 inches at the base to 7½ inches at the top, with a total length of 56 ft. Designed improperly, these poles collapsed under the excessive forces generated by 110 mph winds. Structural calculations indicate that the maximum induced stresses in these poles could have been in excess of twice the ultimate stress. Prestressed reinforced concrete utility poles were also reported collapse due to excessive bending moments generated by wind force. In addition, octagonal hollow-core reinforced concrete lampposts were ruptured near the top at a critical point, as occurred near the international airport at San Juan.

10. HURRICANE INIKI: A CASE STUDY

Hurricane Iniki struck the island of Kauai in Hawaii on September 11 1992, only 18 days after Hurricane Andrew battered Florida and Louisiana. The interest on coastal hurricane damage was still very much the center of attention, which prompted engineering investigations of this event. A study conducted by the American Plywood Association on the damage caused by Iniki is reported in the present writing to reinforce a conclusion drawn from the analysis of Hurricane Andrew.

On September 11, 1992, Hurricane Iniki battered the Hawaiian island of Kauai. Sustained wind speeds of approximately 90 mph with gust up to 143 mph were estimated to have occurred by the local weather bureau in its preliminary evaluation. Kauai is 24 miles long in the north–south direction and 32 miles wide in the east–west direction. Major damage occurred on the north, south, and east coastlines, affecting different kinds of buildings: single-family residential dwellings, multifamily residential structures, condominiums and townhouses, and hotels and other commercial buildings (see Figs. 2.4–2.6). The damage observed throughout the island included that generated by high-velocity winds and, along the south coastline, that caused by high water from an ocean storm surge and wave action.

An investigation by representatives of the American Plywood Association was conducted over several days within two and a half weeks of the hurricane. Their findings were reported in a technical report published in April 1993 (Rose 1993). We report on the basic findings and conclusions drawn from that study in what follows.

Figure 2.4. Impact damage caused by large roof section blown off nearby building. Hurricane Iniki, Kauai, Hawaii. (Photo courtesy of the American Plywood Association (Rose 1993).)

Figure 2.5. Roof section of tongue-and-groove lumber roof decking lifted at large roof overhang over exterior balconies on a resort hotel. Holddown connections for roof framing to posts were also insufficient. Hurricane Iniki, Kauai, Hawaii. (Photo courtesy of the American Plywood Association (Rose 1993).)

Figure 2.6. Plywood roof sheathing panel penetrated wall when blown from roof section of nearby building. Hurricane Iniki, Kauai, Hawaii. (Photo courtesy of the American Plywood Association (Rose 1993).)

Wood Construction Systems

Analysis of damage to lightwood structural construction systems, mainly including single- and multifamily structures, indicated that these kinds of buildings generally performed satisfactory. In fact, many structures of this type escaped with only minor damage to the roofing surface. A major finding was the excellent performance of hip roof systems compared to that of other types of roof systems, such as gable types. Gable roof systems were in fact found to be particularly vulnerable in the studies conducted of Hurricane Andrew wind damage.

Hip roof geometry performs much better than gable roof configurations because of more efficient aerodynamic effects, which can be verified by experimentally obtained shape coefficients. The wind uplift forces on these roof surfaces were much smaller than expected for low-pitched gable structures. Not only are aerodynamic wind forces smaller for hip roofs, but even in structural terms this kind of construction seems to mean higher efficiency compared to that of low-pitched gable roof structures (see Fig. 2.7).

Gable roof structures were not as popular as hip roofs on this island, so that they constituted a minority. Their performance was not as bad as that observed in the damage caused by Hurricane Andrew in southern Florida, which can be attributed to better construction techniques and lower-velocity winds. Typical gable structure constructions on Kauai involved substantial bracings at the terminal parts of gables, so that the end trusses were much less vulnerable. In addition, the overhanging portion of roofs at gable-ends were built using much better engineering criteria than those typically used in southern Florida.

Figure 2.7. Connections of tongue-and-groove lumber roof decking to framing were inadequate to resist wind uplift forces. Hurricane Iniki, Kauai, Hawaii. (Photo courtesy of the American Plywood Association (Rose 1993).)

Figure 2.8. Complete structural failure of house with Hawaii "single wall" construction. Hurricane Iniki, Kauai, Hawaii. (Photo courtesy of the American Plywood Association (Rose 1993).)

Figure 2.9. Roof and wall system failures of residential building with timber-framed post-and-beam construction and in-fill wall panels. Hurricane Iniki, Kauai, Hawaii. (Photo courtesy of the American Plywood Association (Rose 1993).)

The building technology used in Hawaii includes local features that substantially differ from those commonly used in the continental United States. Structural framing and other construction assemblies have particular configurations which would be hard to relate to conventional light-framed American wood systems, such as Hawaiian "single wall" construction typologies (see Fig. 2.8).

Examination of the quality of roofing system performance did not reveal any unexpected outcomes. Asphalt shingles installed according to conventional practices were definitely vulnerable to high-velocity wind uplifts. It may be advisable to consider a dual line of action in roofing to guarantee protection from water penetration, which in itself is a major source of damage independent of structural damage. It is thus advisable to install a base layer consisting of a self-adhering roofing underlaying a rigid roof deck to create a secondary weather-resistant roofing membrane. In addition, as observed in the damage caused by Hurricane Andrew, concrete roof tiles require more dependable anchorage than what is usually provided by conventional installation practices, not only to prevent disruption of the roof itself, but also to prevent the hazard created when these tiles become flying debris which endanger other structures. One aspect of roof decking that did perform satisfactorily was wood shake and single roofing attached over solid roof decks. It is useful to consider that reports on the behavior of roofing material subject to uplift suction forces did not uncover any unexpected behavior, emphasizing that good engineering judgment is the only requirement, provided that negligence is not a factor (see Figs. 2.9 and 2.10).

In terms of damage produced by tidal surges, flooding and wave action damage, including the wave battering and the ramming effect of floating debris against structural and nonstructural building components, we should note the final conclusions fell within the same general

Figure 2.10. Inadequacy of toenailed connection of timber roof trusses to ridge beam for resisting wind uplift forces. Hurricane Iniki, Kauai, Hawaii. (Photo courtesy of the American Plywood Association (Rose 1993).)

criteria common to evaluation of the effects of similar storms. Such factors as the proximity of buildings to the shore, the elevation of the first inhabited floor above the expected maximum high water level, and the structure of the foundation system are all fundamental, and produced the same effects during Hurricane Iniki as they have in other severe storms.

11. HURRICANE ANDREW: A CASE STUDY

The effects of hurricanes on the human habitat are best examined through case studies, for which purpose we include the following analysis of Hurricane Andrew. This hurricane, the last major storm to have occurred prior to the writing of the present book, was one of the most destructive storms to ever batter the southern coast of the United States. By 1992, two other major storms in this century had devastated Florida: Betsy in 1965 and the Labor Day hurricane of 1926. Andrew by far exceeded the strength of these storms and wreaked nearly total devastation (see Figs. 2.11–2.14). It reached Miami in the early morning of August 24, 1992. Moving from the Atlantic across the Florida peninsula, it reached the Gulf of Mexico, where, after regaining strength over open water, it eventually battered Louisiana.

Data on Hurricane Andrew include the following:

- internal diameter of the eye = 12 to 18 miles;
- external diameter of the outer edge of the eye wall = 25 to 30 miles;
- the width of the damage track extended for about 24 miles to the north and 12 miles to the south of the eye center, with the path width varying along the path between 36 and 50 miles;

- translational velocity = 18–20 mph;
- storm strength = category 4 on the Saffir–Simpson scale (see Table 2.3).

Estimated Velocities

An executive summary written and edited by Ronald F. Zollo of the Department of Civil and Architectural Engineering at the University of Miami at Coral Gables, Florida (Zollo 1993) reports on the wind velocities estimated by different research agencies using different methodologies. (These agencies included the Wind Engineering Research Council (WERC), The National Oceanic and Atmospheric Administration (NOAA), and a private research group (Reinhold et al. 1993.) The reports categorically deny the original maximum 200^+ mph sustained wind speed value estimates made by the press in their first reports. A maximum fastest-mile wind speed of 145 mph was recorded on the coast when the hurricane first hit land, which subsided to 120 mph as the storm proceeded overland several miles. More specific values can be found in the Zollo's report.

We should appreciate the clarification in distinguishing the basic methodologies used, so that ambiguities can be avoided. To this end, we must make distinctions between the time periods used to take average speed readings. This includes "peak gusts" (time periods between 2 and 5 seconds), "sustained winds" (1 and 2 minutes), and "fastest-mile" (duration equal to the time a column of air 1 mile long needs to completely pass over a given point). Notice that the values for fastest-mile wind speeds usually fall between those of peak gusts and sustained winds. With this clarification, the preliminary findings of the WERC Post-Disaster Investigation Team reported fastest-mile wind speeds between 110 and 125 mph for Hurricane Andrew. More precise calculations from the same group updated the fastest-mile speed values to 125–130 mph. The values reported by the independent research group were based on anemometer readings include a 179 mph gust at a home in Perrine (located where the northern part of the eye wall hit the shore), and confirmed that, "In much of the area the wind speeds are not substantially higher than a fastest-mile wind speed of 120 mph" (Reinhold et al. 1993).

This report also compares the intensity of Andrew and that of Hugo and found that an estimate of a 140 plus mph fastest-mile wind speed for Andrew at its first encounter with the mainland was smaller than that estimated for Hugo (according to the WERC). In general, Andrew's maximum wind speeds were not higher than those expected for the building structure design in this area. Furthermore, recorded maximum wind gusts were the result of a mechanism typical for hurricanes without the presence of embedded tornadoes.

Learning From Damage Analysis

The damage was originally estimated to be between \$20 and 30 billion for Florida and over \$1.0 billion for Louisiana. The original estimated number of houses destroyed was $60{,}000^+$, a number that eventually doubled after more accurate surveys were completed. The violence of Andrew is reflected in the originally estimated 140 mph sustained wind speed with gusts up to 160 mph reported by the general press.

A written damage evaluation conducted promptly within a few days of the storm is one of the most significant documents for drawing meaningful conclusions for evaluation of engineering practices both in design and construction to protect against the effects of high-velocity winds. The data reported herein on damage caused by Andrew are based upon

Figure 2.11. Total devastation of typical residential structures under hurricane wind forces including impacts from collapsing trees. Hurricane Andrew, Dade County. (Photo courtesy of FEMA (1992a).)

Figure 2.12. Typical damage to boats in the area. Hurricane Andrew, Dade County. (Photo courtesy of FEMA (1992a).)

Figure 2.13. Reinforced concrete utility pole for electric power distribution collapsed under wind loads. Hurricane Andrew, Dade County. (Photo courtesy of FEMA (1992a).)

Figure 2.14. Example of major devastation to public housing. Hurricane Andrew, Dade County. (Photo courtesy of FEMA (1992a).)

Figure 2.15. Lapped joint of splice in wall double top plate inadequate at gable-end wall. Insufficient lateral bracing of wall to roof framing also contributed to the failure shown. Hurricane Andrew, Miami area. (Photo courtesy of the American Plywood Association (Keith and Rose 1992).)

the literature generated by official investigations of the damage. Specific reference is made to the findings of a 45-member investigating team that included structural engineers, scientists, building code officials, and technical representatives of the building construction industry, among whom were Edward L. Keith and John D. Rose, two technical representatives from the American Plywood Association, who prepared a report (Keith and Rose 1992).

The team conducted their investigation within a week of the storm and analyzed a wide spectrum of building typologies: low- and medium-rise buildings in residential and commercial types of applications, light-framed wood structures, concrete masonry structures, modular buildings, etc.. They focused on the relevant construction details that could be isolated as critical causes for building failures. It is beneficial to subdivide the wide spectrum of building construction typologies into a few selected types.

Light-Framed Wood Construction

The vast majority of residential structures in the United States include lightwood framing; indeed, this kind of building construction typology constituted the majority of the buildings damaged in the area struck by Hurricane Andrew. Observation of building failures pointed out that the major culprit responsible for damage was the inadequacy of connections between the major components of building elements (e.g., walls, floors, and gables). Structural or nonstructural failure in this component assembly constituted the first stage in

Figure 2.16. Uplift straps shown at truss bearing. Inadequate connection of straps to trusses at this location resulted in major structural damage. Hurricane Andrew, Miami area. (Photo courtesy of the American Plywood Association (Keith and Rose 1992).)

the collapse mechanism, so that, should these connections have resisted, the second line of failure would still remain to be determined. However, we limit the analysis to the actual damage observed, and the major findings are reported as follows.

Wall-to-wall connections at exterior corners as well as between partitions and exterior walls were observed to fail under resultant wind forces. Roof gables and wall connections were seen to fail under the action of uplift or of other wind forces components. Corner walls were found to often be connected by just a few nails, in some cases only three. The top plates of walls assumed to be lapped joints at corners were often found to be discontinuous, leaving a highly precarious connection (see Fig. 2.15). The bottom plates at such corners were found to be discontinuous, and the connection between the bottom plates of walls and floor platforms depended on the strength of a few occasional nails, creating an extremely weak connection between walls.

The connection between roof gables and exterior walls demonstrated extreme vulnerability. In this structural detail, the beginning of the failure mechanism was found to be detachment of the roof sheathing along the edges up to 12 ft wide, followed by collapse of roof trusses (see Figs. 2.16 and 2.17). These connections eventually collapsed due to a lack of bracing, failing from the roof edge and continuing toward the internal part of the structure, following a domino effect (see Fig. 2.18). Further analysis of connections between roof gables and walls uncovered a major weakness in the connection between trusses and the plates of supporting walls. Such connections often depended on only a few toenails, which

Figure 2.17. Truss tie-downs at supports proved adequate in this concrete block and stucco (CBS) home even though the roof sheathing was lost. Hurricane Andrew, Dade County. (Photo courtesy of Ronald F. Zollo (Zollo 1993).)

Figure 2.18. Typical gable-end wall failure. Roof sheathing blown off adjacent to gable-end wall, resulting in failure in gable-end wall framing and "domino" collapse of adjoining trusses. Hurricane Andrew, Miami area. (Photo courtesy of the American Plywood Association (Keith and Rose 1992).)

were sorely inadequate to provide sufficient shear resistance, both in terms of the inadequacy of toenailing per se and the total number of nails used.

Internal partitions were often poorly connected to other walls, with a few occasional nails which became easy victims to wind pressure once the wind penetrated the interior of a structure. The absence of proper connection within partitions and other walls, whether bearing or nonbearing, constituted double jeopardy. They were vulnerable in and of themselves, not being properly braced, and they failed to provide significant strength for the total structure. It is sadly ironic that a few dollars spent on proper nailing would have strengthened these buildings enough to protect them.

Further along the lines of improper joining, windows and door framing were found to be poorly connected to the rest of the building frame. Only a few nails or screws were used, and the opening in the frame for the window had been made much larger than necessary, so that nails and screws would only penetrate a fraction of an inch into the surrounding frame. Similar weaknesses due to improper connections were in the exterior paneling, which, if they had been fastened to the frame, could have provided the needed stiffening effect for the exterior walls. Exterior panels were often only connected along their exterior edge, with few fasteners connecting panels to the intermediate studs in walls.

Wind uplift forces are a typical phenomenon associated with wind which can be critical in conventional construction design, since they are often neglected. In conventional wood framing, tie-downs are the most effective means to resist uplift forces; in fact, they have been proven effective even when questionably installed. However, damage was frequently caused by improper installation of tie-downs or their absence when necessary. Tie-downs are absolutely necessary in connections between roof gables and exterior walls to prevent uplift, but they are totally ineffective in providing shear resistance.

Other considerations in lightwood framing failure involves nonstructural components, such as ceiling collapse caused by saturation of material by rainwater which penetrated into the attic space after vents or portions of the roof deck were blown off. It was found that, not only did ceiling damage constitute a significant loss by itself, but that it deprived the rest of the structure of a stiffening component. Improper installation of wood trims was also a noticeable nonstructural feature. This allowed contact with the ground, which led to decay due to moisture retention, which extended damage to structural components after a prolonged period of time.

Plywood Roof Decks

Plywood roof decks or oriented strandboard (OSB) roof decks were observed to behave equally in terms of structural performance. Such behavior did not depend on the type of construction typology of the structure. The roof panels constituting decks were usually observed to fail due to a lack of proper mechanical fasteners to supporting structural members. Recommendations for installation of fasteners for plywood and OSB panel sheathing call for either nails or staples of adequate size and with adequate spacing. The recommended nail size for panels up to half an inch thick is not less than 6d; for higher thicknesses the minimum size is not less than 8d. The recommended spacing in both cases is not less than 6 inches c. to c. along the panel edges, while the maximum spacing should not exceed 12 inches c. to c. when nails are installed along intermediate supports. When failures were observed in roof decking, a number of fasteners were inevitably defective, and an erratic nail pattern was typical for those panels detached from the supporting framing. In

Figure 2.19. Typical structural damage to residential buildings. Hurricane Andrew, Dade County. (Photo courtesy of FEMA (1992a).)

some cases, the spacings were as large as 48 inches, indicating extremely poor construction quality. When staples were used in panels that failed, it was noticeable that the fasteners were often placed too close to the edges, greatly reducing their holding capacity. In addition to the critical effects of improper fastening, roof panels were seen to fail more often near the roof edges at gable-ends, as discussed above. Roof failure followed collapse of end-trusses at gable-ends, as this building component is one of the most critical in withstanding wind forces. In fact, gable-ends within the overall surface area of pitched roofs are the zones of maximum wind pressure, as indicated by widely used typical values of the external pressure coefficients.

Roof failure was one the most frequent types of building damage and failure of roof sheathing was a major effect. A later study done about a month after the first one followed the hurricane currents. This survey found that 37% of buildings in the area suffered no loss of roof sheathing panels, while 63% lost one or more roof deck panels. However, even when roof decks did not suffer damage, roofing materials were never left intact. Further, aside from roofs, different types of structural damage to framing affected 37% of buildings (see Fig. 2.19). The buildings inspected in this survey were located in the Country Walk and Lake of the Bay developments and were selected from those that had received medium to severe damage. Of the 34 buildings submitted to detailed analysis, 27 had lost roof deck panels: 14 included plywood and 13 included OSBs. No structural difference between the two types of panels was observed. In both cases, when fasteners failed it was due to insufficient withdrawal resistance. Curiously, either the tip of the fastener was loosened from the framing (the most common occurrence) or the roof panel pulled through the fastener head (rare), leaving the tip still connected to the framing.

Roof failure in this survey was a result of a combination of the effects of external and internal pressure. When windows or doors failed on the windward side allowing air to increase internal pressure, the positive pressure underneath the roof deck added to the negative pressure (suction) above, affecting the structural strength of the roof much more severely than in buildings where internal pressure did not increase.

Asphalt shingles were the most widely used roofing material in this area. They typically failed due to an insufficient number of fasteners. Staples were usually always used in numbers smaller than the minimum required by local building codes, and most failures occurred not due to staple withdrawal but to tears in the shingles themselves, which then pulled through the crown of staples. Flat concrete and barrel-shaped concrete tiles were also used as shingles. The flat type demonstrated higher resistance to uplift than barrel-shaped tiles. We should note that installation of both types of concrete tiles in wooden roof decks is usually accomplished by laying a thin mortar bed over the roof deck and embedding the tiles into the mortar. However, the damage analysis demonstrated that tile uplift occurred due to a lack of bonding between tiles and mortar, but that bonds between mortar and roof deck usually did not fail. No mechanical fasteners had been used to secure concrete shingles to decks, though special perforations were built into the tiles for such purpose. These concrete tiles have become extremely popular in roofing construction in southern Florida and now constitute the majority of roofing systems, replacing the previously popular clay tiles that were traditionally a major component in the architecture of this region. However, the new concrete tiles do not allow the same bond with the mortar bedding as the clay tile had, and are thus a less desirable roofing system with regard to wind uplift. Furthermore, concrete tiles were a major source of flying debris carried aloft by the high-velocity winds. Complete tiles and tile fragments became deadly missiles that struck other structures and damaged whatever was in their path. As mentioned above, it was poor bonding and the absence of mechanical fasteners that generated tile failures. Obviously, special measures can be taken to reduce or eliminate these effects.

Modular Building Structures

Modular structures, including mobile homes, located in the surveyed area demonstrated a strong vulnerability to the wind forces of Hurricane Andrew (see Fig. 2.20). This particular building construction has totally unique inherent structural deficiencies, compared to the other numerous types of traditional building typologies, which make them particularly vulnerable to the effects of lateral forces. Their collapse was usually extremely severe and could only be classified as total losses. The most outstanding feature of their behavior under wind loads is the instability of their connection to external supports. Uplifting and overturning due to the lack of anchorage to a solid foundation was their most obvious deficiency. This is emphasized by the partial benefit that some adequate tie-downs demonstrated in some cases. Tensile ties running over the roof deck and solidly connected to the ground spared some structures from total destruction, demonstrating their effective behavior in resisting uplifting and overturning. However, their action does not have an influence on the effects of the other destructive actions of wind forces. The positive effects of tie-downs were observed in some cases, but only when properly installed. Loosely installed tie-downs had no effect. In addition, tie-downs that did not extend over the roof were also totally insufficient to withstand wind loads, allowing the upper structure to be displaced, only fastening the main chassis to the ground.

Figure 2.20. Damage to mobile homes, one of the most typical housing facilities in Florida. Hurricane Andrew, Dade County. (Photo courtesy of FEMA (1992a).)

Among the other major deficiencies of modular construction is the inadequacy of their primary structural frame, including a particular vulnerability along connections between walls and walls and walls and roofs. The stiffening effects that side panels could add to a structure by creating structural diaphragms is also not obtained because of the extreme flimsiness of the light-gauge panels used, whether these are made of light-gauge steel, aluminum, or plastic. Particle boards used for interiors are also undesirable due to their vulnerability to moisture, which eventually penetrates behind structural components, especially around windows and doors, and constitute a concentrated structural weakness in these areas. Such behavior is further accentuated when the flashing around windows and doors is not properly installed.

In addition to the popular mobile home building methodology, another frequent modular building system involves structures partially prefabricated in a shop and eventually assembled on-site. Typical of this type of building are modular apartment buildings, many of which were studied in the first survey. These include small (800–1100 ft^2) residential single-story units at Homestead Air Force Base. They have precast concrete foundations and a prefabricated wooden upper structure assembled on-site. The anchorage of wood-framed floor systems to concrete foundations by means of lag bolts proved to be inadequate, and some structures were lifted from their foundations and rolled over. This system reflects a similar structural deficiency found in conventional wood-framed building: unsatisfactory connections between walls and walls, walls and floors, and walls and roof systems (see Figs. 2.21–2.23).

Figure 2.21. Connection of wood-framed wall sections to adjacent wall at corners, and wall connections to roof, floor, and cross-connecting interior partition walls, occasionally were inadequate to withstand wind forces. Inadequate attachment of roof sheathing to gable-end truss, and gable-end truss to end wall, often led to this type of building failure. Hurricane Andrew, Dade County. (Photo courtesy of the American Plywood Association (Keith and Rose 1992).)

Figure 2.22. High vulnerability of prefabricated homes. Hurricane Andrew, Dade County. (Photo courtesy of FEMA (1992a).)

Figure 2.23. Prefabricated home totally destroyed. Hurricane Andrew, Dade County. (Photo courtesy of FEMA (1992a).)

Metal Structures

Most metal buildings typically employed for commercial occupancy adopt the structural action of a diaphragm to provide overall static stability. Diaphragm action is usually employed vertically and horizontally, respectively, through efficient utilization of walls and roof decks, providing that adequate stiffness is achieved. Loss of wall panels and roof decking will of course reduce such action when moderate damage occurs and eliminate it completely when extensive damage occurs (see Fig. 2.24). Such effects from reduced or absent diaphragm resistance can be catastrophic to the overall stability of metal buildings (see Fig. 2.25). The following conclusion emphasizes the weakness of metal sheetings and decking with regard to the fastening system that connects them to the main frame: the major causes of failure was due to the pulling through of roof decking from supporting systems below, with either individual mechanical fasteners or spot welding. Even when roof metal decking was installed over plywood decking, the metal roof sheeting was pulled away by the wind uplift force, detaching it from the plywood below, which remained fastened to the structural frame.

Masonry Structures

Observations of the behavior of masonry buildings include some interesting findings. These buildings included bearing and nonbearing masonry walls and a precast T-roof system, in which case the nonbearing masonry walls, parallel to the length of the T, were observed to fail due to wind forces perpendicular to their surfaces, falling inward or outward according to whether the wind pressure was positive or negative (see Figs. 2.26 and 2.27). Additional

Figure 2.24. Typical failure of metal roof decks in commercial type steel-framed buildings. Hurricane Andrew, Miami area. (Photo courtesy of the American Plywood Association (Keith and Rose 1992).)

Figure 2.25. Large low-rise commercial building mostly destroyed when steel roof deck diaphragm failed, allowing concrete tilt-up walls to collapse. Hurricane Andrew, Miami area. (Photo courtesy of the American Plywood Association (Keith and Rose 1992).)

Figure 2.26. Failure of automated rack storage warehouse with precast concrete T-beam roof deck and precast concrete and masonry walls. Hurricane Andrew, Miami area. (Photo courtesy of the American Plywood Association (Keith and Rose 1992).)

Figure 2.27. Masonry wall inadequately attached to concrete T-beam roof deck panels with welded steel connections spaced 10 ft o.c. Hurricane Andrew, Miami area. (Photo courtesy of the American Plywood Association (Keith and Rose 1992).)

findings indicate a weakness in masonry walls at the corner of structures where two walls intersect. It was concluded that this weakness derived from a lack of steel reinforcement within the walls. In addition, failure of nonstructural brick veneer due to wind suction was usually determined by a lack of adequate anchorage of veneer to the wall behind it, due either to inadequate mechanical anchorage or to a lack of bonding between veneer and masonry wall.

Studies

The investigations that followed Hurricane Andrew have prompted various studies aimed at providing engineering recommendations to assist in protecting structures from high-velocity wind forces. One study conducted by the American Plywood Association concerned fasteners and recommended that roof sheathing be secured, considering the uplift and shear forces that roof decks are subjected to during high-velocity wind action.

Hurricane Andrew: A Layman's Approach

In taking a nontechnical approach, it is often rewarding to look at a problem in terms of common sense. We first consider a statement made in the December 20, 1992 issue of *The Miami Herald* in article entitled "What Went Wrong?" about the efficiency of building technology under the devastating force of Hurricane Andrew:

> Hurricane Andrew did not ravage all neighborhoods equally. Wood-framed houses in Country Walk ... were devastated. The more flimsy Dadeland Mobile Home Park ... was leveled. Houses in Mediterranean, below the mobile home park, were badly damaged. But the concrete-block Munne Estates, the houses with the distinctive red-tiled roofs, survived the storm relatively intact. The pattern was repeated throughout South Dade. Damage followed the rigid lines of subdivisions, not the whimsy of wind. Construction quality and design largely determined the degree of hurricane damage.

In addition to these candid remarks, which structuralists should always keep in mind, this article also points out other revealing facts.

The newspaper's investigation covered over 60,000 damage reports through computer mapping of areas of damage and comparing them to estimated wind velocities maps. They found that the two did not coincide, i.e., maximum damage did not necessarily occur where wind velocity was highest. Since the strongest factor affecting damage was identified as building technology rather than wind velocity, the study also highlighted the superior performance of dwellings built before 1980. We in no way mean to minimize the importance of a detailed engineering analysis of damage; nevertheless, it is important to retain a sense of proportion and a sensible perspective, so that which is most obvious is ignored.

12. AFTERMATH RECOMMENDATIONS

After the August 24, 1992 natural disaster of Hurricane Andrew, the Governor's Disaster Planning and Response Review Committee was formed (by Executive Order 92-242) and was tasked to make recommendations on improving Florida's readiness for future disasters. The recommendations of this committee were ready by January 11, 1993, and were presented to the Governor immediately thereafter (State of Florida, 1993). We briefly review them below in the hope that they could serve as a guideline for projects in other parts of the world (see also Lewis 1993).

This document includes an introduction which focuses on improving readiness for future disasters through improved measures in communications, planning, intergovernmental coordination, training, and funding.

The first chapter includes three sections: Communications, Evacuation, and Shelter. Under Communications, the recommendations cover public education, predisaster communication with the public, and communications among emergency response agencies. The topic of Evacuation is subdivided into planning, transportation systems for mass evacuation, and safety measures for people with regard to marinas and bridges. Shelter is subdivided into planning and operations, usage of schools and other public buildings for shelters, reducing shelter demand, registration of buildings for shelter use, and facilitating sheltering operations for handicapped and others in need.

The second chapter includes five sections: Post-Disaster Communications and Public Information; Post-Disaster Response and Recovery Operations; Availability of Damage Assessment Data; Medical Care and Relief; and Coordination of Volunteers, Donations, and Supplies. Post-Disaster Communications and Public Information is subdivided into coordination of emergency communications before and after storms and coordination of information and instructions to the public. Post-Disaster Response and Recovery Operations is subdivided into planning for response and recovery, intergovernmental coordination, security, looting protection, and improvement of federal relief programs. Availability of Damage Assessment Data covers aerial and ground assessment and computerized databanks. Medical Care and Relief covers coordination of medical services, a statewide medical plan, coordination of health care volunteers, and emergency pharmacies. Coordination of Volunteers, Donations, and Supplies is subdivided into collection/distribution of donations and coordination of volunteers.

The third chapter deals entirely with funding and recommends that a state disaster assistance trust fund be established.

Plans of this type which hope to guarantee a program of preparedness for violent storms such as hurricanes and typhoons are obviously necessary, especially for heavily populated regions of the world with large metropolitan areas sited in geographical zones with a climatology favorable for such storms. This briefly outlined set of recommendations in its skeletal structure can furnish a starting point from which other communities around the world can begin to organize themselves, adjusting each item to their specific needs.

An evaluation of these recommendations points out its particular relevance for hurricanes and typhoons, and not to other types of severe storms, such as tornadoes. Much attention is paid to the process of evacuation of a region, and we can note that the nature of hurricanes and typhoons allows for such warning. In addition, the length and width of the path of destruction in hurricanes and typhoons are of such magnitude compared to other types of storms that the damage involved usually reaches extraordinary proportions and affects not just local but national concerns, thus bringing with it social, political, and economic consequences of a special character that has no comparison to the effects generated other kinds of storms. Even the most violent tornadoes, with all their fury and deadly violence, will not generate comparable impact on a national level as destructive hurricanes, which can bring about large-scale government regulation and executive/legislative action.

13. BUILDING CODE IMPROVEMENTS TO DEAL WITH THE FORCE OF HURRICANES

Evaluation of damage wreaked behind by Hurricane Andrew has produced positive and constructive recommendations that could be applied to improve building codes in areas where hurricane-velocity winds are likely to occur. A group of experts published a report on a workshop conducted in October 1992 in Ft. Lauderdale, Florida, which included professionals working in wind engineering, meteorology, and building design/construction. Their efforts generated a significant amount of data that will be briefly discussed below (see Cook 1993). The specific categories addressed in the workshop included wind speeds, wind loads, building codes, building enclosures, fenestration, doors, building inspection, dissemination of information, and prefabricated housing.

The workshop generated 28 specific recommendations. which, though not unanimously approved by workshop members, did represent the view of a substantial majority. We present some of these recommendations as follows:

They decided that designs related to wind loads and residential structures should conform with the specifications of the ACSE 7 (American Society of Civil Engineers 1990) documents, except for structures below 60 ft in height, to which should be applied §1205.2 of the American Standard Building Code. They also recommended that standards for high-velocity wind regions be applied to single- and two-family dwellings.

Special attention was given to clarification of the widely different definitions of wind speed. The recommended definitions are those contained in ACSE 7: for basic wind speed use 3- to 5-second peak gust instead of the fastest-mile wind speeds. Using this definition would mean that all the wind speed values in the ACSE 7 documents need to be converted.

They recommended that residential construction within 20 miles of a coastline be designed according to ACSE specifications, which take into account the convective gustiness of hurricane wind flows by assuming a value of the Importance Factor (I) equal to 1.05 (§1205.2 of the American Standard Building Code a value of 1.10). When Southern Building Code Congress International (SBCCI) standards are used, the value of the fastest-mile wind speed should be multiplied by 1.05, but the maximum value should not exceed 110 mph.

They also recommended that a set of regulations within the building codes be enforced to guarantee adequate safety for typical residential structures designed without adoption of a wind engineering analysis. Wind analysis for more critical buildings would not be superseded because of the minimum standards included in the building codes, since they would be only be applied to noncritical engineering structures (which would otherwise be bypassed in a rigorous wind analysis). In establishing design and construction criteria for residential structures within this context, they strongly adopted wind engineering, as opposed to the empirical trial-and-error approach.

The fundamental objective of the design of building envelopes to guard against the effects of wind loads is based on providing adequate resistance against wind pressure and protection from the impact of flying missiles carried aloft. This point has been reiterated over and over again in the many reports on hurricane damage. The same reports have continually pointed out that additional damage results when the building envelope is violated, so that wind and rain penetrate into the interior. In fact, when a strong wind hits the exterior of a building, its pressure may cause additional damage, acting on walls and especially on the underside of roof decks, thereby contributing to uplift of roof structures. Significant damage can also result to the exterior when the wind penetrates into the building, an event which can exponentiate the cost of resulting damage. Based on the foregoing, the workshop recommended that particular attention be paid to the integrity of the building envelope in terms of fenestration and doors, the most vulnerable envelope components, and that special consideration be given to protection of glass surfaces through shutters as well as connection of window and door frames to the main structure.

They further recommended that the design of single- and multifamily dwellings located in high-wind regions meet SBCCI standards.

Based on the numerous observations of gable-end wall failures, they recommended that vulnerable structural component be strongly reinforced. Trusses at the ends of a gable roof should be especially braced to the other trusses, especially at the level of bottom chords. Special consideration should also be given to the ceiling system, where the bottom part of trusses should be independent of each other, so that failure of one will not compromise the integrity of others. Further within the context of gable-end failures and roof integrity, a rational structural framing of roof overhangs was strongly recommended.

To increase the safety standards of existing buildings, whether residential or commercial, it was recommended that shutters be applied to doors and windows not as mandatory requirements but as recommended practice with appropriate incentives. To provide such an incentive, the recommendations point to special credits that insurance companies could supply under the sponsorship of the State Board of Codes and Standards.

Based on the vulnerability of metal edge flashings, roof coverings, and windows and doors under wind loads, it was recommended that research in this area be expanded by the building industry and that it follow the recommendations of the State Board of Codes and Standards.

CHAPTER THREE

Tornadoes

1. INTRODUCTION

The three chapters of the present work deal with high-velocity wind in order of increasing intensity. At the peak stand those violent storms referred to as tornadoes in English-speaking countries. Several other terms are used in other languages which present a different pictorial image of these storms. The Italian language uses *tromba d'aria* (i.e., "air trumpets") because of shape of the storm resembles that of these musical instruments, which start with a conical-shaped funnel near its parent clouds and extend down in a slanted tube. The Latin term used to identify this phenomenon was *tonare*, which can be used in formulation of a more general terminology which would include tornadoes with different intensity, size, wind speed, pressure drop, and other mechanical characteristics, as well as waterspouts.

Tornadoes form a category of storms that drastically distinguish themselves in terms of wind speed from large tropical cyclones such as hurricanes and typhoons. They are extremely concentrated phenomena that in practical terms do not compare with others in terms of overall damage, and therefore commanded less attention within engineering circles until the advent of such supercritical structures as nuclear power plants. One could say that these structures were the main reason that has prompted so many studies of these storms and brought them to the attention of the world in much more conspicuous terms. Even if they do not cause damage comparable to that produced by hurricanes, the feelings they induce in the population of tornado-prone areas comprise a terrifying fact of life for millions. It is well known that the United States constitutes a geographical zone where tornadoes occur with the highest frequency. Because of the high standard of living, the fear of constant exposure to tornado-induced damage, and the casualties incurred, needless to say, tornadoes constitute a major concern.

The following account reports on a major tornado system that illustrates typical tornado activity which strikes vulnerable areas across the United States.

On Sunday, March 27, 1994, a series of tornadoes developed within a weather system over in the Gulf of Mexico. Moist unstable air rose, began moving, then moved northeasterly under jet stream action and collided with a low-pressure cold front. Tornadoes hit Alabama, Georgia, Tennessee, and the Carolinas that caused several fatalities, numerous injuries, and severe property damage.

The damage in Charlotte caused by more several tornadoes that hit the west side of town amounted to $20 million. More than 300 residential buildings were affected, 10% of which were destroyed. Other typical damage included uprooting of trees and collisions between trees and buildings. Also typical were damage to vehicles, damage to light storage buildings, and falling powerlines (see Figs. 3.1–3.9).

An episode that caused the greatest number of fatalities during this storm system was the structural collapse of a church in Piedmont, 20 miles southeast of Gadsden, in northeastern Alabama at 11:30 a.m. CST. The tornado hit the Goshen Methodist Church and caused uplifting of the roof structure followed by inward collapse of the supporting walls, which crushed the interior. With detachment of the roof, the concrete block masonry walls were no longer stabilized by the diaphragm action of the roof, thus loosing their ability to resist horizontal forces. Due to the high-density occupancy during a service, casualties included 19 dead and 90 injured. It is important to note that religious buildings are involved in such tragedies.

The 44 fatalities recorded during this storm system was a low number for such high-intensity tornado activity. However, this number is equal to normal annual total. Of note in terms of the high electrical activity in the atmosphere associated with tornadoes and the vulnerability of aerial utility lines are two fatalities, one caused by lightning on the front porch of a residential building (Charlotte, NC) and the other due to electrocution by accidental contact with a downed powerline (Boger, NC).

In this chapter tornadoes are given highest priority, reflecting the terror they cause for millions of people and the critical role they play in the design of nuclear power plants because of the high-intensity forces they generate.

Before getting into a detailed treatment of the subject, we present a brief overview. The most common tornadoes in the Northern Hemisphere are weak, of short duration (approximately 1–2 minutes at ground level), and move in a northeastern direction at 30 mph with an approximate path of 1 mi × 150 ft. A rare few extremely violent tornadoes can last over an hour, move at speeds up to 68 mph, and include a path up to 100 miles wide and 2.5 miles long. The overall distribution of storm types in terms of violence includes 62% minimum intensity, 36% strong, and 2% extremely violent tornadoes.

The threat posed by tornadoes to the containment structures at nuclear reactors cannot be overemphasized especially. Some of the effects of the disaster at Chernobyl in the Ukraine will last for thousands of years and tens of thousands of lives will eventually be cut short. In terms of the United States, considering the effects that tornadoes have on the population by themselves, the horror of a catastrophic nuclear accident is still hard to image.

Official government records allow us to compare the strength of tornadoes and other natural atmospheric phenomena. The National Oceanic and Atmospheric Administration (NOAA) (1966–80) publishes an annual *Climatological Data: National Summary* along

Figure 3.1. Tornado damage, March 27, 1994, Charlotte, NC. This interior photograph of a building on Jay Street clearly shows the path of a tornado, with its twisted sheet metal and steel. The front and right walls of this brick building remained standing, while the rear and half of the left wall were destroyed. (Photo courtesy of Deborah Melaragno.)

Figure 3.2. Tornado damage, March 27, 1994, Charlotte, NC. In the Tuckaseegee Road area, on the corner of Jay and Thrift Streets, a building exploded and completely collapsed. The buildings across the street were hardly affected or not affected at all. (Photo courtesy of Deborah Melaragno.)

Figure 3.3. Side view of building in Fig. 3.2. (Photo courtesy of Deborah Melaragno.)

Figure 3.4. Tornado damage, March 27, 1994, Charlotte, NC. Just off Berryhill Road, this large uprooted tree smashed through the roof of a house. (Photo courtesy of Deborah Melaragno.)

Figure 3.5. Side view of Fig. 3.4. (Photo courtesy of Deborah Melaragno.)

Figure 3.6. Tornado damage, March 27, 1994, Charlotte, NC. Typical damage along the path of the tornado between Tuckaseegee Road and Jay Streets. (Photo courtesy of Deborah Melaragno.)

Figure 3.7. Tornado damage, March 27, 1994, Charlotte, NC. Typical scene of tree trunks on roof, with gutters ripped from overhangs, broken glass, torn shingles, and dislodged doors. Just off Berryhill Road. (Photo courtesy of Deborah Melaragno.)

Figure 3.8. Tornado damage, March 27, 1994, Charlotte, NC. Total devastation to the interior of a commercial building with destruction of roof and facade. (Photo courtesy of Deborah Melaragno.)

Figure 3.9. Tornado damage, March 27, 1994, Charlotte, NC. Clearing efforts after the storm. (Photo courtesy of Deborah Melaragno.)

with monthly summaries (the *Storm Data* monthly). Table 3.1 presents the major tornadoes, floods, and tropical cyclones that occurred between 1983 and 1993, presented in terms of their destructive effects in terms of damage and casualties.

The numbers in the table for U.S. fatalities do not indicate any substantial difference in order of magnitude among tornadoes, floods, and tropical cyclones. In fact, the numbers are relatively small considering the continental expanse involved. However, when considering Asia, these numbers assume disproportionately high values that a first glance escape comprehension: the 70,000 deaths caused by the April 30, 1991 cyclone, the 10,000 deaths caused by the May 25, 1985 cyclone, and the 300,000 deaths caused by the November 13, 1970 cyclone, all in Bangladesh. Reflecting on these data, it is logical to assume that such a disparity in loss of human life could not be completely accounted for by a higher intensity of the wind force of these storms and must involve the nature of the structures which housed the victims. The vulnerability of the structures involved makes the context of civil engineering much more important.

Catastrophes generated by the action of high-velocity winds bring into perspective the importance of the scientific basis for the building arts and the absolute necessity of regulating this ancient art with a set of ethics that reflects the dignity of mankind. Traditional building technologies, often as old as the origins of the written language of the residents, have proved inadequate when tested by forces that go beyond an intuitive understanding of conventional loads, such as what is typical during earthquakes. This situation is severely exacerbated when the structures involved are substandard within their own cultural context due to the restrictions caused by poverty, corruption, negligence, etc. This region of Asia, where humans live in precarious shelters, brings the topic of natural disasters back into the

Table 3.1. Tornadoes, Floods, and Tropical Storms, 1983–93

Item	1983	1984	1985	1986	1987	1988	1989	1990	1991	1992	1993 (prel.)
Tornadoes[1]											
Number	931	907	684	764	656	702	856	1133	1132	1303	NA
Lives lost, total	32	122	94	15	59	32	50	53	39	39	NA
Most in one tornado	3	16	18	3	30	5	21	29	13	10	NA
Property loss >$500,000	95	125	69	75	38	48	60	91	64	NA	NA
Floods											
Lives lost	200	126	304	80	82	29	81	147	63	87	NA
Property losses (mill.$)	4100	4000	3000	4000	1490	114	415	2058	1416	800	NA
North Atlantic tropical storms and hurricanes[2]											
No. reaching U.S. coast	4	12	11	6	7	12	11	14	8	6	8
Hurricanes only	1	1	6	2	1	1	3	–	1	1	1
Lives lost in U.S.	22	4	30	9	–	6	56	13	15	24	3
Property losses (1990 mill.$)[3]	2392	77	4457	18	8	9	7840	57	1500	25,000	35

– = 0; NA = not available.

[1]A violent rotating column of air descending from a cumulonimbus cloud in the form of a tubular- or funnel-shaped cloud, usually characterized by movements along a narrow path and wind speeds from 100 to 300$^+$ mph. Also known as a "twister" or "waterspout." [2]Source: National Hurricane Center, Coral Gables, FL, unpublished data. Tropical storms have maximum winds of 39–73 mph; hurricanes have maximum winds of 74$^+$ mph. [3]Source: Hebert, Jarrell, and Mayfield, "The Deadliest, Costliest, and Most Intense U.S. Hurricanes of This Century," NOAA Technical Memo NHC-31, February 1993. Source: Except as noted, U.S. National Oceanic and Atmospheric Administration, Storm Data (monthly).

proper perspective of providing for basic human needs on a global scale using the most current knowledge of the building sciences.

2. TORNADOES: GENERAL CHARACTERISTICS

Scientific description and analysis of the tornado phenomenon date back to the late seventeenth century, when François Lamy (1689) published a work on the physical interpretation of air vortices. Five years later, a posthumous treatise by Geminiano Montanari on the physical and mathematical aspects of tornadoes was published, which provided a detailed description of a storm that struck the environs of Venice in northern Italy on July 29, 1686. Another early work on was Boscovich's (1749) manuscript about the tornado which damaged much of Rome on June 11–12, 1749. A century later Peltier (1840) published his work on tornadoes in Paris. Much has been published on the meteorological and physical aspects of tornadoes since these early studies, but very little about the forces that produce destructive effects on buildings.

The so-called tornado season begins with the approach of each spring in the United States, bringing death and damage to somewhere in the country. Statistical data collected during the 16-year period of 1953–69 indicate an average number of 640 tornadoes per year, with a peak of 912 in 1967 and a minimum of 437 in 1953 (data provided by the NOAA).

Subsequent data give us a more precise understanding of U.S. tornado frequency due to the much more sophisticated detection system network which leaves few storms undetected. 1992 data indicate a total of 1293 tornadoes, with two temporal concentrations: 123 tornadoes in mid-June and 93 in November the week before Thanksgiving. The maximum monthly number was 399 in June. A very large number of tornadoes was also reported in 1990 (1133) and 1991 (1125).

Focusing on the storms that caused the most deaths, Table 3.2 furnishes a comprehensive view of killer storms within the period 1925 to 1994. The table indicates only the most violent storms and the respective number of deaths. The highest number of lives taken (689) was caused by the storm of March 18, 1925. The next highest total (455) was caused by the storm of April 5, 1936 in Mississippi and Georgia, followed by the storm of April 3–4, 1974 (315). The annual average losses for this period are $75 million and 71 lives.

Scientific information on the qualitative and quantitative characteristics of tornadoes was for many years incomplete and inconclusive. Public opinion (shared by design professionals) held that nothing could be done to prevent or alleviate losses and human suffering.

The present situation is quite different. Though the problem persists, the body of scientific knowledge on the subject has increased considerably in the last few years, especially in the fields of meteorology and fluid dynamics. The majority of research clearly indicates that the forces generated by tornadoes are much lower than previous estimates. Wind velocity estimates which at one time exceeded 300 mph are now on the order of 200 mph. These new numbers give structural designers a completely different perspective, so that tornado-resistant structures now seem feasible, not only from a technical but also from an economic point of view.

For the first time in the history of construction, tornado-resistant structures were designed and built to resist tornadic forces based systematic scientific approach in the late 1960s. These structures included nuclear power plants built in the United States and elsewhere. It is feasible to predict that the design criteria used by future architects and engineers for other

Table 3.2. Notable U.S. Tornadoes, 1925–94

Date	Location	Deaths
1925 Mar 18	Missouri, Illinois, and Indiana	689
1927 Apr 12	Rock Springs, Texas	74
1927 May 9	Arkansas, and Poplar Bluff, Missouri	92
1927 Sep 29	St. Louis, Missouri	90
1930 May 6	Hill, Navarro, and Ellis counties, Texas	41
1932 Mar 21	Alabama (series of tornadoes)	268
1936 Apr 5	Mississippi and Georgia	455
1936 Apr 6	Gainesville, Georgia	203
1938 Sep 29	Charleston, South Carolina	32
1942 Mar 16	Central to NE Mississippi	75
1942 Apr 27	Rogers and Mayes counties, Oklahoma	52
1944 Jun 23	Ohio, Pennsylvania, West Virginia, and Maryland	150
1945 Apr 12	Oklahoma, Arkansas	102
1947 Apr 9	Texas, Oklahoma, and Kansas	169
1948 Mar 19	Bunker Hill and Gillespie, Illinois	33
1949 Jan 3	Louisiana and Arkansas	58
1952 Mar 21	Arkansas, Missouri, and Tennessee (series)	208
1953 May 11	Waco, Texas	114
1953 Jun 8	Michigan, Ohio	142
1953 Jun 9	Worcester and vicinity, Massachusetts	90
1953 Dec 5	Vicksburg, Mississippi	38
1955 May 25	Kansas, Missouri, Oklahoma, and Texas	115
1957 May 20	Kansas, Missouri	48
1958 Jun 4	Northwestern Wisconsin	30
1959 Feb 10	St. Louis, Missouri	21
1960 May 5–6	Southeast Oklahoma and Arkansas	30
1965 Apr 11	Indiana, Illinois, Ohio, Michigan, and Wisconsin	271
1966 Mar 3	Jackson, Mississippi	57
1966 Mar 3	Mississippi and Alabama	61
1967 Apr 21	Illinois and Michigan	33
1968 May 15	Midwest	71
1969 Jan 23	Mississippi	32
1971 Feb 21	Mississippi Delta	110
1973 May 26–27	South and Midwest (series)	47
1974 Apr 3–4	Alabama, Georgia, Tennessee, Kentucky, and Ohio	315
1977 Apr 4	Alabama, Mississippi, and Georgia	22
1979 Apr 10	Texas and Oklahoma	60
1980 Jun 3	Grand Island, Nebraska (series)	4
1982 Mar 2–4	South and Midwest (series)	17
1982 May 29	Southern Illinois	10
1983 May 18–22	Texas	12
1984 Mar 28	North Carolina and South Carolina	57
1984 Apr 21–22	Mississippi	15
1984 Apr 26	Oklahoma to Minnesota (series)	17
1985 May 31	New York, Pennsylvania, Ohio, and Ontario (series)	75
1987 May 22	Saragosa, Texas	29
1989 Nov 15	Huntsville, Alabama	18
1989 Nov 16	Newburgh, New York	9
1990 Jun 2–3	Midwest and Great Lakes	13
1990 Aug 28	Northern Illinois	25
1991 Apr 26	Kansas and Oklahoma	23
1992 Nov 21–23	South and Midwest	26
1993 Aug 6	Petersburg, Virginia	4
1994 Mar 27–28	Alabama, Tennessee, Georgia North Carolina, and South Carolina (series)	52
1994 Apr 26	Central Texas	4

types of buildings will include specifications to withstand expected tornadic forces when the region has a high probability of their occurrence.

A couple of trends corroborate such a prediction. The present expansion of housing centers into rural areas makes it much more likely that tornadoes will strike urban centers, and this trend is constantly increasing. In addition, the standard of living in the United States and other countries is such that the present toll in terms of lost lives and property destruction will be increasingly unacceptable.

The present state of the art in tornado engineering includes a general understanding of the atmospheric phenomenon and the magnitude of forces involved, a methodology with which to compute the probability of their occurrence, and some existing storm-resistant structures. It is therefore essential to refine data and to fill in gaps where necessary. In addition, information and data from other scientific disciplines need to be made available to members of the design profession (e.g., architects and engineers) and to the general public.

Tornadoes are not limited to the North America, so that tornado engineering is an international concern. The worldwide proliferation of nuclear power plants makes it absolutely necessary to prepare tornado probability studies for proposed plant sites and to include specific design criteria. Many countries, including the United Kingdom, have no official policy on protection from tornado forces at plant sites. Since the consequences of structural failure and radioactive contamination are undeniably catastrophic, dissemination of safety information is extremely important.

Tornado Formation

Tornadoes usually form during warm and humid weather and are quite often associated with severe thunderstorms. More than one tornado can be formed from a parent thunderstorm.

Theories on the origins of tornadoes vary widely. Some are based on thermal phenomena, some on mechanical or electrical principles, and some on a combination of these factors. One of the most popular is that tornadoes form during unstable conditions when layers of cold air lie above warm air near the ground. As cold air at high altitudes descends, warm air rises in a spiral motion that initiates a vortex (see Fig. 3.10). During their relatively short lifespan, tornado vortices descend from the mother cloud, reach the ground, and occasionally rise, as they move erratically in their translational motion (see Fig. 3.11).

Tornado Parent Storms

The occurrence of tornadoes is not an isolated weather phenomenon but is associated with wide range of atmospheric disturbances. One such disturbance is referred to a the "mesocyclone," which has larger-scale circulation. Mesocyclones are usually on the order of 2–6 miles in diameter (3–9 km). A mesocyclone in turn develops from a larger system, termed a "supercell," which is a particularly virulent thunderstorm. The typical position of the tornado itself within a storm is at the edge of the main updraft area and close to the downdraft. An individual supercell can include more than one mesocyclone (up to six have been observed). These mesocyclones and their corresponding tornadoes are usually distant from one another. Not all tornadoes derive from supercells, but those that do are the most violent ones.

SINKING COLD AIR

RISING WARM AIR

Figure 3.10. According to one theory, tornadoes form when cold air at higher altitudes descends as warm air near the ground rises. The warm air acquires spiral motion as it rises and thus originates the typical tornado vortex.

Tornadoes can also be fathered by hurricanes, after these storms which form over open water reach mainland. Tornadoes are formed within the hurricane circulation zones which constitute the rainbands.

The weather conditions usually observed in association with the birth of tornadoes includes a layer of warm moist air near the ground approximately 3000 ft (1 km) thick, overlain by a layer of dry air. This condition is quite unstable: the more buoyant warmer layer spontaneously tends to rise (convection) but is prevented from doing so by the cooler layer above. These conditions are further intensified when the temperature in the upper layer drops at a rapid rate. Then tornadoes will definitely form when certain disturbances in the upper layer break the inversion state, setting off updraft convective motion. At this point, horizontal wind paths are deflected vertically, creating a vertical wind shear as an upper draft is formed, while a rotatory (anticyclonic) path at speeds higher than 20 knots is formed within a layer up to 9000 ft thick but lower than the jet streams.

In this pretornadic stage, great disturbances (cyclones) form below the jet streams and slowly translate from the west at high altitudes. A layer of dry air proceeds at the same time at medium altitudes from the west, while warm moist ocean air lies near the ground. Simultaneously, low-level and vertical wind shears near the ground complete the scenario, and tornadoes are formed.

The Shape of Tornadoes

Tornado vortices extend downward from their parent thunderstorm cloud to the earth's surface. The dark thunderstorm cloud, several thousand feet tall, is a cumulonimbus (i.e., a

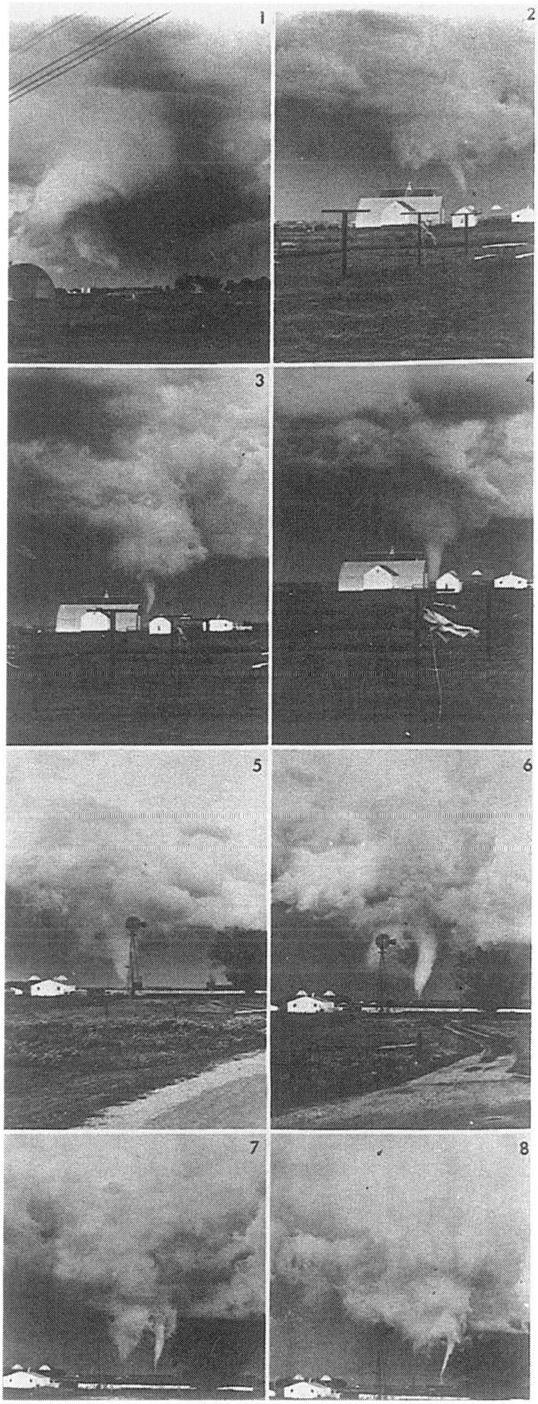

Figure 3.11. Sequential evolution of a tornado: 1) formation of a thunderstorm mother cloud; 2) formation of the tornado funnel; 3) extension of the funnel down from the cloud to the ground; 4) lifting of the funnel and subsequent descent; 5) rising again; 6) descent of the funnel and formation of a second funnel (behind the windmill); 7) further development of the second funnel; 8) regression and extinction of the second funnel while the first is still fully developed.

Figure 3.12. Vortex configuration of the tornado of June 27, 1955, near Scotts Bluff, Nebraska. (Photo by J.E. Jarrell, courtesy of the National Weather Service.)

massive cloud typically shaped like a cauliflower and associated with thunder and heavy precipitation). The shape of the vortex of rotating air is not the same for all tornadoes. The general funnel shape sometimes assumes the appearance of a long slender column that can be either vertical or curvilinear. The shape of large-diameter vortices is usually conical (see Figs. 3.12 and 3.13).

Fire-Generated Vortices

A convective air column rising over high-intensity fires can cause two types of air vortices to form: one resembles a whirlwind, while the other is similar to a tornado. The latter type originates about 1000 ft over the fire and descends to the ground, where it picks up debris, making the funnel visible. It also picks up burning debris if the vortex descends within the fire. This combination of strong wind and burning particles makes these vortices particularly dangerous. Typical in forest fires, tornado-like whirlwinds originating from long-lasting oil fires have also been reported in the literature. In an 1925 oil fire that occurred in San Luis Obispo, California, a great number of high-speed vortices were observed to move horizontally over the land throughout the several days that the fire raged. Whirlwinds of this type were also generated during the Great Chicago Fire of 1871, with lethal consequences.

Tornado Detection

Vonnegut (1960, 1966) demonstrated that ~150 kHz electromagnetic waves are produced during electrical discharges in the atmosphere, which provides evidence for the theory that tornado formation and its energy mechanism are based on electrical energy. The components in the 54–60 kHz band can be captured on a television screen, which shows a luminous signal that lasts for a few minutes, which has a practical application in tornado detection (Gates 1972). Turn on a television and set the contrast control at maximum. Turn the dial to

Figure 3.13. Conical configuration of tornado vortex. Tornado of June 8, 1953, at Hoytville, Ohio. (Photo courtesy of the National Weather Service.)

the highest numbered channel. Then adjust the brightness of the screen until the contrast is at the threshold of blackness. Turn the selector back to channel 2 without resetting the brightness. Lightning will appear on the screen as flashes. If the screen then becomes bright or if a picture is visible on the darkened screen, a tornado is within 20 miles of you.

Of course, the most dependable tornado detection technique, the one used by weather stations, is radar. A familiar hook-shaped signal on a radar screen indicates the presence of a tornado.

Definitions of Terms

Boscovich (1749) used the term *tromba* (Italian for "trumpet") to describe the shape of the mother cloud of a tornado. The use of this word spread to other lands, becoming *tromben* in German and *trombe* in French. The word *tornado* is Spanish in origin. It appeared in English usage in the nineteenth century.

To avoid misinterpretations, we now clarify some of the terms used in meteorology, according to the International Meteorological Vocabulary (World Meteorological Organization):

1. **Spout**: a phenomenon often comprised of a violent whirlwind revealed by the presence of a cloud column or inverted cloud cone (funnel cloud) protruding from the base of a cumulonimbus or of a "bush" composed of water droplets raised from the surface of the sea, or dust, sand, or litter lifted from the ground.

2. **Tornado**: a) the term given in North America to an intense spout with a large diameter; and b) the term incorrectly given to a violent thunder squall in West and Equatorial Africa.

3. **Dust whirl, sand whirl, dust devil**: an ensemble of dust or sand particles, sometimes accompanied by small litter, lifted from the ground in the form of a whirling column of which varies in height and has a small diameter and an approximately vertical axis.

4. **Waterspout**: a spout above a water surface.

Most European countries use these definitions and refer to such wind phenomena that occur in Europe as "spouts." However, they do use "tornado" in the case of exceptionally large or destructive storms. The term "spout" is not used at all in English and Australian scientific papers. "Tornado" is commonly used regardless of size or location (Evesson 1969; Gilbert and Walker 1963; Kert and Dean 1963; Lacy 1968; Lamb 1957, 1965).

In the United States, as in other English-speaking countries, "tornado" is not limited to storms on North American territory, nor to large vortices, but includes any phenomenon described as a "spout" in the International Meteorological Vocabulary.

Though American tornadoes are often assumed to be much larger than those in other areas, many reports do not validate this assumption. Wegener (1917) reported a maximum diameter for American tornadoes of about 3 km, compared to 2.3 km for European tornadoes. Thus, "tornado" is applicable to American, European, Asian, and similar storms throughout the world. The TORRO (Tornado and Storm Research Organization) Scale devised by Meaden (1976) (see Table 3.4 below) evaluates average tornadoes in Great Britain, France, Germany, and the United States with T values of 8, 9, 9, and 11, respectively.

Waterspouts

The International Vocabulary and many professionals in the field assume that waterspouts are similar to tornadoes, except that they occur on water surfaces. However, there are more differences than similarities between the two. Waterspouts usually originate in developing cumulus congestus clouds (which have a maximum top elevation of 12,000 ft). They may also form in shallow stratocumulus clouds. These phenomena are less intense, are of shorter duration, and move at slower speeds than tornadoes. Unlike tornadoes, they do not necessarily develop within severe thunderstorms, which implies that they have a lower-energy origination source. In addition, their tangential velocity seems to increase linearly from the center upwards a certain distance, as in rotation of a solid, and then decreases. This correlates with the latest theories about tornadoes and was corroborated by measurements reported by Kessler (1970).

Much official data has recently been collected on waterspouts. Segman (1971) reported on work conducted by Golden and others supported by several agencies, including the National Severe Storms Laboratory (NSSL) of the NOAA. One study measured the temperature in the core directly by passing a towed probe through the vortex of five waterspouts. Other data were obtained by dropping smoke flares onto the water and firing flares through waterspouts while movies and still photos were taken. In other experiments, the wing of an airplane was flown through a core, penetrating about 5 ft within. The wing was lifted during approach and dipped during the flyby. These observations seem to imply a rising vertical circulation at the exterior boundary and a downward flow along the interior axis. The same circulation has been proposed for tornadoes, but has not yet been proven. Another interesting observation that could lead to important developments is that the waterspout in this experiment dissipated shortly after the plane passed through it. The vortex of waterspouts dissipated within a short time on four other occasions when an airplane flew within 50 ft them. So there is hope that a mechanism will be discovered to dissipate these phenomena.

Probability of Occurrence

To determine whether tornado forces should be considered in the design of a specific building, it is necessary to know the probability of tornado occurrence in the proposed location. For the United States, it is feasible to use a formula proposed by Thom (1963):

$$P = \bar{z}\, t/A,$$

where P is the probability of occurrence (probable number of tornadoes per year) at the location in question; $\bar{z}$ is the average area of the damage path (Thom calculated this value at 2.8209 mi^2 for the U.S.), t is the mean annual frequency of tornadoes in a zone whose area is A; and A is an area equal to a 1° latitude × 1° longitude square that includes the location in question. The interval of time between two consecutive tornadoes that can strike the same location is denoted by $1/P$, where P is the probable frequency (so that $1/P$ is the probable period). Some critics of the Thom formula state that $\bar{z}$ was calculated based on Kansas and Iowa tornadoes, whose size do not reflect those of the rest of the country. Still, this formula has been the most practical available since 1962, and it is the one used in design of American nuclear power plants. A tabulation of values of t for any location in the United States can be found in Thom (1963).

Thom also studied several tornadoes recorded in Iowa from 1953 to 1962. His findings, which determined the average area of tornado path to be equal to 2.82 mi^2, were found to be valid for the Kansas but not for other parts of the United States, as other researchers have pointed out. In analyzing the data used to determine path length, Thom rejected all reports of paths exceeding 100 mi, since he believed these values to be doubtful. He derived an average length of 3.94 miles based on 106 cases. Similarly, he arrived at an average path width value of 154.5 yd by rejecting values larger than 1000 yd as unreliable.

Skaggs (1970) analyzed tornado occurrences in 12 eastern states. He demonstrated that an average tornado area for the entire United States would be an oversimplification, since a great variety of tornado sizes exists for various states. For example, Skaggs computed the average tornado path area to be 0.59 mi^2 for Texas and 5.38 mi^2 for Minnesota (the minimum and maximum values found for the country). He also found a value less than 0.75 mi^2 for the eastern coastal states. Reevaluating the 1950–64 data for Iowa, he was able to verify Thom's value of 2.82 mi^2.

Howe (1975) also examined the path area of many tornadoes. His work (as well as Skaggs') is quite important, because it renders more accurate path area values when Thom's formula is applied. Howe examined 4400 tornado reports from 1953 to 1972 for all states east of the Rocky Mountains. He followed acceptance criteria similar to those proposed by Thom. He rejected width values smaller than 4 yd or larger than 1000 yd. He also rejected length values that exceeded 100 mi or were smaller than 40 yd, and lengths shorter than widths. His calculations yielded an average area of 0.96 mi^2 for all the states examined. The largest value he found was 1.49 mi^2 for Oklahoma, and the smallest was 0.58 mi^2 for Minnesota (note the discrepancy between Skaggs's and Howe's values).

Based on 8900 observations, the National Severe Storm Forecast Center (NSSFC) calculated the average tornado path area for the United States at 0.25 mi^2 (Schafer et al. 1979).

As mentioned above, Thom's formula is still valid, and it can be updated by using specific values for various locations determined after his work was published. Notice also that Thom's formula was derived as a simplification of a more general formula, which is expressed as follows:

$$P = 1 - [(1 - \overline{z})/A]^t.$$

The terms in this formula are as defined above. Further discussion of Thom's formula must emphasize the importance of the values assumed for A, which influence the final values of P (probability of occurrence). Thom specified A as an area 1° latitude × 1° longitude square that includes the location in question. McDonald et al. (1975) pointed out that the probability value could vary considerably in order of magnitude as A increases (e.g., from 1° to 9° of latitude and longitude).

Tornado Ratings

Several methods have been proposed to rate various tornado characteristics, similar to the scales used to rate earthquakes. These methods include the intensity, overall effects, and size of storms. Each parameter can be related to the number of deaths or the magnitude of damage, or both. In discussing intensity, the number of deaths and the amount of damage can be referred to a certain unit — for instance, a 1 mi^2 surface area. Further specifications can distinguish the unit area in terms of geological conditions (e.g., open country or built-up areas of various density). One can consider absolute number of deaths and damage values when rating in terms of magnitude. Obviously, there are many different ways in which a rating system can be structured.

Various rating systems have been proposed: Hazen (1890), Seelye (1945), Fujita (1971, revised in 1973), Meaden (1976), and Dames and Moore (1975). Hazen (1890) proposed the first tornado rating system based on storm violence. He analyzed 2221 tornadoes that occurred between 1873 and 1888. The scale included nine categories, ranging from 3+ to 1–. The categories were based on a monetary assessment of tornado damage.

Seelye (1945) classified tornadoes in terms of intensity, using a scale from 0 to 5 to describe the expected type of damage for each class. In class 5 storms, well-constructed buildings are demolished. In class 3 storms, outbuildings, verandas, and roofs are carried away. No damage is done because the tornadic funnel cloud does not touch the ground in class 0 storms.

Fujita (1971) proposed a rating scale which describes the damage expected in each of several classes (see Table 3.3). Each class was defined by minimum and maximum expected wind speeds and type of damage produced, described in words and by photographs. He modified the scale in 1973 by substituting new photographs. The classes in the scale range from F– to F12: F– includes wind speeds less than 40 mph which correspond to 12 on the Beaufort Scale, while F12 is equal to the speed of sound in the atmosphere (Mach 1). In reality, class F6 (maximum speed 319 mph) is the highest used because no tornadoes have had higher speeds according to official records. The equation which relates wind speed to class in the Fujita scale is written as follows:

$$\text{Wind speed} = 14.1\,(F + 2)^{3/2}\ \text{(mph)}.$$

Table 3.3. Fujita Scale Classification of Tornado Wind intensity (1976)

Class	Characteristics	Verbal description of damage
F–	Doubtful tornado less than 40 mph	40 mph speed corresponds to Beaufort 8 or "Fresh Gale." Beaufort specification for use on land is "Breaks twigs off trees." Little damage is expected.
F0	Very weak tornado 40–72 mph	This speed range corresponds to Beaufort 9 through 11. Some damage to chimneys or TV antennae; breaks branches off trees: pushes over shallow-rooted trees; old trees with hollow inside break or fall; sign boards damaged.
F1	Weak tornado 73–112 mph	73 mph is the beginning of hurricane wind speed or Beaufort 12. Peels surface off roofs; windows broken; trailer houses pushed or overturned; trees on soft ground uprooted; some trees snapped; moving autos pushed off the road.
F2	Strong tornado 113–157 mph	Roof torn off frame houses, leaving strong upright walls standing; weak structure or outbuildings demolished; trailer houses demolished; railroad boxcars pushed over, large trees snapped or up-rooted; light-object missiles generated; cars blown off highway; block structures and walls badly damaged.
F3	Severe tornado 158–206 mph	Roofs and some walls torn off well-constructed frame houses; some rural buildings completely demolished or flattened; trains overturned; steel-framed hangar warehouse-type structures torn; cars lifted off the ground and may roll some distance; most trees in a forest uprooted, snapped, or leveled; block structures often leveled.
F4	Devastating tornado 207–260 mph	Well-constructed frame houses leveled, leaving piles of debris; structures with weak foundation lifted, torn, and blown off some distance; trees debarked by small flying debris; sandy soil eroded and gravels fly in high winds; cars thrown some distance or rolled considerable distance, finally to disintegrate; large missiles generated.
F5	Incredible tornado 261–318 mph	Strong frame houses lifted clear off foundation and carried considerable distance to disintegrate; steel-reinforced concrete structures badly damaged; automobile-sized missiles fly through distances of 100 yds or more; trees debarked completely; incredible phenomena can occur.
F6–12	Inconceivable tornado 319 mph to sonic speed	Should a tornado with a maximum wind speed in excess of F6 occur, the extent and types of damage may not be conceivable. A number of missiles, such as ice boxes, water heaters, storage tanks, automobiles, etc., will fly a long distance, creating serious secondary damage on structures. Assessment of tornadoes in these categories is feasible only through detailed survey involving engineering and aerodynamical calculations as well as meteorological models of tornadoes.

Source: Fujita (1976).

Meaden (1976) proposed a rating system for tornadoes that refined the concepts adopted by Fujita. The scale he adopted is called the TORRO Scale. The various categories in this scale correspond to certain wind speeds and specific types of damage. Each category is identified by FC (funnel cloud) or a tornado force number from 0 to 12 (see Table 3.4). The

Table 3.4. TORRO Tornado Intensity Scale

Force (T)	Type	Verbal description of damage[1]
FC	Funnel cloud or incipient tornado	No damage to structures, unless to tops of tallest towers, or to radiosondes, balloons, and aircraft. No damage in the country, except possibly agitation to highest treetops and effect on birds and smoke. Record FC when tornado spout seen aloft but not known to have reached ground level. A whistling or rushing sound may be noticed.
0	Light tornado 42–58 mph	A. Loose, light litter raised from the ground in spirals. B. Temporary structures such as marquees seriously affected. C. Slight dislodging of the least secure and most exposed tiles, slates, chimney pots, or television aerials may occur. D. Trees severely disturbed, some twigs snapped off. Bushes may be damaged. Hay, straw, and some growing plants and flowers raised in spirals.
1	Mild tornado 59–77 mph	A. Heavier matter levitated include planks, corrugated iron, deck chairs, light garden furniture, etc. B. Minor to major damage to sheds, outhouses, locksheds, and other wooden structures such as henhouses and outhouses. C. Some dislodging of tiles, slates, and chimney pots. D. Hayricks seriously disarranged; shrubs and trees damage to hedgerows, crops, trees, etc. Trees may be uprooted.
2	Moderate tornado 78–97 mph	A. Exposed, heavy mobile homes displaced; light caravans damaged. B. Minor to major damage to sheds, outhouses, lockup garages, etc. C. Considerable damage to slates, tiles, and chimney stacks. D. General damage to trees, big branches torn off, some trees uprooted. Tornado track easily followed by damage to hedgerows, crops, trees, etc.
3	Strong tornado 98–119 mph	A. Mobile home displaced, damaged, or overturned; caravans badly damaged. B. Sheds, lockup garages, outbuildings torn from supports/foundations. C. Severe roof damage to houses, exposing much of roof timbers; thatched roofs stripped. Some serious window and door damage. D. Considerable damage (including twisted tops) to strong trees. A few strong trees uprooted or snapped.

cont'd

Force (T)	Type	Verbal description of damage[1]
4	Severe tornado 120–142 mph	A. Caravans and mobile home destroyed or gravely damaged. C. Entire roofs torn off some frame/wooden houses and small/medium brick or stone houses and light industrial buildings, leaving strong upright walls. D. Large well-rooted trees uprooted, snapped, or twisted apart. The ground possibly furrowed by tornado spout to a depth of about 1 meter.
5	Intense tornado 143–166 mph	C. More extensive failure of roofs than for force 4, yet with housewalls remaining. Small weak buildings, as in some rural areas (or as existed in medieval towns), may collapse. D. Trees carried through the air.
6	Moderately devastating tornado 167–192 mph	A. Motor vehicles over 1 ton lifted well off the ground. C. Most residences lose roofs and some a wall or two; also some heavier roofs torn off (public and industrial buildings, churches). More of the less strong buildings collapse; some totally ruined. D. Across the breadth of the tornado track, every trace in mature woodland or forest uprooted, snapped, twisted, or debranched.
7	Strongly devastating tornado 193–219 mph	C. Walls of frame wooden houses and buildings torn away; some walls of stone or brickhouses and buildings collapsed or are partly beaten down. Steel-framed industrial buildings buckled. Locomotives and trains thrown over.
8	Severely devastating tornado 220–247 mph	C. Entire frame houses leveled; most other houses collapse in part or whole. Some steel structures quite badly damaged. Motor cars hurled great distances.
9	248–276 mph	C. Many steel structures badly damaged. Locomotives and trains hurled some distances.
10	Intensely devastating tornado 277–306 mph	C. Entire frame/wooden houses hurled from foundations.
11	307–337 mph	C. Steel-reinforced concrete buildings severely damaged.
12	Super tornado 338+ mph	

[1]A, B, and C apply broadly to urban situations; D to rural situations.

Source: Meaden (1976).

relationship between tornado force T and the wind speed v (mph) is derived as follows, starting with the Beaufort Scale equation:

$$v = 1.870(B)^{1.5},$$

Then substituting B,

$$B = 2(T + 4),$$

into the Beaufort Scale equation, Meaden obtained

$$v = 5.288(T + 4)^{1.5}.$$

3. THE MECHANICS OF THE STORM: MEASUREMENTS

Scientific investigation of tornado-induced loads for building design largely depends on measurement of the physical forces involved in storms. Here we are particularly concerned with the destructive forces that cause the damages left in the wake of a tornado. These forces include the wind pressure itself and the pressure on the exterior walls of structures that derives from the sudden change in atmospheric pressure. The measurements discussed herein are thus those that define these forces: wind velocity, barometric pressure, pressure gradient, and rate of pressure drop.

There are two types of measurements: direct and indirect. Direct measurements include data obtained by direct readings on calibrated instruments, which are very rare, as it is quite difficult to set up the proper instrumentation. Instruments must be designed and constructed to operate properly under severe tornado conditions, and they must be located in the path of the tornado. The few numbers available are related to barometric pressure and were recorded by chance on rare occasions when a tornado happened to pass over or near a barometer. In recent years, additional direct measurements have been taken, as reported by Kessler and Lee (1976). These measurements were taken by some of the 30–60 stations located in zones of high tornado probability. Another series of direct measurements was taken by penetrating the core of the vortex of waterspouts with probes, as mentioned above.

The value indirect measurements is derived from other available data when the mathematical relationships between them and the measurements are known. The mathematical relationship sometimes is based on a subjective assumption, which, of course, reduces the reliability of results. In other instances, the results can be considered equivalent to direct measurements because of the direct connection between available data and measurements.

Most values of tornado force have been computed indirectly using one of the following means:

1. Structural analysis of damage.
2. Analysis of ground marks.
3. Examination of motion pictures of tornado funnels.
4. Examination of splinters that penetrated other objects.
5. Analysis of the shape of the funnel.

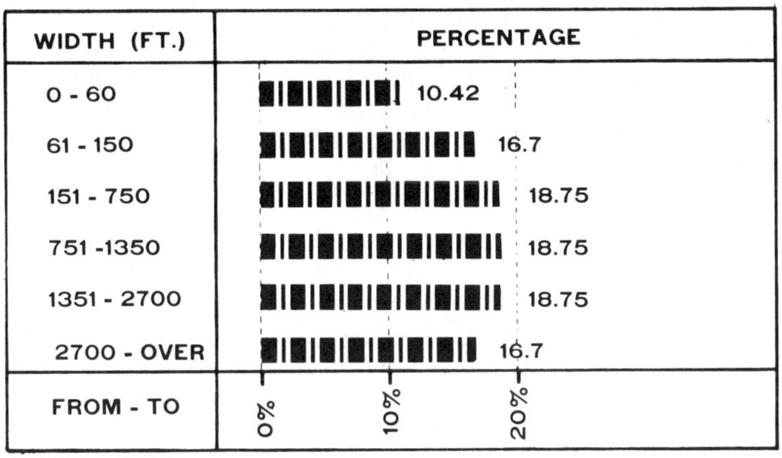

Figure 3.14. Distribution of tornadoes in Australia according to size (width of damage path). Damgage width does not represent the funnel diameter on the ground, since more than one tornado could translate simultaneously.

Derivation of wind velocity from analysis of film (i.e., studying frames, pinpointing airborne particles and observing their change of position from one frame to the next, scaling the distance traveled in the frame, and determining the time elapsed from the frame speed) seems to be very dependable. The results of such analysis can probably be classified as direct measurements because of the precise calculations they allow. Other indirect measurements, however, must be evaluated according to the reliability of the method employed.

Characteristics of the Tornado Path

Because the path cut by tornadoes remains visible for a considerable period of time, it can be measured without much difficulty. Records of virtually thousands of storms have been collected and constitute the basis for several studies, as reported by Flora (1953).

Asp (1963) analyzed 924 tornadoes in Oklahoma and concluded that the average path length was close to 10 mi and the average width 440 yd, from which he derived an approximate area of 2.5 mi^2.

Martin (1940) studied 1000 tornadoes chosen at random from around the United States and found an average path length of 13.4 mi with an average width of 396 yd; the average area would thus be approximately 3.0 mi^2.

In a study of 209 tornadoes in 1950 reported by Flora (1953), it was found that the average path length was 12 mi and the average width 438 yd; the average area was thus about 3.0 mi^2.

The reader will recall that Thom's study (1963), based on tornadoes in Iowa and Kansas, indicated an average path area of 2.8209 mi^2.

Data collected on Australian tornadoes show a considerable variation in size, with diameter maximum values greater than 2700 ft. The category with diameters of 2700 ft and greater includes more than 16% of the storms studied. This seems to disprove the common belief that tornadoes in other parts of the world are smaller than typical North American tornadoes. Figure 3.14 illustrates the distribution of tornadoes in Australia according to size.

Wolford (1960) indicated that the average path length was 16 mi, while Battan (1959) calculated it to be only 2 mi. However, to arrive at this conclusion he rejected all accounts of tornadoes that presented discontinuities. He also indicated that the path length in 95% of the cases was less than 8 miles. Thom (1963) concluded that the average path length for tornadoes in Iowa and Kansas was 3.94 mi and that the average path width was 154.5 yd. Howe (1975) calculated the average path length and width of tornadoes east of the Rocky Mountains were 4.7 mi and 165 yd, respectively.

The prevailing direction of a tornado path is usually from southwest to northeast. Paths are normally curved, and the ends tend to veer northward. Wegener (1917) observed that European usually turn to the right when emerging from wooded areas into open fields or passing from land to water, and turn left when passing from water to land. The edges of the path are not continuous and smooth but reflect rather erratic motion, with sudden changes in direction and diameter; they also at times rise from the ground and then fall.

It is important to notice that the path is indicated by the extent of visible damage, and does not necessarily correspond to the edges of the vortex.

There have been many tornadoes with lengths greater than 100 mi. The longest tornado ever recorded (293 mi) struck Mattoon, Illinois, on May 26, 1917. Widths up to 2 mi have been measured. The May 11. 1970 tornado that hit Lubbock, Texas, was estimated to have been 2 mi wide at some points.

Violence as a Function of Path Length

We define the violence or intensity of a tornado as the absolute value of energy involved per unit path area. However, we can substitute wind speed, wind pressure, or amount of destruction (i.e., injuries, fatalities, and damage) per unit area into this definition in place of absolute energy value, since these are expressions of that energy. Since many factors seem to be associated with the intensity level of tornadoes, either as causes or effects, we should point out that path length is one of the effects that could be used as an index of tornado violence; in other words, tornadoes with longer paths appear to be more violent.

To support this supposition, we can cite the work of several investigators. Wilson and Morgan (1971) indicated that tornadoes with long paths cause more deaths and damage per mile of path length than tornadoes with shorter ones. Pearson (1971) supported this thesis. From an analysis of 740 tornadoes that occurred in 1971–72, Fujita and Pearson (1973) found that length and size were proportional to storm violence. They calculated that higher intensity was associated with path length varying from 3 to 30 mi, and path widths varying from 175 to 556 yd.

Forward Velocity

The tornado system as a whole, from the mother cloud to the funnel bottom, moves horizontally over the earth's surface with a velocity that varies in both magnitude and direction from instant to instant and from storm to storm. Although the general direction of forward motion is usually from southwest to northeast, the path is sometimes very erratic. There are accounts of storms that turned 180°. Because translational velocity is not constant, we must distinguish two different velocities: average and instantaneous.

There are countless records of average velocity since such measurements are easy to make; conversely, little is known about instantaneous velocity, except that it is very erratic, at times falling to 0 mph and then increasing at an undetermined rate.

There are available minimum and maximum average forward velocity values computed from recorded cases, and mean values published by different authors. The lowest recorded average speed is 5 mph for the storm of May 24, 1930, in Pratt, Kansas. The maximum recorded average speed is 65 mph during a storm that hit Kansas from Grenola to Uniontown on May 25, 1917. Flora (1953) mentioned the tornado of March 15, 1938, in Batesville, Illinois, which moved at speeds of 60–65 mph, and the tristate tornado of March 18, 1925, which moved at speeds higher than 59 mph in Indiana. The U.S. Department of Commerce (1965) indicated that the highest recorded translational speed was 68 mph. According to Flora (1953), Martin (1940) calculated the mean to be 45 mph based on 1000 tornadoes selected at random. Tepper (1958) assumed the mean to be between 25 and 40 mph, with an average of 32.5 mph. Kessler (1970) determined the average path length to be 10 miles and the average duration 20 minutes, yielding a mean of 33 mph.Since there are no data to indicate what the maximum instantaneous velocity is, we can only speculate. Since we know that the forward velocity can drop to 0 mph and can also change or reverse direction, it is obvious that there are peak velocity values above the average forward velocity. We therefore know that we cannot use average values in design problems and be safe — some allowance must be made.

Knowledge of instantaneous translational velocity is essential in order to compute the following:

1. The total horizontal component of the velocity, which is the vector sum of forward and tangential velocity.
2. The loading rate due to wind and pressure change.

A direct consequence of computing values 1 and 2 would be to establish whether such loads act dynamically or statically.

3. The rate of pressure change, so as to design adequate venting systems.

Tangential Velocity

Tangential air velocity in the funnel is perpendicular to the vertical axis of the tornado and to the radius of the orbits in which air particles travel. At given altitude, tangential velocity varies with radial distance according to a law that resembles that of Rankine's vortex. This flow consists of a forced vortex inserted in the core of a free vortex. In the forced vortex, which extends distance R from the axis, the fluid particles have viscosity, which makes the flow rotational. The fluid thus acts as a solid, and, as in a solid, velocity varies linearly from zero at the center to the maximum at the edges.

In rotational flow, each particle rotates around its own centroidal axis as it revolves around an axis common to all other particles. To visualize this phenomenon, imagine that each particle has an arrow. If at a given moment all arrows point toward the common axis, they will always be pointing toward it. If we compare them to a stationary arrow which points to the north, they are constantly changing direction, which indicates that the particles are rotating. On the other hand, if the arrows at a certain instant all point north and continue to do so constantly during their revolution around the common axis, the flow is called irrotational.

The flow is rotational in the forced vortex, as in solids. The equation of motion in this case is

$$V = \xi r/2 = (K/2\pi)(r/R^2), \qquad (1)$$

where ξ is vorticity, r is the radius at any point, R is the maximum or minimum (critical) radius at the forced vortex, and $K = \xi\pi R^2$ is the strength of the forced vortex. In the free vortex, the flow is irrotational, which is typical of ideal fluids. The equation of flow here is

$$\begin{aligned} V &= K/2\pi r, \\ Vr &= K/2\pi = \text{constant}. \end{aligned} \qquad (2)$$

This formula is represented by a hyperbola where $V = 0$ and $r = 0$ are the axis and asymptote, respectively.

This airflow pattern was corroborated by a study of a tornado which hit Dallas on April 2, 1957. Hoecker (1960) found that tangential velocity increased linearly from 0 mph at the axis to a maximum of 170 mph at a radial distance of 125 ft at an altitude of 150 ft. It then continued to decrease up to a distance of 250 ft following the following equation:

$$Vr^{1.6} = \text{constant}.$$

From a 250-ft radius outward, the velocity continued to decrease according to

$$Vr^{0.8} = \text{constant}.$$

This variation from (2) indicates that the flow at this altitude was not irrotational; however, the data do indicate irrotational flow at an elevation of 1200 ft. Kessler (1970) agrees with a distribution of tangential velocity similar to that of Rankine's combined vortex.

The transition from rotational to irrotational flow at distance R from the axis is not abrupt, but follows a transitional curve with a concomitant reduction in maximum velocity (see Fig. 3.15).

Horizontal Component of Resultant Velocity

The vector sum of the forward and tangential velocities is a horizontal projection of the resultant velocity. This sum is a vector that, for a given air particle, changes continuously in magnitude and direction as the particle rotates and moves forward. For counterclockwise rotation of the vortex (typical in the Northern Hemisphere), the vector is maximal on the right side of the path, minimal on the left, and intermediate in between.

General Estimates of Maximum Wind Velocity

Several statements about estimated maximum wind velocity in tornadoes have been made by different authors. These statements have not been directly substantiated by measurement; rather, they represent expert opinion. They are mentioned in order to give a representative illustration of what is known today about tornadoes and to clarify misunderstandings.

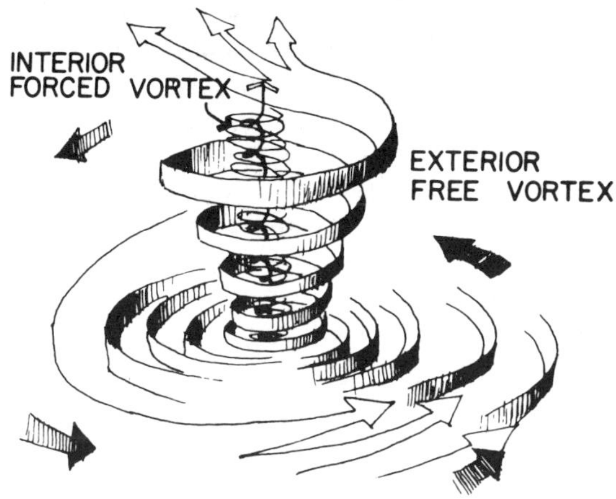

Figure 3.15. There is an almost general consensus that a tornado vortex is a combined Rankine vortex consisting of a free vortex on the periphery and a forced vortex at the core. In the free vortex, typical of ideal fluids, the velocity increases exponentially as the air particles get closer to the center. In the forced vortex at the core, the velocity decreases linearly as the air particles get closer to the center, just as it does with solid bodies.

In 1961, the American Society of Civil Engineers Task Committee on Wind Forces expressed its conclusions about tornado forces as follows:

> The peripheral wind velocities have been estimated to be in excess of 300 mph, on the basis of damage done; and the dynamic pressure due to the velocity is 230 psf, which corresponds to an average design pressure of 260 psf.

Flora (1953) reported much indicative data on wind velocities. He estimated the maximum rotary wind velocity at 454 mph for the Texas Panhandle tornado of April 6, 1947. He also mentioned that, based on observation of damage done by many tornadoes, wind velocities ranged from 450 to 500 mph. He also suggested that much evidence indicates that the velocity in some parts of the vortex is on the order of the speed of sound. A similar conclusion was derived from a theory proposed by Abdullah (1955) on supersonic wind speed in the "area limited by the 'limiting circle' and the 'critical circle.'"

Chang (1969) described a tornado as a long vortex core with maximum circumferential velocity ranging from 100 to 400 mph, while Tepper (1978) reported that whirling speeds in tornadoes have been estimated up to 500 mph.

A report by the U.S. Department of Housing and Urban Development (1966) presented several recommendations for stormproofing houses against hurricanes and tornadoes. This report quotes an engineer who headed a group of technicians investigating storm damage in Florida and Mississippi: "We know now that the forces of a tornado are not irresistible. We once thought that their velocity was 600–700 miles per hour, but actually it ranges from 100 to 300 miles per hour and probably approximates 200–250."

Edwin Kessler (1967), director of the NSSL, stated that his personal belief (in his words, "little better than a guess") was that the damage observed in tornadoes could have been produced by wind speeds of 300 knots (150 m/sec, or 335 mph).

A pamphlet from the U.S. Weather Bureau (1965) (now called the National Weather Service) presents a concise list of information on tornadoes, which states that the estimated wind velocity within a tornado is more than 300 mph. Estimates of wind velocities for tornadoes to be used in design of nuclear power plants were originally presented in 1967 by Swanson et al. More specific estimates of these velocities were published by the U.S. Atomic Energy Commission in 1974.

Walter H. Hoecker (1967) of the Atmospheric Trajectory Branch of the Environmental Science Services Administration pointed out that indirectly measured wind speeds have approached 300 mph.

Swanson et al. (1967) reported that wind velocities estimated from structural damages have ranged below 250 mph. They also stated:

> A few estimates have been made of speeds-not-exceeded for structures that have withstood the full force of tornadoes. These range from about 200 to 500 mph. It appears that a maximum wind speed of 300 mph is conservative for the "average" tornado, and that a speed of 500 mph has a probability in any given storm that is at or below the level of 2%.

Fujita et al. (1967) discuss several investigations made by different researchers on the structural damage caused by tornadoes and proposed minimum wind velocities estimates. They concluded that, "It seems reasonable to assume that minimum speeds ranging between 55 and 217 mph are required to result in the typical damage left behind by Midwestern tornadoes."

Maximum Wind Velocity Estimated from Structural Damage

Three transmission towers were destroyed in a tornado in Worcester, Massachusetts, on June 9, 1953. Booker (1954) reported that the towers had been tested to failure in the factory and that their strength was known. This was quite, important because most conclusions based on structural analysis are affected by the validity of the assumptions made; in this case, the test to failure was unequivocally certain.

The towers were identified by numbers 113, 115, and 116. The direction of the powerline supported by the towers was almost at 90° to the direction of the tornado. Tower 113 was to the left of the centerline path, while towers 115 and 116 were to the right. Tower 114, which resisted the tornado winds, was almost in the center. Based on the counterclockwise circular motion of the storm, we can say that the three towers were exposed to transverse winds and that tower 114 was also subjected to frontal winds. According to Booker, towers 115 and 116 fell forward and tower 113 fell backward, as a result of the counterclockwise direction of the rotating winds. The minimum wind velocities that would have caused the collapse of the towers were determined to be 148 mph for tower 113 and 170 mph for tower 116. Engineering calculations showed that the velocity near tower 114 was about 343 mph.

Bigler (1975) determined estimated wind speed by evaluation of structural damage in connection with small tornadoes that occurred on March 17, 1975, which struck Avim, Port Bolivar, and Texas City, Texas. Analysis of a damaged water storage tank in Texas City provided the necessary information for computation of the minimum wind speed required to produce observed damage. The tank was 13.58 ft high and had a 6.05-ft diameter. In the computations, the tank was assumed to have been filled with water. The minimum wind forces required to overturn the tank and to shear the tank off a wooden base were then computed, and the minimum wind forces required to produce these effects were determined to be 303 mph for overturning and 285 mph for slippage. Considering that at 285 mph slippage would occur first, this is the assumed wind speed value.

Important wind velocity data can be found in a study conducted by Segner (1960). These data were obtained from computations of the minimum wind speed required to produce damage to various structures during the Dallas tornado of April 2, 1957. An analysis was conducted for each structure, considering several reasonable failure modes. The results were influenced by so many factors that they have different degrees of dependability. Unfortunately, one point raises some doubt about the validity of the results: the impossibility of determining whether the damage was caused only by wind forces or by flying objects as well. The results of the study follow:

1. Two 8" concrete block walls in a service station were blown outward. The walls were 12 ft high and perpendicular to each other. The calculated minimum wind speeds for the walls were 91.6 and 92.2 mph.

2. A small empty steel storage tank, 4 ft in diameter and 5.29 ft tall, was overturned. The tank was 0.125" thick, weighed 472 lb when empty, and was supported on four legs braced horizontally. The supporting structure consisted of $1\frac{1}{2} \times 1\frac{1}{2} \times \frac{3}{6}$" steel angles. Eyewitnesses saw the tank roll for about 100 ft after it overturned. The author estimated a minimum wind speed of 55–65 mph for the failure of the supporting structure and greater speeds to account for the rolling action of the tank.

3. A flat roof covering a two-story structure was lifted. The roof deck, estimated to weigh 13 lb/ft^2, was nailed to 2 × 10 supporting joists. Neither the joists nor the ceiling attached below were damaged, but the deck was blown off. Segner assumed that only half of the common nails anchoring the deck to the joist were effective, and concluded that the lift was caused by a minimum wind speed of 179 mph.

4. Railroad freight cars were overturned. Segner considered eight of them and computed the required wind velocity for each as follows: 128, 144, 144, 217, 83, 144, 105, and 143 mph. The velocities for each car were different due to such factors as weight, amount of load, mode of overturning, and position of doors (open or closed).

5. A 45-ft elevated signboard collapsed. This structure was built with 13 yellow pine poles, each 45 ft high with an approximate diameter of 15". The poles, spaced approximately 8 ft on center, were braced together with bracers starting 28 ft from the ground and extending to the top. The signboard itself was 56 ft long and 17 ft high. The minimum wind velocity was calculated at 302 mph. Among the several failure modes, the most severe that Segner assumed possible was that the billboard failed first and then each pile collapsed individually under the force of the wind acting on the pile surface. The velocity computed in this case was the highest calculated among all of Segner's analyses. Noticing the uniqueness of this value, Segner suggested that, instead of direct wind action, impact from flying objects might have caused the structural failure.

6. Perimeter walls of a 200 × 160-ft rectangular building collapsed. The exterior walls were 12" thick and consisted of 8" lightweight concrete block and 4" brick

interlaced. The wall clearance height was 18′, 5". Of the four exterior walls, two collapsed outward, one collapsed inward, and one resisted the tornado. The minimum wind speed for the two walls falling inward was calculated at 107 mph; the velocity calculated for the one falling outward was 109 mph.

7. An empty truckbed overturned. He assumed that a combination of vertical uplift and horizontal wind action overturned the truckbed and the hydraulic hoist connected to it. He estimated the effects of the shape factor and considered the result to be approximate.

8. A 42-ft flagpole comprised of three steel pipes of different diameter was bent. Starting from the base, the first section was 3.5" OD with a 15.5-ft height from the ground. The second section, 17.7 ft long, was 3" OD; the third section was 8.8 ft long and 2.5" OD. The pole extended 6 ft into the ground and was embedded in concrete. The pole remained straight along its entire length, but it yielded at the connection with the concrete, as shown by a permanent deformation confined to that point. Segner considered a minimum yield point of 35,000 psi and estimated the minimum wind speed at 115 mph assuming an elastic action, or 133 mph for plastic behavior.

9. A roof section was torn from a warehouse. The portion of the roof analyzed was 25 × 21 ft and consisted of 3" longleaf pine sheathing with four-ply tar and gravel roofings. Segner could not determine whether the lifting was due to the typical pressure drop or to suction produced by wind action. The estimated wind velocity was 189 mph. However, he advised that the accuracy of the result should be assumed to be ±25% due to the approximate assumptions about joint failure.

Maximum Wind Velocity Estimated from Ground Marks

In a study of the North Platte Valley, Nebraska, tornado of June 27, 1955, Van Tassel (1955) drew interesting conclusions and derived a wind velocity of 484 mph in the funnel. On aerial photographs taken about 40 hr after the tornado, ground marks were clearly visible on some cultivated fields. These marks were actually ridges or dikes, from 0.25 to 0.5" high, arranged in waves. They were systematic and elliptical in shape, and their dimensions were calculated by scaling the photographs. The major axis was calculated to be 230 ft, the minor axis 152 ft. The forward speed determined by radar observations was estimated at 12 mph; the average distance between rings at the center of the leading edges was determined to be 15′, 4". The basic assumption was that the marks were produced by some object carried by the tornado within the funnel.

Wind speed was determined as follows:

$$V = eNS,$$

where V is the wind speed (mph), e is the approximate perimeter of the ellipse (mi), N is the number of rings per unit length (per mile), and S is the forward speed of the tornado (mph).

In the formula that follows, a is half the major axis and b is half the minor axis:

$$
\begin{aligned}
e &= 2\pi \sqrt{(a^2 + b^2)/2} \\
&= 2\pi \sqrt{(115^2 + 76^2)/2} = 2\pi\sqrt{9500} \\
&= 612 \text{ ft} \\
&= 612 \text{ ft}/5280 \text{ ft/mi} = 0.116 \text{ mi.}
\end{aligned}
$$

Number of orbits in 1 mi: $N = 5280/15.3 = 345$.

S was assumed to be 12 mph, and

$$V = 0.116 \times 345 \times 12 = 480 \text{ mph.}$$

Several wind velocities have been estimated by analysis of ground marks visible along tornado paths. These marks are usually cycloidal in shape, indicating two component paths — one rotational, the other rectilinear — representing the rotational and forward motion of a point moving with the tornado.

Interpretations of the mechanics that produce these marks vary among the researchers who investigate this phenomenon. Van Tassel (1955) assumed that the marks are produced by an object picked up by the tornado winds and moved according to wind direction. Fujita et al. (1967), in contrast, assumed that among the different types of marks left on the ground by tornadoes, one, cycloidal in shape, is produced by suction force. Around the tornado center, there are several spots in which suction force is concentrated, spots that are not larger than 50 ft in diameter and which rotate around the tornado center. The suction power at these points is smaller than that of a household vacuum cleaner.

In analyzing cycloidal suction marks left by a tornado on April 21, 1967, which extended southwest of Belvidere and north to Woodstock, Illinois, Fujita et al. concluded that rotational speeds on two occasions were 150 to 160 mph. The forward speed of the tornado was measured at 50 mph. The resultant velocities, found by vectorial addition of the rotational and forward speeds, were 210 and 200 mph, respectively.

Using the same method, Fujita et al. computed several velocity values for suction spots observed after the Fargo, North Dakota, tornadoes of June 20, 1967. The resultants of rotational and forward velocities were 172, 176, 173, 180, 180, 173, and 166 mph. In this case, the forward speed was 62.5 mph. These values are very close to those calculated by Fujita et al. when they analyzed motion pictures of the Fargo tornadoes. It is very encouraging when two completely different methods of analysis agree on relatively low maximum velocities.

Wind Velocity Derived from Scaling Motion Pictures

Hoecker (1960) presented a large number of tornado wind velocity values for the Dallas tornado of April 2, 1957. High-quality tornado movies were available for study of the Dallas tornado, including some scenes taken with a 10X telephoto lens. Fragments of clouds and debris in circular motion were clearly visible within the funnel, and their location during different time intervals could be identified quite accurately because the time periods between pictures were known.

Hoecker also presented a distribution of wind velocities in mph at several points located at different radial distances from the axis of the tornado funnel and at different elevations from the ground. The wind velocities were determined only at those points where tracer particles were located by chance. However, it is probable that higher wind velocities could have existed at points different from the ones analyzed. He assumed that the moving tracer elements, cloud fragments, debris, and dust particles were moving at the same speed as the air in which they were imbedded, neglecting aerodynamic shape factors.

The largest velocity value obtained by Hoecker was 170 mph, which was a tangential velocity component. The point in the tornado where this velocity was calculated was located 225 ft above ground and at a 130-ft radial distance from the center of the funnel.

Fujita et al. (1967) derived tangential wind velocities for the Fargo tornadoes of June 20, 1957, from a study made of motion pictures showing the tornado in action. Fujita analyzed seven cloud pendants appearing at the base of a sheared-off funnel just above the ground. These cloud pendants were located at radial distances from the funnel center of 296, 196, 98, 196, 360, 328, and 296 ft. Their velocities were 112, 112, 93, 100, 111, 100, and 102 mph, respectively.

Maximum Wind Velocity Estimated from the Shape of a Funnel

Glaser (1959) computed a maximum tangential wind speed of 230 mph for the Texas tornado of March 21, 1956. He used a method based on evaluation of the shape of the tornado funnel. The tangential wind speed was computed for an altitude of 820 m at a radial distance of 61 m from the center.

Vertical Velocity

Though tornadoes have often been described as nature's giant vacuum cleaners, Fujita et al. (1967) concluded that their suction power near the ground smaller than that of conventional household vacuum cleaners. This is confirmed by the quick decrease in vertical velocity with decreasing altitude. Hoecker's work (1960) on the Dallas tornado includes calculations of vertical wind velocity, which he found to vary with altitude. Along the vertical axis of the tornado, upward velocity was calculated at 20, 50, 110, 152, and 0 mph at respective altitudes of 0, 6, 50, 135, and 800 ft. These findings indicate a constant increase in speed up to an altitude of 135 ft and then a linear decrease to 0 mph at an elevation of 800 ft. This change in velocity appears to result from the friction effect of the ground, which extends upward to a substantial altitude.

Table 3.5 provides a summary of wind velocity estimates. The values are grouped according to the criteria used to determine them.

Pressure Change

The relationship between air pressure and velocity in the tornado vortex can be expressed by a cyclostrophic equation:

$$\frac{\partial p}{\partial r} = \frac{\partial V^2}{\rho},$$

where P is the pressure, r is the radial distance, ρ is the density, and V is the velocity.

Since the relationship between tangential velocity and radial distance is known, as shown above, it is possible to integrate the cyclostrophic equation over the radius starting at a distance where the pressure is undisturbed. After integrating we obtain the radial distribution of the pressure as theoretically expected in a tornado. Hoecker (1960) used this method in his study of the Dallas tornado. He obtained the pressure field from the velocity field, which were derived from motion pictures of the vortex. The pressure distribution in a vertical plane through a tornado axis is represented by plotting the isobar lines as functions of height and radial distance. For the Dallas tornado, the spacing of isobars decreased up to a certain radius and then increased toward the axis. Obviously, spacing of the isobars was inversely proportional to wind velocity. In fact, wind speed increased as the spacing decreased.

Since the vertical component of wind force can be estimated from the pressure gradient, the distribution of this component can be derived for any point. It follows that there is a zone near the axis where the force is directed downward, while in the outer region the force is upward. Note that such double vertical circulation is in agreement with the latest observations made on waterspouts.

Estimates of the Maximum Pressure Differential

Several authors have expressed opinions about the variation of pressure within tornadoes. Finley (1884) considered the possibility that the center of a tornado vortex is a near vacuum because of the centrifugal force of the tornado cloud, thus implying a pressure drop of about 2000 psf. Teesdale (1928) gave an estimated maximum pressure drop value of up to 0.5 atm, about 1000 psf. C.F. Brooks (1949) assumed the pressure in the funnel to vary between 0.8 and 0.5 atm (1600–1000 psf).

Ferrell (1889) assumed the pressure in the funnel to be 0.75 atm, which means a 0.25 atm (500 psf) reduction. Logie (1919) claimed that the reduction of pressure can exceed 50 mb (104 psf) and added that many authorities assume that 500 mb (1040 psf) could be the maximum value. Humphreys (1940) believed pressure reduction to be about 1/10 of an atmosphere (~200 psf). E.M. Brooks (1951) stated that the probable value might be greater than 200 mb (418 psf).

Abdullah (1955) arrived at a minimum pressure of 500 mb at the limiting circle of the tornado, starting from a pressure of 1000 mb at infinity. The pressure drop would thus be 470 mb at the limiting circle, and he assumed that this value could also be considered valid for the center of the funnel. A pressure drop of 470 mb would be equal to a pressure of 981 psf.

Kessler (1970) suggested a maximum wind speed of 300 knots, or an equivalent pressure drop of about 200 mb, according to Bernoulli's equation:

$$200 \times 2.088 = 418 \text{ psf.}$$

Unofficial Pressure Differential Measurements

One of the largest drops in atmospheric pressure recorded during a tornado is an unofficial measurement of questionable reliability. Several authors have mentioned this exceptionally high reading, with slight differences among them. In reporting on the tornado that passed through St. Louis, Missouri, on the afternoon of May 27, 1896, Frankenfield (1896) had been given an aneroid barometer with a metric scale to be reset. The barometer belonged to a Mrs. Klemm and had been in her house when the tornado struck. Her son had read the barometer at 680 mm (26.78"). Frankenfield compared this reading to the official readings of the U.S. Weather Bureau office in St. Louis and corrected the value to 27.30" to account

Table 3.5. Wind Velocity Estimates[1]

a. Maximum speeds (mph)

Chang (1966)	400
Flora (1953)	500
Fujita (1967)	217
Hoecker (1967)	300
Kessler (1967)	335
Melaragno (1968b)	300
Swanson et al. (1967)	500
Task Commission on Wind Forces (1961)	300
Tepper (1978)	500
U.S. Weather Bureau (1964)	300^+

b. Average maximum speeds (mph)

Flora (1953)	450–500
Chang (1966)	100–400
Melaragno (1968b)	200–250
Swanson et al. (1967)	200–500 (300 most probable)

c. Maximum speeds estimated from structural damage (mph)*

Bigler (1957)	Water tank	285
Booker (1954)	Transmission tower	148
	Transmission tower	170
	Transmission tower	343
Segner (1960)	Concrete block wall	91.6
	Concrete-block wall	92.2
	Storage tank	65
	Roof	179
	Railroad freight car	128
	Railroad freight car	144
	Railroad freight car	217
	Railroad freight car	83
	Railroad freight car	144
	Railroad freight car	105
	Railroad freight car	143
	Signboard	302 (doubtful)
	Brick-and-block wall	107
	Brick-and-block wall	109
	Truckbed	133
	Roof	189

cont'd

d. Maximum speeds estimated from cycloidal ground marks (mph)

Fujita (1967)	176
	180
	180
	173
	166
Fujita et al. (1967)	210
Van Tassel (1955)	484

e. Maximum tangential speeds estimated from scaling motion pictures (mph)

Fujita (1960)	112
	93
	100
	100
	111
	100
	102
Hoecker (1960)	170

f. Maximum speeds estimated from funnel shape (mph)

Dergarabedian and Fendell (1971)	208
Glaser (1959)	230

[1]These values represent speeds required to cause the damage observed. Actual maximum speeds may of course be higher.

for the difference in elevation between the Klemm house and the weather bureau office. The difference he reported was 2.05".

The U.S. Weather Bureau (1965) reported the same pressure differential reading to be 2.42" of mercury. Reynolds (1958) indicated that the variance between the two readings was 2–2.5". Flora (1953) also mentioned this series of events.

Baier (1896) provided additional information on the subject. He performed a complete investigation of the circumstances surrounding the reading of the pressure done by young Klemm. He also checked the accuracy of the instrument. He concluded that the original value of 27.30 in, representing the corrected reading computed by Frankenfield, should be corrected to 26.94". This additional 0.36" brings the pressure differential from 2.05 to 2.41" of mercury (170.47 psf).

Another unofficial pressure differential reading is mentioned in a report on the August 20, 1904 tornado in Minneapolis, Minnesota (Outram 1904). The reading is exceptionally high and therefore questionable; however, Outram seemed to be convinced of the dependability of the two men who measured the barometric pressure. They observed the aneroid barometer dropping to 23" and then rising almost immediately to its original position. Assuming that the initial reading was the same as that observed by the local forecaster (28.67"), the drop was 5.67" (401 psf). Outram noted that the aneroid barometer had been checked at the local weather station and compared with the other instruments not very long before the storm and had been found to be correct.

Official Pressure Measurements

The first recorded official measurement of pressure change at the eye of a tornado is probably the one reported in the *Monthly Weather Review* for the that hit Little Rock, Arkansas, on October 2, 1894. The tornado center seemed to have passed directly over the telegraph office where the local weather bureau office was also located. George S. Harkness (1894), an observer at the office, reported the events. The tornado struck a building on the south side of the telegraph office, throwing the second story over the office roof. The windows in the office were then blown in. The wires connecting the instruments on the roof with the recording apparatus inside were broken, and many instruments were damaged. This occurred at 8:28 p.m. It appeared to Harkness that the tornado passed over the office in about 1 minute, during which time the self-recording Richards barograph recorded a 0.38" drop. The diagram that furnished the record of the barometer variation indicated that, after a slight oscillation, the pressure dropped in a linearly from 29.31 to 28.93" and returned so quickly that it was difficult to tell the time elapsed during the oscillation. The horizontal movement of the paper in the barograph was such that the recorded oscillation could have occurred in a minute or less, but it is not possible to establish the time period exactly. The trace of the pen would have been practically the same whether the time period had been a minute or a second. Even if the pressure gradient cannot be defined, this piece of data has great significance because of its dependability.

Another reliable pressure reading was taken during the tornado that passed through Dyersburg, Tennessee, on March 21–22, 1952. A weather station equipped with a barograph was operated by the Civil Aeronautic Authority at the local airport. The tornado path at that point was 258 yd wide, and the barograph was in the path, located exactly 41 yd from the centerline. The reading indicated a pressure drop of 0.65" (45.98 psf). Carr (1952) indicated that the real drop may have been higher than the value indicated on the barogram trace because of a probable lag in response of the instrument.

On August 20, 1904, a severe storm which seemed to be a tornado struck Minneapolis, Minnesota. According to Outram (1904), the official pressure reading was 28.82", which gradually fell to 28.67" during the afternoon:

> Just about the time of the greatest severity of the storm, the barograph dropped suddenly to 28.25, returning immediately and rising to 28.80, then dropping back quickly to 28.70.

It then fell gradually until the next day. The 0.55" jump is an official reading.

Pressure Values Estimated from Scaling Motion Pictures

Probably the most interesting and valuable data on pressure change are presented by Hoecker (1960). He estimated wind speeds in the Dallas tornado of April 2, 1957, from high-quality motion pictures that showed the funnel and movement of the cloud fragments and debris circulating around the tornado. The study includes a diagram of wind velocity in mph at several points located at different altitudes and at different radial distances from the storm center, along with curves connecting points of equal velocity (isotachs). It should be pointed out that, though velocities are indirect measurements, they are a logical consequence of direct measurements. If the measurements from which the velocities are derived are correct, then the velocities are correct as well. In case of motion pictures, the time interval between one frame and the next is definitely known. The distance between two different locations of the same object at two different instants can thus be accurately scaled.

We must therefore conclude that these measurements are nearly direct and are not just based on theoretical assumptions, as are many others.

In a second paper, Hoecker (1961) elaborated on wind velocity values already derived and computed pressure gradient values at the same points at which he had previously determined wind velocity. To obtain the pressure gradients in millibars per meter (mb/m), he used the cyclostrophic wind equation introduced in Table 3.5. We should note that some assumptions are made when using this equation, so that even if they are theoretically justified, the resulting values cannot be considered as close to direct measurements as previously derived velocity values. The reader is referred to Long (1960) and Glaser (1959) for the justification for using the cyclostrophic equation.

Hoecker (1961) also presented pressure distribution with respect to distance from tornado axis and altitude. Assuming an average forward velocity of 27 mph for the Dallas tornado, he also included the pressure distribution at ground level. His data are very useful for design of venting systems, because calculations are based on pressure and time factors when determining vent sizes. The maximum pressure drop computed value was 60 mb (1.8" of Hg), which corresponds to a pressure of 125 psf. Table 3.6 contains some values of pressure drop, including estimated values, unofficial records, official records, and some measurements derived from scaling motion pictures of tornadoes.

Rate of Pressure Change

The pressure change rate over time or with respect to distance from the tornado center is an essential element in the design of tornado-resistant buildings. If the pressure differential between the inside and outside of a building must be balanced, a venting system must be designed. Its basic function would be to make a certain volume of air escape from the building as the pressure on the outside dropped. The size of outlets would be determined according to the rate of pressure drop. Most important also would be to establish whether the pressure load could be a dynamic load, which is possible if the time interval in which the pressure drops were equal to the fundamental period of vibration of the structure.

The absence of data on peak forward speed makes accurate estimates of the rate of pressure change impossible. Reynolds (1958) suggested an empirical value for the pressure gradient of 1" for 120 ft. Assuming a diameter of 0.25 mi and a constant forward motion of 40 mph, he derived that a point at the edge of the limit of destruction would be in the center of the storm after 11 sec. This procedure assumed that the pressure varies linearly from the undisturbed area to a minimum value at the center of the tornado, which is not probable for the entire cross-section through the vortex.

Hoecker's 1961 analysis of the Dallas tornado indicated a maximum time rate of pressure change of 26 mb/sec. Note that such a value was not constant and was observed for only 0.5 sec. Further, this value was derived at a certain distance from the center of the storm (4 sec from the center, which corresponds to approximately 159 ft, for the translational speed of 27 mph used by Hoecker). It is thus not necessarily the highest value.

Table 3.6. Pressure Drop

Source	Pressure drop estimates (in Hg)	Difference between atmospheric and reduced pressure
Abdullah (1955)	–	981 psf
C.F. Brooks (1948)	–	1000–1600 psf
E.M. Brooks (1951)	–	418 psf
Ferrell (1889)	3.35	500 psf
Finley (1882)	13	2000 psf
Humphreys (1940)	–	200 psf
Kessler (1970)	–	418 psf
Logie (1919)	–	1000 psf
Teesdale (1928)	6.7	1000 p@f
Unofficial records		
Baier (1896)		–
Flora (1954)		–
Frankenfield (1896)		–
Outram (1904)		401 psf (5.67 in Hg)
Reynolds (1958)		–
U.S. Weather Bureau (1894)		170.47 psf (2.41 in Hg)
Official records		
Carr (1952)		0.65 in (45.99 psf)
Harkness (1894)		0.38 in (26.88 psf)
Outram (1904)		0.55 in (38.9 psf)
Pressure values estimated from scaling motion pictures		
Hoecker (1961)		60 mb (125 psf)

Case studies of criteria adopted for the design of some nuclear power plants indicate the following:

1. A pressure differential of 3 psi in 3 sec were adopted for: Peach Bottom Atomic Power Station, Peach Bottom, Pennsylvania; Limerick Generating Station, near Philadelphia, Pennsylvania; Fort St. Vrain Nuclear Generating Station, Platteville, Colorado; Surry Power Station, Surry, Virginia; Forked River Nuclear Generating Station, Forked River, New Jersey; Davis–Besse Nuclear Power Station, Oak Harbor, Ohio; Trojan Nuclear Power Plant, Prescott, Oregon; and others.

2. A pressure differential of 3 psi in 3 sec (1 psi/sec) followed by calm for 2 sec and repressurization was adopted for: Arkansas Nuclear One, Russellville, Arkansas; Millstone Nuclear Power Station Unit No. 2, Waterford, Connecticut; and others.

3. A pressure differential of 3 psi in 5 sec was adopted for: Oconee Nuclear Station, Seneca, South Carolina; Vermont Yankee Generating Station, Vernon, Vermont; and others.

4. EFFECTS ON STRUCTURES

Horizontal Torque

When the resultant of the wind force and the reaction from the foundation do not lie on the same vertical plane, a horizontal moment is applied to a building. This often occurs because the foundation's center of gravity and the wind resultant do not coincide with the windward face of the building's center of gravity due to the hyperbolic distribution of the wind force.

During the tornado that hit Lubbock, Texas, on May 11, 1970, the 21-story Great Plains Life Building received severe structural damage due to a horizontal torque (Fig. 3.16). A service shaft at one end of the building offset the horizontal resultant of the foundation from the geometrical centroid. A plastic deformation made the top of the steel structure rotate 12" to the east with respect to the base, forcing the building to be evacuated. This very clearly indicates a suggested design procedure to avoid such torque. The residual horizontal moment due to hyperbolic distribution of wind force can probably be resisted by the residual strength furnished by redundant members. Some strengthening methods can be used for such small structures as dwellings. Rotation of the structure can be prevented by eliminating translations and using tension members at corners anchored in the concrete foundation, by using steel bracings notched in the studs for wood-framed structures, or by using concrete piers that act as stiffeners.

Foundations

We can see that for small-diameter tornadoes with high forward speed it is more probable that loading rate can be high enough to produce magnification of dynamic forces, with more severe consequences to the structure and foundation. Studies on the resistance of soil under dynamic conditions have indicated that the shear strength and bearing capacity for sandy granular soils decrease as loading rate increases. However, this is only true up to a value of 0.002"/sec for dry or saturated soil conditions. It has also been found that the wedge of soil under shear stresses is smaller as loading rate increases (Casagrande and Shannon 1949; Seed and Lundgren 1954; Whitman 1957; Whitman and Healy 1962). Further studies conducted at the Georgia Institute of Technology (Banks 1963; Woodard 1964) indicated that, as loading rate is increased over 10"/sec, the bearing capacity of dry sand increases up to a static value and then remains constant. For saturated sand, not only did the bearing capacity increase sharply, but the ultimate strength became several times greater than its value under static loading. However, it would found that settlement can reach 4% of the footing diameter at ultimate strength, which cannot be considered acceptable for practical purposes.

Horizontal forces applied to foundations are opposed by reactions generated either by passive soil pressure or by friction between foundation and soil. Research in this area indicates that displacements are small in the case of large dynamic horizontal loads, and that they are negligible in the case of frictional reactions acting alone. On the other hand, when reactions from friction and passive pressure act together, they expected that lateral displacement could be as high as 4% of foundation depth. In view of their findings, they recommend

Figure 3.16. The 22-story Great Plains Life Building in Lubbock, Texas, was damaged by the tornado of May 11, 1970. Although the steel frame was permanently deformed, the fully engineered building was restored to useful service. (Photo courtesy of the Institute for Disaster Research, Texas Tech University.)

reducing such displacement by designing foundations to provide a reaction due to pressure that is twice the required minimum.

Missiles

The literature on tornado accidents includes a large number of reports on objects that have penetrated solid materials driven by tornado winds. Frankenfield (1896) reported that a 2 × 10" wood plank was driven by the wind into the 0.5" web of a wrought-iron girder on the Eads Bridge during the St. Louis tornado of May 17, 1896. Documented by a photograph, this incident provides irrefutable proof of the effect of the wind (see Fig. 3.17).

Figure 3.17. 2 × 10" board penetrating the half-inch wrought-iron web of a bridge girder after the May 17, 1896 tornado in St. Louis, Missouri. (Photo courtesy of the U.S. Weather Bureau.)

Splinters of wood penetrating trees, straw penetrating solid wood, and the like are common occurrences during tornadoes. One of the most incredible reported incidents is that of a flimsy piece of wood that perforated the metal brake drum of an automobile during the May 31, 1947 tornado that hit Leedy, Oklahoma. A similar incident was reported following the April 30, 1949 tornado in Norman, Oklahoma, where a wood splint pierced the metal casing of an electric drill (Fig. 3.18).

Experimental work since 1967 has attempted to reproduce penetrations similar to that which occurs during tornadoes (see Keller and Vonnegut 1976). A pneumatic gun designed by W. Sykes (1967) of Arthur D. Little can accelerate small particles such as straws and splinters to known speeds and reproduce this effect. This gun has been helpful in evaluating the wind speeds required to produce penetration of objects into different materials (see Table 3.7).

Tornadoes carry a large amount of solids and condensed fluid particles aloft that can move at speed comparable to that of the air in the vortex. Thus, the characteristic dark color of the funnel-shaped column of air is actually caused by material in suspension, of course including condensation.

The types of debris that can be flying in the orbits of a tornado vortex are incredible in their variety. When a storm moves over the countryside, the most probable suspended materials include loose soil particles (such as dust, sand, gravel, and rocks) and fragments of vegetation (such as tree limbs that break off). When a storm passes over built-up urban

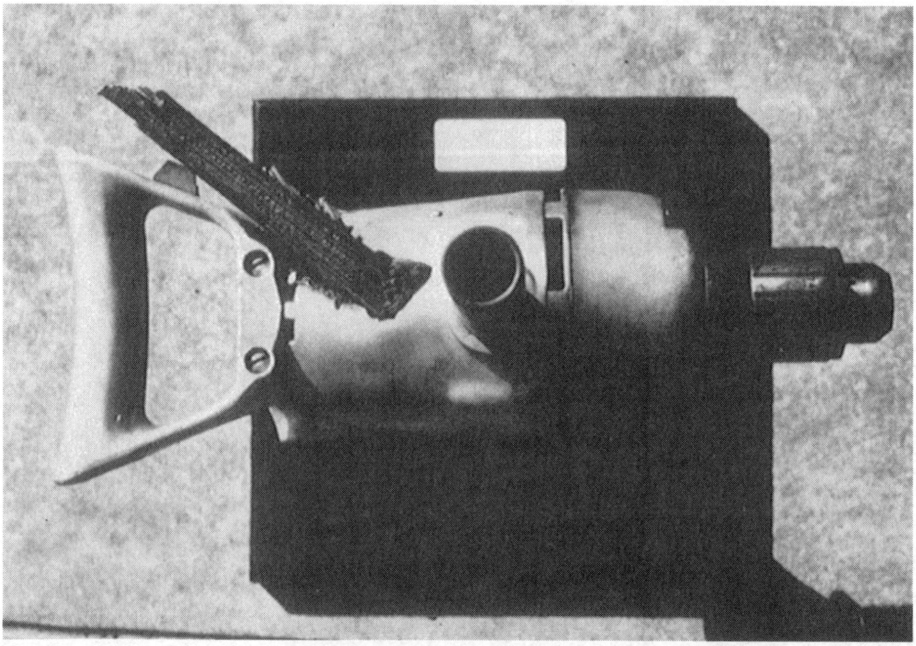

Figure 3.18. Metal casing of a drill penetrated by a piece of wood during April 30, 1949 tornado in Norman, Oklahoma. (Photo courtesy of the National Severe Storms Laboratory, NOAA.)

areas, the variety of debris is much greater since it can include debris from streets, damaged or destroyed buildings, the interior contents of buildings, and a myriad of other sources.

Many objects are picked whose weight is small compared to the area exposed to the winds. Vehicles generally fit this category. They can become windborne, but without reaching the high speeds of air particles. Automobiles are also rolled over (and over), and are dragged along the ground for short distances until they crash into an obstacle or are ejected from the vortex by centrifugal force (Fig. 3.19). The damage they suffer is usually very severe, which makes them very unsafe refuges during tornadoes. In addition, they can inflict severe damage to structures upon collision. Other types of windborne material from urban areas can include such items as streetlights, utility poles, signs, fences, manhole covers, electrical transformers, and telephone/electrical poles.

Whether completely destroyed or partially damaged, buildings are probably the largest contributors to the amount of tornado debris. Roofs and walls produce such debris as shingles, boards, nails, plaster, bricks, blocks, and metal sheets. Metal sheets are extremely dangerous. In the case of the April 14, 1969 tornado that hit Dacca, Bangladesh, which killed and injured more than 5000, the major individual cause of death and injury was corrugated sheetmetal panels popularly used as roofing material. The were deadly missiles during the storm, taking on the character of flying razor blades. There are similar examples of metal sheets used in buildings for general cladding or roofing in the United States which became windborne when panels were detached or ripped from their anchors.

Table 3.7. Missile Penetration During Tornadoes

a. Threshold speeds required for penetration by toothpicks

Target	Threshold speed (m/sec)	Threshold speed (mph)
Ponderosa-pine board (seasoned)	30	67
Fir plywood	70	157
Oak, dark, soft area (seasoned)	200	447
Maple board (seasoned)	215	481
Oak, light, hard area (seasoned)	>230	>515
Glass, 6 mm thick	>300	>671
Cement block	>310	>693

b. Threshold speed required for penetration by broom straws

Target	Threshold speed (m/sec)	Threshold speed (mph)
Ponderosa pine (seasoned)	65	145
Fir, soft rings (seasoned)	75	168
Oak, dark, soft area (seasoned)	140	313
Dead, barkless elm-tree branch	170	380
Fir, hard rings (seasoned)	>200	>447
Maple (seasoned)	>210	>470
Oak, hard, light area (seasoned)	>215	>481

c. Maximum observed projectile penetration and associated speed

Projectile	Target[1]	Penetration (mm)	Penetration (in)	Speed (m/sec)	Speed (mph)
Toothpick	Pine board	Complete, 19	0.75	190+	425
Toothpick	Maple board	3.5	0.14	300	671
Toothpick	Maple board, wet	15	0.59	300	671
Toothpick	Elm branch	6	0.24	220	492
Toothpick	Soft oak	2	0.08	235	526
Toothpick	Plywood	Complete, 10	0.39	200+	447
Toothpick	Glass, 2.5 mm thick	Glass shattered	270	604	
Broom straw	Pine	6	0.24	190	425
Broom straw	Pine, wet	6	0.24	190	425
Broom straw	Soft fir	9	0.35	170	380
Broom straw	Elm branch	2	0.08	180	403
Broom straw	Soft oak	2.5	0.1	195	414
Broom straw	Plywood	3.5	0.14	170	380
Bamboo straw, pointed	Pine board	Complete, 19	0.75	180+	403
Bamboo straw, blunt	Pine board	16	0.63	190	425
Bamboo straw, pointed	Maple	7.5	0.30	215	481
Bamboo straw, blunt	Maple	5.5	0.22	180	403

[1]All wood targets were seasoned wood.

Figure 3.19. Damaged car cleary indicates the vulnerability of automobiles during tornadoes. (Photo courtesy of the National Weather Service.)

Other types of building debris include doors, window frames and glass, chimneys, fans, and air conditioners. Many missiles are produced by the interior furnishings of damaged buildings and all types of stored goods, e.g., materials from warehouses and outdoor storage areas (Fig. 3.20).

Once an object is in motion by wind force, it behaves according to its own aerodynamic characteristics. Whether the object is injected into the tornadic vortex, the extent to which it is accelerated, and the velocity it finally reaches are all matters that depend on the drag-and-lift coefficients of the object and its orientation with respect to the direction of the wind.

Due to the great variety, it is convenient to divide the spectrum of potential missiles into three broad categories:

1. At one extreme are those objects with low weight-to-surface area ratios assumed to be injected into the vortex and to reach the same velocity as the air particles around them. These missiles have good aerodynamic shapes, are expected to remain in orbit for a prolonged time, and can be at any altitude within the vortex.

2. At the other extreme are heavy objects with poor aerodynamic shapes, such as spherical bodies. They never achieve the velocity of the air nor are they injected into the vortex; rather, they roll, tumble, and bounce over the ground.

3. The last category includes those objects with intermediate characteristics that place them between the two other categories.

Figure 3.20. Pine board driven through a tree by the violent force of the tornado that hit Pryor Creek, Oklahoma, April 24, 1942. (Photo courtesy of the Oklahoma Historical Society.)

Impact Loading

All the various missiles mentioned above differ considerably from each other in terms of weight and area of contact when they collide with an obstacle. Some types, whose weight and contact area are particularly critical, can be selected for design purposes and their effects evaluated. We now examine the design specifications of some U.S. nuclear power plants. The criteria adopted to satisfy the requirements of the Nuclear Regulatory Commission with respect to tornado forces have produced basic models that most designers follow.

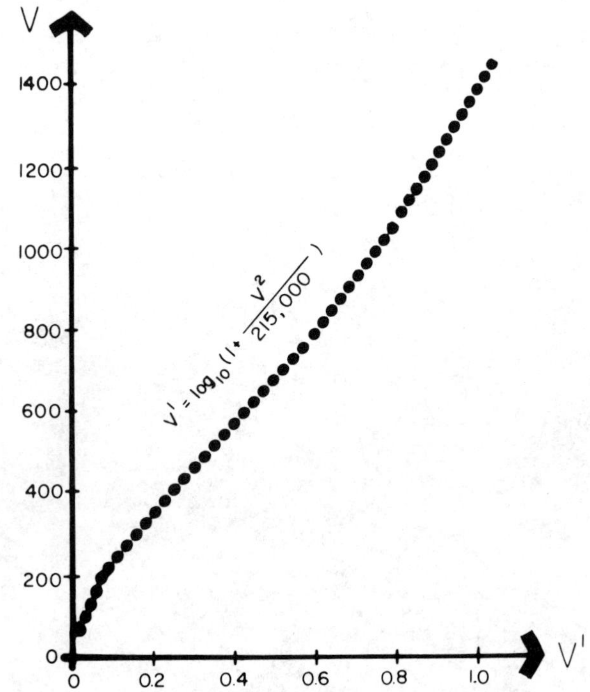

Figure 3.21. Velocity factor in the modified Petry formula expressed as a function of impact velocity *V*.

Missiles were divided into three categories. Type 1 missiles include those missiles expected to reach maximum possible velocity (assumed as the sum of the maximum tangential and forward velocity). These values vary up to 360 mph (considered a very conservative estimate). Type 2 missiles are usually longitudinal, slender, travel head-on, and have a small impact area (e.g., heavy and slow missiles such as automobiles and similar objects, which are not injected into the vortex). Of these, the most common is the 4000-lb automobile traveling at 50 mph. Type 3 objects include those that travel at intermediate speeds (e.g., a 3" diameter pipe traveling at 100 mph).

Simplified Analysis of Penetration

When the mass of a missile is small compared to that of the struck structure, the energy absorbed and dissipated by the structure, so that verification of structural stability can be neglected. The important analysis to performed is penetration of the missile into the structure, which depends on the following factors: impact velocity, propagation of stress waves, impact cratering, friction, heating, spalling of the stricken structure, and crushing of the missile. We can identify two types of impacts: high- and low-velocity. The dividing line in terms of impact speed is 500 ft/sec (341 mph). Certain aerodynamic tornado-driven missiles do fall within the high-velocity impact category. In this case, we can neglect several factors: hypervelocity effects, friction, heating, impact crater, the nose shape of the missile,

Table 3.8. Coefficient K_1 for the Modified Petry Formula

Material	ft^3/lb
Limestone	5.38×10^{-3}
Concrete $f'_c = 2200$ psi	7.99×10^{-3}
Reinforced concrete: $f'_c = 3200$ psi and 1.4 percentage of reinforcement	4.76×10^{-3}
Reinforced concrete: $f'_c = 5700$ psi and 1.4 percentage of reinforcement	2.82×10^{-3}
Stone masonry	11.72×10^{-3}
Brickwork	20.78×10^{-3}
Sandy soil	36.70×10^{-3}
Soil with vegetation	48.20×10^{-3}
Soft soil	73.20×10^{-3}

and missile crushing. There are several empirical formulas based on these assumptions, the most popular of which is the modified Petry formula:

$$\begin{aligned} D &= K_1 \frac{W}{A} [\log_{10} (1 + \frac{V^2}{215{,}000})] \\ &= K_1 \frac{V}{A} V', \end{aligned} \tag{3}$$

where V' is the velocity factor, equal to $\log_{10}[1 + (V^2/215{,}000)]$; D is the depth of penetration into infinite thickness (in ft); W is the weight of the missile (in lb); A is the minimum effective cross-sectional area of the missile (in ft^2); V is the impact velocity (in ft^2/sec) (see Fig. 3.21); and K_1 is the experimentally obtained coefficient typical for each material (see Table 3.8).

The effective penetration, D_1, into a slab of finite thickness T varies with ratio T/D as follows:

$$D_1 = D[1 + e^{-4[(T/D)-2]}], \tag{4}$$

where D_1 is either equal to D (when the slab thickness is twice D) or larger (when the slab thickness is smaller than D). The minimum thickness required to avoid total penetration is thus $T = 2D$. In practice, D must be computed first; then a thickness larger than $2D$ is to be selected (see Fig. 3.22). To illustrate the use of equation (3), several missiles at different speeds were selected and their penetration into various materials calculated. The results are presented in Table 3.9.

Local Effects

Observing penetration at closer range, we can see that a certain amount of material in the impact area is crushed when the stresses generated during collision exceed the ultimate strength of the material. Some of the energy is spent in crushing the missile, with a

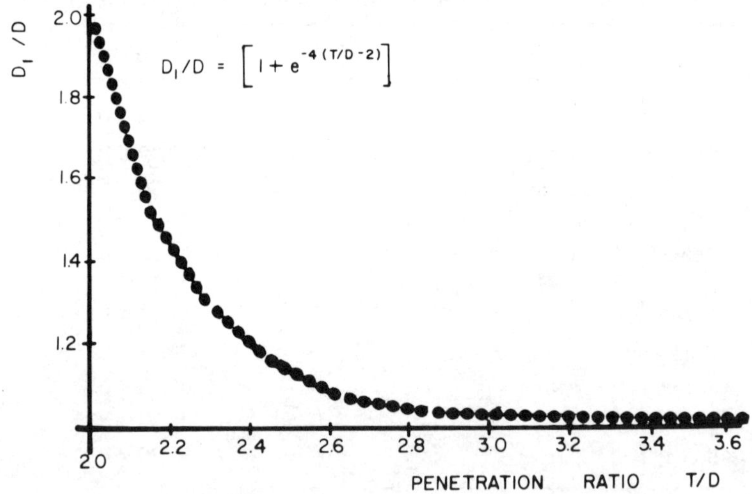

Figure 3.22. D_1/D as a function of T/D in the penetration formula.

consequent reduction of penetration. For instance, in the case of a 12 ft long 4 × 12" wood plank traveling at 200 mph, calculations show that depth of penetration in a concrete wall is about 4". However, in reality, this number will be smaller. In the case of a 4000-lb automobile traveling at 50 mph striking a reinforced concrete wall, the depth of crushed concrete is 0.125" according to calculations. Formulas used to calculate penetration of missiles into targets, though they take the characteristics of the target material into account, ignore those of the missile material. If missiles consist of lightweight low-strength materials (e.g., wood) or are low-weight (compared to their volume) hollow objects, their impact resistance is poor and they crush as they collide. The energy spent in crushing reduces impact energy, thereby lessening the severity of the effects on the target.

Still using a concrete wall as an example and examining the impact phenomenon further, we can see that, when a missile strikes the exterior face of a wall, a compressive shockwave originates at the area of impact and travels through the wall toward the interior (free) face. Once there, the shockwave is reflected and becomes a tension pulse that travels back toward the exterior face where the impact occurred.

If the material has poor tensile strength, as in the case of concrete, the tensile force causes spalling of the material starting from the interior face. The spalling proceeds toward the impact surface, gradually decreasing in intensity as the tensile pulse decreases. Finally, it stops when the tensile strength of the concrete is greater than its tensile stress. Spalling produces a number of small pieces of the material, which in turn become missiles when they travel away from the structure at a certain velocity. This velocity is generated by the energy the wall receives from the missile. The depth of spalling in reinforced concrete is reduced by reinforcement and usually does not exceed 3". Fortunately, the velocity that these fragments acquire is low, i.e., less than 20 ft/sec (13.63 ft/sec, as determined in several studies analyzing the spalling effects of high-velocity missiles).

Table 3.9. Missile Penetration Into Various Materials

Missile	Weight w (lb)	Cross sectional area (ft^2)	w/A	Impact velocity v (mph)	Impact velocity v (ft/sec)	$\log_{10}(1+v^2/215{,}000)$	Material coefficient K	Penetration D (ft)
a. Brick Masonry								
4" × 12" × 12" timber	131	0.274	480	200	294	0.146	20.48×10^{-3}	1.44
4" × 12 × 12" timber	131	0.274	480	250	367	0.212	20.48×10^{-3}	2.08
35-ft utility pole	1370	0.785	1745	200	294	0.146	20.48×10^{-3}	5.23
4000-lb automobile	4000	20	200	50	73	0.01	20.48×10^{-3}	0.041
4000-lb automobile	4000	30	133	100	147	0.04	20.48×10^{-3}	0.109
2000-lb automobile	2000	25	80	100	147	0.04	20.48×10^{-3}	0.065
Common brick	4	0.048	82	200	294	0.146	20.48×10^{-3}	0.246
b. Sandy soil								
4" × 12" × 12" timber	131	0.274	480	200	294	0.146	36.7×10^{-3}	2.57
4" × 12" × 12" timber	131	0.274	480	250	367	0.212	36.7×10^{-3}	3.73
35-ft utility pole	1370	0.785	1745	200	294	0.146	36.7×10^{-3}	9.26
4000-lb automobile	4000	20	200	50	73	0.01	36.7×10^{-3}	0.0734
4000-lb automobile	4000	30	133	100	147	0.04	36.7×10^{-3}	0.195
2000-lb automobile	2000	25	80	100	147	0.04	36.7×10^{-3}	0.117
Common Brick	4	0.048	82	200	294	0.146	36.7×10^{-3}	0.44
c. Reinforced concrete								
4" × 12" × 12" timber	131	0.274	480	200	294	0.146	4.76×10^{-3}	0.333
4" × 12" × 12" timber	131	0.274	480	250	367	0.212	4.76×10^{-3}	0.484
35-ft utility pole	1370	0.785	1745	200	294	0.146	4.76×10^{-3}	1.21
4000-lb automobile	4000	20	200	50	73	0.01	4.76×10^{-3}	0.0095
4000-lb automobile	4000	30	133	100	147	0.04	4.76×10^{-3}	0.024
2000-lb automobile	2000	25	80	100	147	0.04	4.76×10^{-3}	0.0152
Common brick	4	0.048	82	200	294	0.146	4.76×10^{-3}	0.0572

cont'd

Missile	Weight w (lb)	Cross sec- tional area (ft^2)	w/A	Impact velocity v (mph)	Impact veloc- ity v (ft/sec)	$\log_{10}$ $(1 + v^2 /215{,}000)$	Material co- effi- cient K	Pene- tra- tion D (ft)
d. Stone masonry								
4" × 12" × 12" timber	131	0.274	480	200	294	0.146	11.72×10^{-3}	0.82
4" × 12" × 12" timber	131	0.274	480	250	367	0.212	11.72×10^{-3}	1.19
35-ft utility pole	1370	0.785	1745	200	294	0.146	11.72×10^{-3}	2.99
4000-lb automobile	4000	20	200	50	73	0.01	11.72×10^{-3}	0.023
4000-lb automobile	4000	30	133	100	147	0.04	11.72×10^{-3}	0.062
2000-lb automobile	2000	25	80	100	147	0.04	11.72×10^{-3}	0.037
Common brick	4	0.048	82	200	294	0.146	11.72×10^{-3}	0.141
e. Soil with vegetation								
4" × 12" × 12" timber	131	0.274	480	200	294	0.146	48.2×10^{-3}	3.36
4" × 12" × 12" timber	131	0.274	480	250	367	0.212	48.2×10^{-3}	4.9
35-ft utility pole	1370	0.785	1745	200	294	0.146	48.2×10^{-3}	12.25
4000-lb automobile	4000	20	200	50	73	0.01	48.2×10^{-3}	0.096
4000-lb automobile	4000	30	133	100	147	0.04	48.2×10^{-3}	0.256
2000-lb automobile	2000	25	80	100	147	0.04	48.2×10^{-3}	0.154
Common brick	4	0.048	82	200	294	0.146	48.2×10^{-3}	0.577
f. Soft Soil								
4" × 12" × 12" timber	131	0.274	480	200	294	0.146	73.2×10^{-3}	5.13
4" × 12" × 12" timber	131	0.274	480	250	367	0.212	73.2×10^{-3}	7.45
35-ft utility pole	1370	0.785	1745	200	294	0.146	73.2×10^{-3}	18.70
4000-lb automobile	4000	20	200	50	73	0.01	73.2×10^{-3}	0.1464
4000-lb automobile	4000	30	133	100	147	0.04	73.2×10^{-3}	0.39
2000-lb automobile	2000	25	80	100	147	0.04	73.2×10^{-3}	0.234
Common brick	4	0.048	82	200	294	0.146	73.2×10^{-3}	0.88

Materials

In very broad terms, we can say that the impact resistance of a structural member is measured by the amount of (strain) energy that the member can absorb during deformation caused by a load. If we consider a static load, *P*, gradually applied to a member, and draw the load deflection curve, the area under the curve represents the energy absorbed by the member and its impact strength. The triangular area represents the elastic-energy-absorbed resilience, and the total area represents the elastic plus plastic energy absorbed (toughness). In other words, the former is the impact strength at the elastic limit, whereas the latter is the impact strength at rupture (see Table 3.9).

If we allow only elastic deformation in the structure, we must consider the resilience as the impact capacity. If we allow some permanent deformations to occur, we can consider the toughness of the material as a strength characteristic.

The impact resistance of prestressed concrete members has been proven by experimental testing to reach an optimum level at a specific percentage of reinforcement, amount of prestress, and concrete strength. This is so critical that changing the proportions (e.g., even by increasing the amount of steel) will cause a reduced impact strength. Is prestressed concrete better than concrete (or vice versa) in resisting impact loads? By testing reinforced versus prestressed concrete beams with equal static strength, it was concluded that there is no real difference between the two and that the only index of strength is beam resilience. This can be made to vary, though the beams still have the same static strength. This can be accomplished by changing the percentage of reinforcement or prestress and allowing various deflections of the beams under load.

Dwelling Structures

How can we evaluate the effects of tornado forces on different types of buildings? It is particularly helpful in this respect to use test results collected for structures exposed to nuclear blasts. A certain analogy which exists between tornado forces and nuclear explosions makes this possible. Indeed, though nuclear explosions and tornadoes are different phenomena, several aspects of the forces involved are comparable, including the drag force of the wind and the pressure differential on opposite sides of walls, floor slabs, and roofs.

Nuclear tests were conducted in Nevada on March 17, 1953 and May 5, 1955 (see Fig. 3.23). Various types of buildings were located at different distances from the ground zero so as to evaluate the damage caused by a variety of loadings. From the damage analysis, it was found that wood-framed dwellings collapse at a pressure of 720 psf (5 psi) (see Fig. 3.23); they do not collapse at 245 psf (1.7 psi), but they are damaged beyond repair. Ordinary masonry construction differs little from wood-framed ones in overall strength. On the other hand, seismic-resistant structures made of reinforced concrete panels and reinforced concrete block walls can tolerate pressures of 245 psf (1.2 psi) with damage limited to doors and windows. It is very useful to establish that, for average buildings, broken doors and windows constituted an adequate venting system to balance pressure in most cases. Interestingly, the seismic-resistant structure could resist loads up to 720 psf (5 psi) with only negligible additional damage (Melaragno 1982).

The Nevada tests clearly indicate that buildings using ordinary construction technology are vulnerable to forces similar to those unleashed by tornadoes. Nevertheless, using such special materials as concrete and adopting such special techniques as those used for earthquake-resistant structures tremendously improve building strength. Typical dwelling

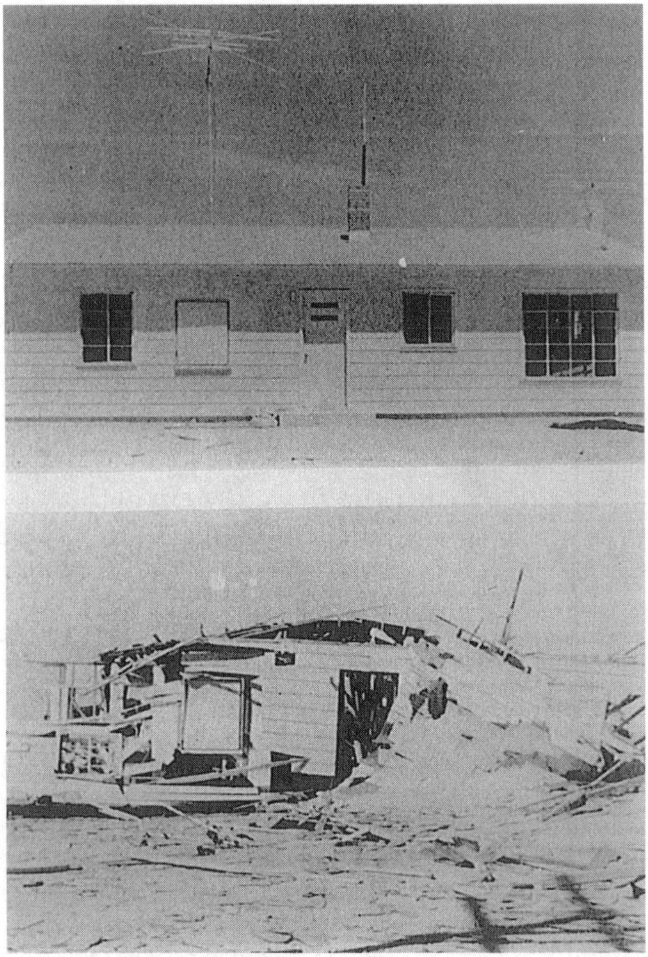

Figure 3.23. Test on structure subjected to nuclear explosions at the Nevada Testing Grounds in 1955. The photo above indicates the structure before testing; below, after testing under 5 psi load. (Photos courtesy of the U.S. Atomic Energy Commission.)

structures are usually designed by empirical methods without the benefit of structural analysis. Building codes are generally sufficient to guarantee their stability under conventional loads, so that engineering verifications are omitted. Economic factors also usually rule out professional fees. Tornado-resistant criteria are rare for almost all buildings, except for nuclear power plants. However, for buildings that require structural design, storm-resistant criteria are established by the engineer after the forces involved are defined. On the other hand, structures used to house people are left in an unresolved situation.

For small buildings in those areas where tornado frequency is significant, the following criteria should be considered. The basic structure is assumed to be divided into two zones: a central one (referred to as the primary zone) and the rest of the building (the secondary zone). The primary zone built in reinforced concrete should satisfy three basic structural requirements:

1. To be able to resist the various forces (conventional loads and tornado forces) applied to the structure so that its stability is guaranteed with a high probability of safety.

2. To function as a resistant element for the secondary structure against the action of horizontal forces.

3. To serve as a structural support against conventional vertical loads.

The wind force directly applied to the exterior surface generates a vertical overturning moment that must be balanced by a larger resisting moment. The use of grade beams is an obvious method to increase the resisting moment. Protection from the tendency of a structure to slide must be made assuming that a frictional force acts between the soil and the foundation, and that passive soil pressure acts on the vertical projection of the underground part of the foundation. Note that these two forces are reactions generated by the horizontal wind force. Maximum soil pressure must be checked against the soil's allowable bearing capacity. The structure must be analyzed for shear and properly reinforced. Consideration must also be given to the above-mentioned horizontal moment.

A venting system for pressure balance is necessary; however, since the propagation of a pressure wave travels at the speed of sound, the location of vents is not critical in small structures. Vents can thus be located where most convenient, i.e., on the roof slab rather than on the walls, which would provide better shielding action with fewer openings.

Sizing of vents should be based on conservative assumptions. The Nevada nuclear tests confirmed that size is a critical factor. In some cases, it was found that airflow through first-floor windows was more rapid than through small basement windows, which produced a pressure differential between the basement and first floor that caused the floor to collapse (see Fig. 3.24).

For impact loads from flying debris, reinforced concrete should be used for the primary zone. As mentioned above, stiffening of the whole structure should not endanger the stability of the primary zone. To satisfy this condition, it is necessary that supports connecting the primary to the secondary structure be designed so that they fail when the reactions reach certain preestablished limits. This criterion should be applied to the horizontal and vertical components of reactions.

The floorplan of the primary zone is illustrated by several schemes that compare different wall arrangements in terms of the shielded area (see Fig. 3.25). The shielding action is provided by reinforced concrete walls, assuming that doors have no protective capacity. The doors should be solid, reinforced, and be able to slide so that they can be retracted behind the walls; however, when closed, they should bear against the walls, acting as slabs supported on three sides.

The secondary zone, which constitutes the largest part of the building, is designed based on the concept of controlled damage to avoid total collapse, but it is not expected to totally withstand the full fury of a tornado. To keep additional building costs as low as possible, the secondary structure should be built (with few exceptions) using conventional materials and standard construction techniques. For instance, additional reinforcing elements from the foundations to the upper plate should be installed. The most important factor with respect to the stability of the secondary structure, however, is not the stability of the structure itself but the contribution offered by the primary structure when coupled with it. For further discussion on building technology applicable to tornado-prone areas, see Melaragno (1973a, 1973b, 1973c).

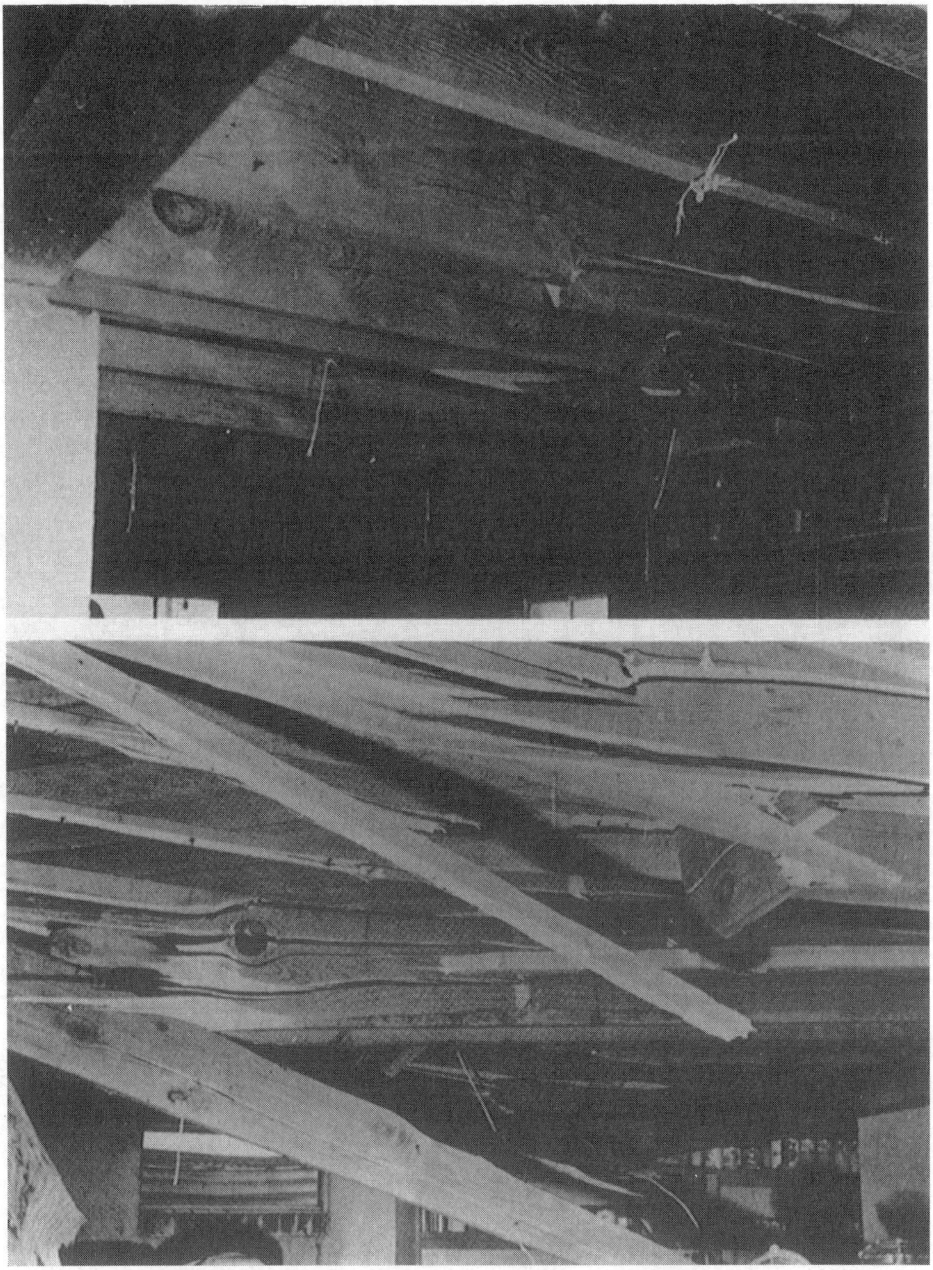

Figure 3.24. Damage by pressure differential above and below floor in building at the Nevada Testing Grounds in 1955. The difference in window size between the upper floor and the basement allowed more air to move in the upper floor; therefore, the pressure in the upper area was different from that in the basement. This pressure differential was applied to the floor, causing the damage shown. (Photos courtesy of the U.S. Atomic Energy Commission.)

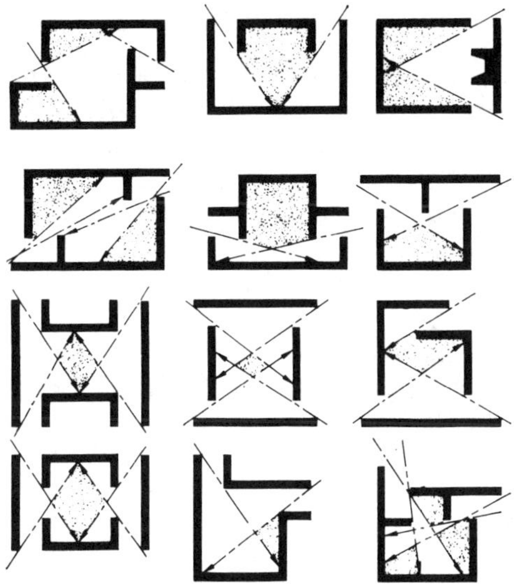

a

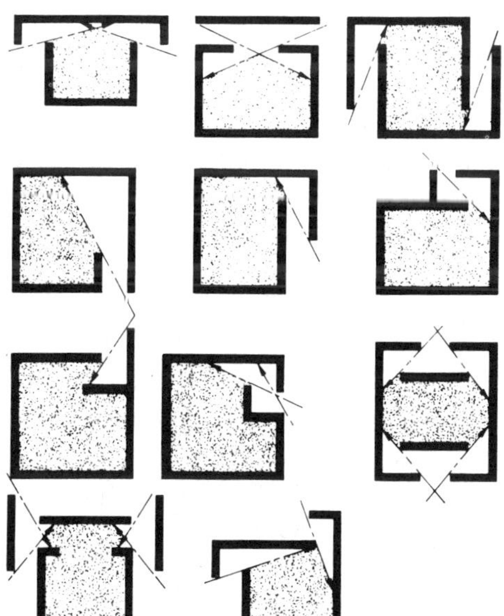

b

Figure 3.25. The central core. At first glance, there seems to be no problem in designing the arrangement of the core: designating the area, enclosing it with concrete walls, and arranging baffles to protect door and window openings from flying debris. However, as we start sketching general schemes, we realize that some work and many do not, i.e., the sheltered area is extremely small for the same amount of wall area. After sketching a number of alternatives, we divided them into two groups: *a*, acceptable, and *b*, unacceptable.

Tornado Shelters

Fear (indeed, terror) is a painful reality for those who live in parts of the world where tornado probability is high or where past tragedies caused by tornadoes cannot be forgotten. In some areas of the Midwest in the United States, tornado watches are a daily occurrence during tornado season, and tornado warnings (i.e., when tornadoes are actually detected in the vicinity) are frequent. There is a constant state of insecurity and anxiety. The people of Dacca in Bangladesh will never forget the tornado that battered their city on April 14, 1969. Decapitating metal sheets shot through the air, 600 died, and 5000 were injured. In view of the terrible danger to human life, the need for some sort of tornado protection cannot be ignored. Even the rugged nineteenth-century settlers of the Great Plains recognized the need for tornado shelters and built so-called "storm cellars" near their homes.

Since the 1950s, several plans for tornado shelters in residential areas have been prepared by such agencies as the American Red Cross, the Portland Cement Association, and Kansas State University. However, their architectural components would disrupt rather than enhance the appearance of conventional city dwellings. We include some schematic proposals made by the present author (Melaragno 1968a) (see Figs. 3.26–3.29). Types A–D are made out of reinforced concrete, have pressure balancing vents, and use a top slab configuration that could be used as a landscape feature. These structures are not completely underground, which alleviates feelings of claustrophobia.

Safety Measures

The latest evaluations of tornado-caused deaths and property damage related to many tornadoes indicate the existence of new correlations between safety measures and their effectiveness. The U.S. Department of Health, Education, and Welfare (1979) emphasized some safety points and disproved some commonly accepted assumptions. The validity of this study was corroborated by agencies whose representatives participated in the survey, including the Centers for Disease Control, the American Red Cross, and the U.S. Air Force School of Health Care Sciences.

The object of study was the disaster caused by a tornado that struck the southwest corner of Wichita Falls, Texas, in the late afternoon of April 10, 1979. The following data were compiled:

- City population: 100,000
- Damage area: 7 mi long, 1 mi wide
- Deaths: 44
- Injuries: 1871 individuals
- Damage: \$300 million$^+$
- Destroyed residential units: 6000$^+$
- Warning system: operational and efficient, with sirens sounding twice (1 hr and 5 min) before tornado touchdown

Relating the number of deaths to the other parameters, we note the following:

- Deaths were 0.044% of the city population
- Deaths were 2.3% of the total number (1915) of people physically struck

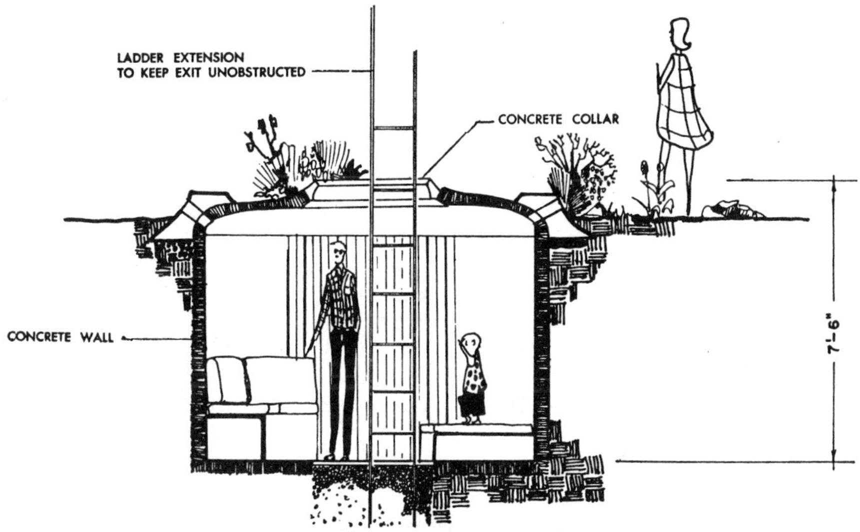

Figure 3.26. Type A outdoor tornado shelter.

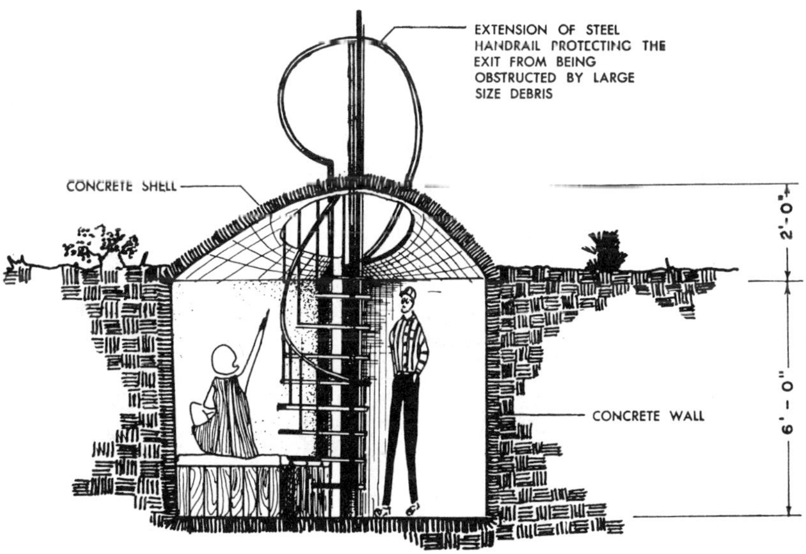

Figure 3.27. Type B outdoor tornado shelter.

When the number of accidents and fatalities among the people who stayed indoors during the tornado was compared to that for people who looked for shelter outdoors, it was obvious that conventionally constructed dwellings were much safer. This was true not only for dwellings on the fringe of the tornado path but also for those in the most severely damaged part of path. Even when structures were totally demolished, the people inside were protected. The components that were particularly effective in shielding people included base-

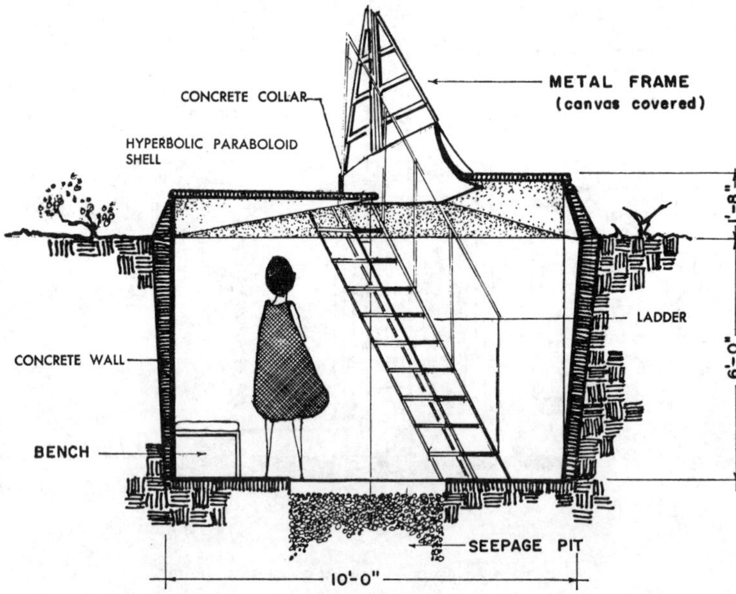

Figure 3.28. Type C outdoor tornado shelter.

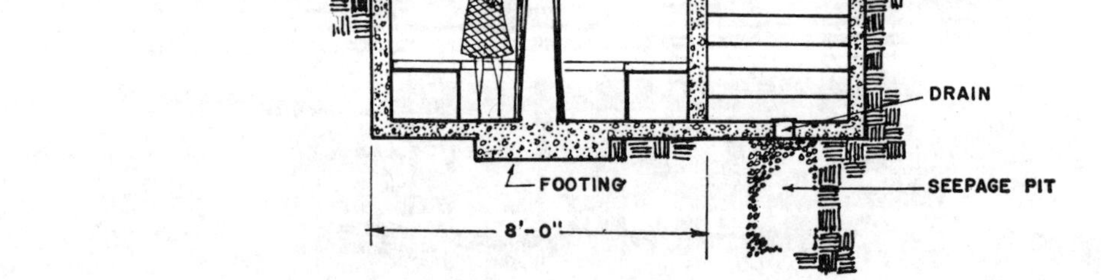

Figure 3.29. Type D outdoor tornado shelter.

ments, closets, hallways, and heavy furniture. Outdoor features (e.g., culverts, underpasses, and ravines) were less effective.

Summary

The forces with which tornadoes act upon building structures in their path can be summarized as follows: wind forces per se, air pressure changes between building interior and exterior, and impacts by missiles. The wind forces are the result of rotational and translational velocities, and their intensity is inversely proportional to distance from the tornado eye. Pressure changes are proportional to wind velocity, so that pressure varies with distance from the eye. The pressure change rate, which is a function of diameter, is also related to the translational velocity.

The structural response of buildings varies accordingly to the forces involved (wind pressure, pressure change, and impact). The lateral resistance of buildings to wind pressure can be based on such structural systems as vertical diaphragms, horizontal diaphragms, shear walls, frame action, and vertical trussing.

Pressure change can be dealt with, either by reducing it or by resisting it. Pressure reduction can be obtained by exterior and interior venting. Pressure resistance implies design of floors, roofs, and walls according to specific pressure loads.

Resistance against the impact of flying debris is most conveniently based on protective structures, which, for conventional buildings, can include baffled walls, special window design, and protective shelters.

As mentioned above, tornado forces were not considered in building design until the late 1960s, when construction of nuclear power plants required that tornado forces be included in structural design. Since then, quite a few such critical structures have been built in the United States and around the world, creating a great amount of documentation on design data, as illustrated below.

High-Rise Buildings

Most records of tornado forces acting on buildings involve residential structures, since these are the most numerous, while high-rise buildings are almost never included. We consider the case of a 20-story building exposed to a tornado in Lubbock, Texas in 1970. This event is still a major relevance for the area of wind engineering. The structure in question was the Great Plains Life Building in downtown Lubbock, Texas, which has since been renamed after being rehabilitated by new ownership (see Fig. 3.16 above). It is still the tallest structure to have ever been reported on with regard to tornado forces. The only other high-rise reported in the technical literature is a 5-story building in Waco, Texas, hit by a tornado in 1953.

The behavior of high-rise buildings with respect to lateral forces is obviously a critical aspect for structural design. When these forces are extremely accentuated, as in the case of tornadoes, they become even more relevant. The role that this case study plays in the literature on severe storm wind engineering cannot be overemphasized. We report it in its generalities, leaving the reader responsible for a more detailed analysis of the record concerning this structure.

The Building

The building in question was located at 1220 Broadway at Avenue L and was called the Great Plains Life Building (after the company that owned and operated it). It is now called the Metro Tower and is owned and operated by a different company. After the May 11, 1970 storm, during which the building escaped total destruction, it was eventually rehabilitated after structural modification and repairs.

The 20-story building included a rectangular floorplan that was 125 × 75 ft at the base, which was comprised of the first three stories. The plan included 32 bays — 25 in the short direction and 7 in the long direction. The 125-ft long side of the building was oriented north to south, with its short side oriented east to west. Only the first three stories occupied the full area, with a tapered portion extending up to the 20th story, which had a tower. The structure had a steel frame made of A-7 steel with riveted connections, typical of this kind of steel, which was well suited to welding. These connections followed standard American Institute of Steel Construction (AISC) practice and were considered moment-resistant, yet they behaved more like semi-rigid connections.

The floor system included an 18-gauge corrugated metal deck integral with a 4" lightweight concrete slab. The exterior walls varied, i.e., on the north side, where the wall had no major openings, the construction was different from that on the other sides, where windows occupied the full width of the base. On the north side, there was an 8" clay tile wall carried on the spandrel beams, and in front of it there was an 8" thick brick wall structurally independent and acting as a veneer. The two walls were tied together by a header brick every seven courses, creating lateral stability. The external wall was supported at the base by the first three stories alone, while from the third story all the way to the top the brick veneer rested entirely on a shelf angle at the third story. In the walls where 15" windows occupied most of the bays, only the portion below the windows consisted of masonry. These masonry portions consisted of 8" clay tiles, walls carried by spandrel beams, and 8" brick veneer external walls carried by shelf angles. The columns, consisting of 14" wide flange profiles, were encased in gypsum blocks for fireproofing and covered by brick veneer in the portion below the windows. On the interior face, a 1" thick plaster layer covered the tile wall and interior face of the column.

Completed in 1955 at a cost of $2.5 million, this building rose to a height of 271 ft, included 20 stories, and had a 3-story 125 × 56 ft base. In addition to the 20 stories, there was a penthouse which contained the mechanical equipment and a 60-ft antenna, adding additional surface area exposed to the wind.

The Storm

On the evening of May 11, 1970, two tornadoes struck downtown Lubbock, Texas. The first one, smaller in size and intensity, was followed by a much wider and more destructive tornado. The second touched down at 9:45 p.m., creating severe damage along an erratic 8-mi path, killing 26, and injuring more than 1500. The Great Plains Life Building was at least six blocks away from the direct path of the storm. Eyewitnesses furnished an invaluable description of the building's behavior under load. A group of people were in a restaurant on the 20th story when the storm struck. They descended to the ground floor during the storm, and thus experienced the behavior of the building throughout the storm and felt it swaying on each floor. They estimated the motion of the building on the upper floors as on the order of 4–5 ft. Their comments described these motions as similar to that of a ship in a storm. All 30 some occupants of the building were able to reach exits and were able to leave safely.

Building Damage

An inspection of the structure of the building the next day confirmed its structural soundness. A general inspection that provided preliminary data on the building's vulnerability established that two of the four elevators were still operational, as were the air conditioning system and the cooling tower.

An engineering survey of the building distinguished two categories of overall damage: superficial and structural. The superficial damage included destruction of 60% of the glass area, cracking of internal partitions, collapse of some plaster surfaces, collapse of portions of the external brick veneer, and bending of the 60-ft roof antenna. A major amount of damage to the interior and exterior was directly connected to the sway of the building and to torsional deflection (see below). The structural damage originated due to elastic and plastic deformations.

Measurement by transit to establish the overall deformation of the building revealed that the two top corners on the northern side remained unaffected, while the southern corner exhibited maximum deformation in the eastern direction. At the southwest corner, column deflection amounted to 12" from vertical. After the first measurements by transit, a more complete structural survey of the steel frame was conducted by an engineering firm after removal of the exterior cladding on the south of the building between the 5th and 6th stories, where maximum deflection had occurred. This survey found that a deformation had occurred at this level which amounted to 2.25" and was more than 12 ft tall. This deformation was eventually found to be plastic because the deformations remained practically unchanged after removal of partitions and portions of the wall.

Observation of the frame action of steel bends revealed that the stiffness of beams was much greater than that of the columns. Deflection of the frame was mostly due to deformation of columns.

The torque action that the storm winds inflicted on the structure was fundamentally important in the deformation of this building. Because of its structural asymmetry, the center of resistance at the foundation level was not aligned vertically with the joint of application of the resultant of the wind forces on the building, so that a horizontal movement about a vertical axis was begun. It should be noted that the northern end of the building was much stiffer than the southern end, which shifted the center of resistance at the foundation level toward the southern end.

Reclaiming the Structure

From the assessment of damage done to the steel structure of the building, it was established that the stresses had reached the yield point, which created permanent deformations. This became evident when the structure was freed from all secondary building components. Had the deflection been elastic, the steel frame would have gone back to its original configuration. It was obvious that straightening would be necessary to bring the structure back into service. In other words, an opposite permanent deformation had to be induced by applying adequate forces at appropriate points. These points were not along the beams and the columns which constituted the frame, but at the joints where the beams and columns met.

We will consider each individual bend, consisting of multistory and multi-bay portals. When a bend was distorted, each portal was distorted. Therefore, each portal had to be straightened to straighten each bend. Distorted portals could be easily straightened by applying a diagonal tensile force which would permanently deform each portal. A straightening portal would acquire its original shape with 90° corners. Obviously, straightening of each individual portal would result in straightening of the whole structure in a series of simple individual operations. As mentioned above, most of the deformations in each portal were due to deflection of columns rather than the deflection of beams, since the latter were much stiffer than the former.

Interestingly, observation of the exposed steel frame determined that the paint on members provided excellent indicator of the stresses exerted on the frame. The paint acted as a brittle coating that cracked under stress in a clear manner that could recorded easily and calibrated easily.

Reclamation of the building included quite an extensive amount of demolition and reconstruction, which can be assessed independent of the present context (which focuses on structural aspects of the damage). In this respect, to obtain access to the steel frames, it was necessary to demolish portions of the exterior walls, internal partitions, suspended ceilings, etc. These demolitions did not extend throughout the building, but only involved those locations where straightening of the steel frame was required. Four transverse bays, starting on the south side, required full exposure of the steel frame between the 4th and 10th stories. Once the steel frame was exposed and straightened, it was important to strengthen the frames with additional bracings before reopening the building. Beam-to-column connections and column splices did not fail in the frame, so they are above all criticism. Some structural reports on the behavior of the building indicate that the connections acted in a semi-rigid behavior, allowing large-amplitude lateral swaying that could have been reduced with additional bracing. However, plastic failure was located within columns not adequately sized to resist the bending moments induced by lateral forces. It is important to note that the structural design of the frame was adequate to handle the wind loads anticipated by the building codes, but obviously inadequate when faced with the loads that actually occurred.

We can learn important wind engineering lessons from this structure, both from a structural point of view and from the perspective of overall building design. The steel frame acted particularly well in that it allowed easy restoration of its integrity, further indicating that major lateral stiffening could have easily been obtained with adequate bracing if the building had originally been designed for the forces that hit it during storm. Steel frames have not been at fault for high-rise building disasters under severe storm wind forces, but the flexibility of the structure and that of the secondary components have to be compatible so as to minimize relative interference. Other lessons are self-explanatory, including the vulnerability of the glass surfaces that constituted a large percentage of the building cladding.

5. TORNADOES AROUND THE WORLD

Such a natural phenomenon as tornadoes, which in the minds of many laymen are confined to the New World, have been observed in other parts of the world since ancient times. Historical accounts describe these storms occurring over land and at the sea. These accounts were authored by such illustrious historians as Lucretius (99–55 B.C.), Seneca (A.D. 4–65), and Pliny the Elder (A.D. 23–79). Niccolo Macchiavelli described such storms centuries later. His vivid pictorial narratives include a report on an extremely violent storm which struck Italian territory on August 24, 1456. From his account, this storm definitely displayed the typical characteristics of a North American tornado because of the extreme violence demonstrated by the damage it caused. The intensity of the storm was further corroborated by the extreme length of its path, which extended from the Adriatic to the Tyrrhenian Sea, crossing both the Marche and Tuscany. It was first sighted near Ancona about an hour before dawn and was later sighted near Pisa, reflecting its exceptionally long itinerary compared to that of conventional *tromba d'aria*. It was described as about 2 miles in diameter, which also supports the supposition of an extremely energetic storm. Also in agreement with a description of a North American tornado is the fact that this storm was accompanied by high electrical activity and the astounding level of sound reported by eyewitnesses. However, its most significant physical characteristic was the intensity of damage inflicted to masonry structures, which included not only destruction of roofs and chimneys but also absolute and total demolition of structures.

Macchiavelli makes specific mention of destruction caused at Castello di San Caseiano and in the village of Santo Andrea, only 8 miles from Florence. Other localities mentioned included San Martino a Bagnolo and Santa Maria della Pace. Reports of loss of human life were included with awe-filled remarks about the terrifying and supernatural power of the storm, which gave one the feeling of witnessing divine intervention. For many, this was a demonstration of God's wrath against sinful mankind. It could also have been considered as a warning due to the fact that its destruction was mainly confined to sparsely inhabited areas, as the tornado avoided high-density urban areas where the damage would have been catastrophic. The overall effect that one takes away this Macchiavelli's account is that of a particularly violent storm, one quite unusual for the climatology of the territory of Italy. However, one is left with little doubt that a tornado did indeed occur.

Tornadoes Outside the United States

The frequency of tornadoes cited in the literature is based on storms that have been observed and recorded, and not on the absolute number of actual occurrences. It is obvious that a great many factors can affect the number of recorded observations. The population density, the efficiency of news media, the public interest, and the efficiency of the weather stations, among others, are all factors that influence the number of recorded cases. This is confirmed by comparative analysis, if one considers a certain zone and examines the number of cases reported for different years. A sudden increase in the number of tornadoes beginning in 1953 is clearly noticeable In the United States. This date coincides with the beginning of tornado-forecasting programs and the use of radar in weather stations. The number of storms recorded in Europe from 1650 to 1916, as reported by Wegener (1917), varies sharply with political and social factors, including war (see Fig. 3.30). The annual frequency under stable conditions demonstrates considerable variations regardless of the influence of exterior parameters. The graph in Figure 3.31 indicates that from 1953 to 1969 the recorded frequency in the United States varied widely, with maximum and minimum values having a ratio greater than 2:1.

The meaning of the frequency data is clearer when they are collected under different circumstances in various geographical locations. The European statistics given in Figure 3.32 (Wegener 1917) have informative and historical value. They show very clearly that atmospheric vortices are not limited to the plains of North America. The ratio of number of tornadoes in North America to that in Europe was estimated by Wegener to be about 20:1. According to data collected by the Swiss Meteorological Institute, only three vortices were officially recorded in Switzerland in one 80-year period (August 19, 1890, June 12, 1926, and August 26, 1971). This is a very low number compared to the data collected in Italy. According to Wegener, they contradict the pattern of the percentage distribution among European countries. This indicates the subjectivity of the various data collecting sources.

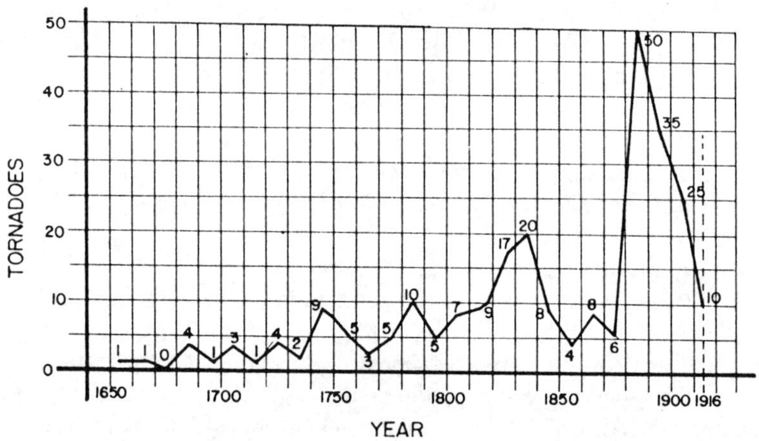

Figure 3.30. Tornadoes in Europe, 1650–1916 (Wegener 1917).

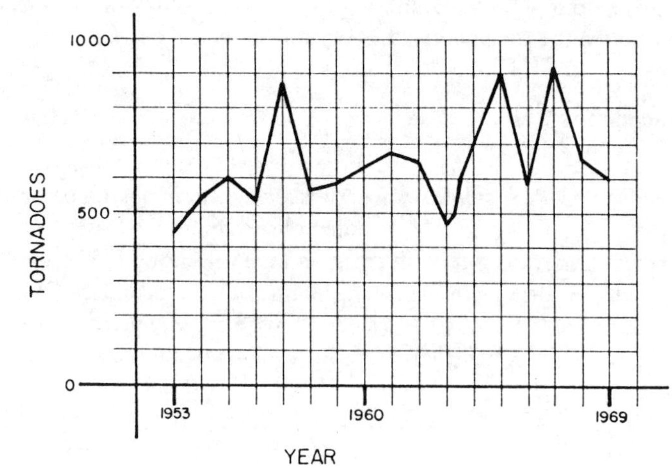

Figure 3.31. Tornadoes in the United States, 1953–69 (NOAA).

Germany has the highest percentage of vortices in Europe (Italy has only 20% of the cases reported in Germany).

A more recent annual frequency distribution (see Fig. 3.33) indicates a growing number of cases in Italy, beginning from the end of World War II. This coincides with a generally revival of observations of meteorological phenomena. The reason for the absence of data for 1966 and 1967 is not clear; it may be an oversight. The Italian frequency data are also very variable. In 1964, there was a maximum of 25 cases. A monthly Italian distribution done by Wegener (1917), indicates that, as with other European countries, the largest number of cases occurs in July (see Fig. 3.34). (The highest frequency in the United States is in May.) The regional distribution in Italy is extended over the whole peninsula and includes Sicily and Sardinia. Some zones have much higher frequencies than others. The 11% frequency in

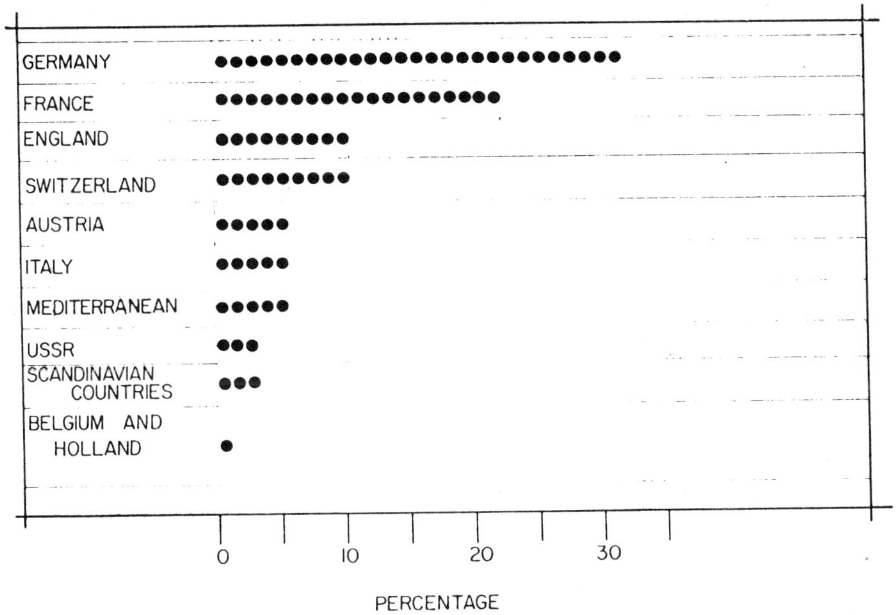

Figure 3.32. Percentage distribution of tornadoes in Europe (Wegener 1917).

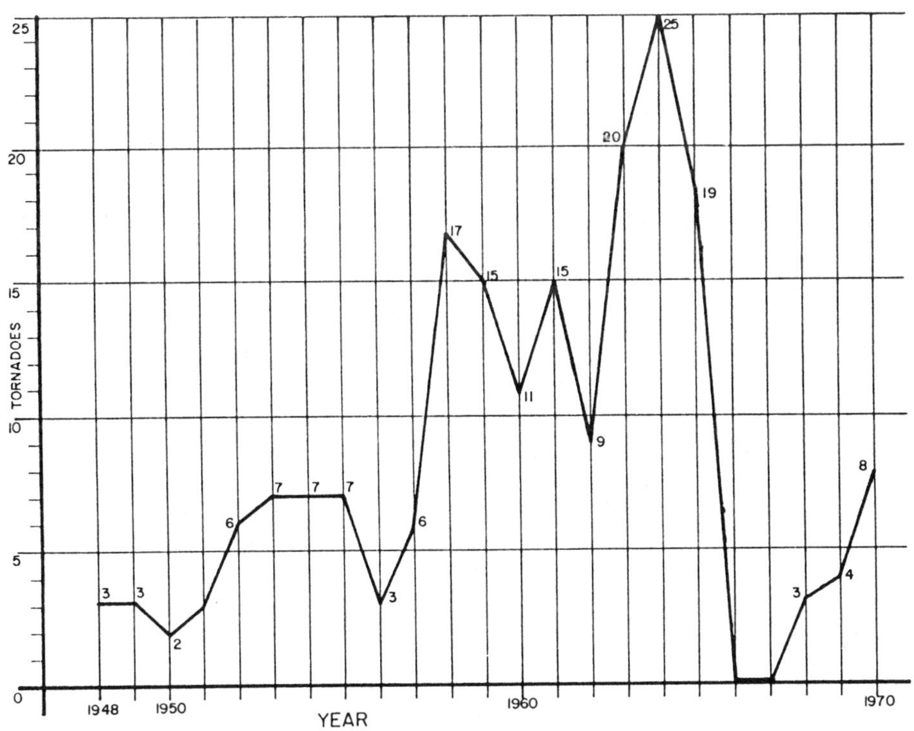

Figure 3.33. Annual tornado frequency in Italy, 1948–70.

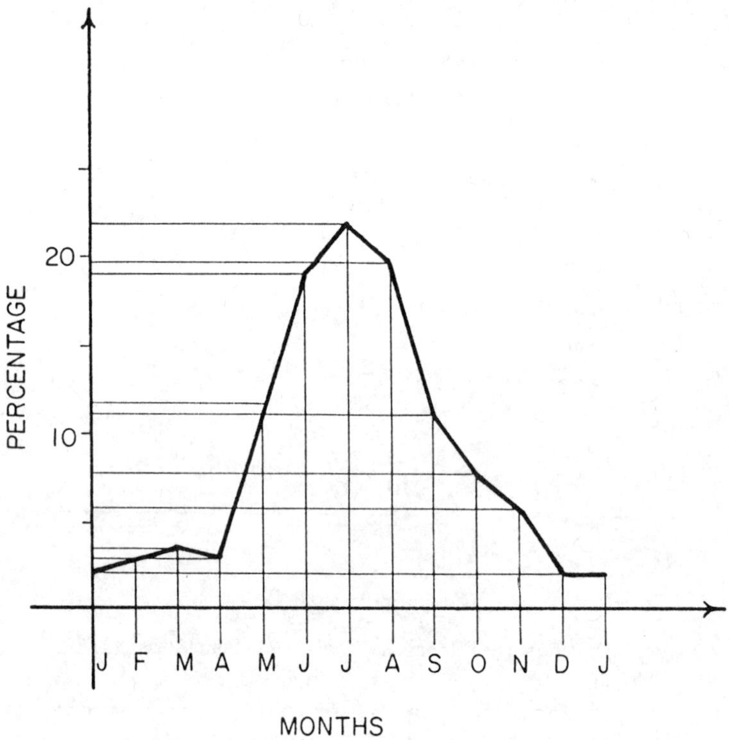

Figure 3.34. Monthly frequency of tornadoes in Europe (Wegener 1917).

the province around Rome clearly is highest, followed by Genoa (4.6%), Turin and Milan (4.15%), and Verona and Palermo (1.56%). It is interesting to note that the first studies on air vortices were performed in Italy.

Montanari (1694) discussed the theory of air vortices in connection with observations of one reported in the territory near Venice in 1686. Boscovich (1749) dealt with the action of an air vortex over Rome (see Wegener 1917). Other frequency data are given in Figure 3.35. The hourly distributions in Holland and the United States indicate that the afternoon is the part of the day when vortices are most likely to occur: 3 p.m. in Holland and 5 p.m. in the U.S. (Wegener 1917).

The first report of a tornado in Australia appears to be one dated October 20, 1826. More methodical records were begun in 1887, which included newspaper descriptions given by eyewitnesses and other unofficial documents whose reliability is questionable since they were not verified by qualified experts. Several types of local Australian storms have similar manifestations that can lead observers to draw erroneous conclusions. Additional confusion arises from the misuse of such terms as cyclone, tornadic squall, hurricane, cockeyed bob, willy-willy, and tornado. Clarke (1962, p. 56) states:

> A search of files on the occurrence of several local windstorms in Australia has made possible some tentative conclusions concerning the frequency, intensity and geographical distribution of these tornado-like storms. They are on the whole about as frequent as, but much less intense than American tornadoes.

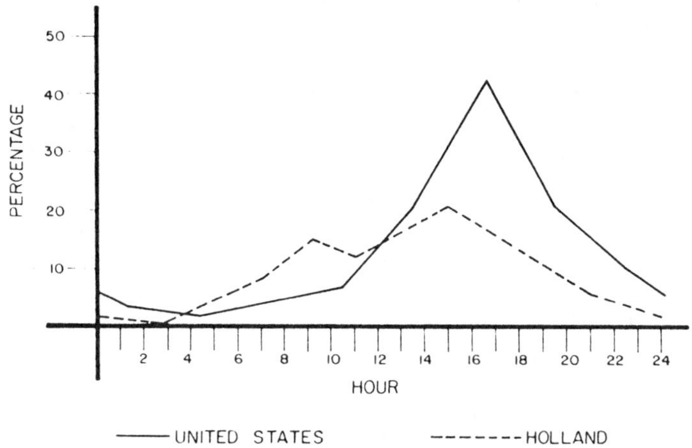

Figure 3.35. Hourly frequency of tornadoes in the United States and Holland.

The size and violence of tornadoes seem to be related to local geographical conditions. It is thus possible to differentiate tornadoes in terms of these parameters. Recent work in this area supports this theory. American tornadoes, for example, can be distinguished from tornadoes in other parts of the world; they can even be differentiated according to state.

Violence can be measured in terms of wind speed and pressure differentials or by their manifestation expressed in terms of deaths and damage per unit area of tornado path. It is obvious, however, that other such factors as population density and the effectiveness of warning systems can confuse an analysis of storm violence.

The size of a tornado path affects the number of deaths and the amount of damage as absolute values, so that a larger but less violent tornado could be assessed as being more destructive than a smaller higher-intensity storm. (In other words, it is critical not to confuse intensity with overall destructiveness in assessing tornadic effects.) However, newer studies support the theory that tornadoes with longer paths are usually not only more destructive but also more violent.

In terms of geographical occurrence, we can thus see that tornadoes vary as a function of the above-described characteristics: size, violence, and overall destructiveness. To confirm this differentiation in practical terms, the Meaden (1976) tornado rating system classifies tornadoes in Great Britain as having an average *T* value of 8. This implies wind velocities of 225–47 mph. Such storms are described as "severely devastating tornadoes." Meaden classifies French and German tornadoes as having an average *T* value of 9, which implies wind velocities of 248–76 mph, with a description similar to that for Great Britain. He sets the average *T* value for U.S. tornadoes at 11, with wind speeds of 307–37 mph and describes the storms in this category as "intensely devastating."

According to Evesson (1969, 1970), the average number of reported tornado days per year in New South Wales is 15.2: 11.1 in the inland area and 4.1 in the coastal area. The areas of the United States with the highest frequency are Kansas (24.9), Oklahoma (23.7), and Nebraska (15.4). The average annual reported tornado days per population density in New

South Wales is 0.16 (0.12 inland and 0.04 coastal). Equivalent U.S. data include: Kansas (1.13), Nebraska (1.10), Oklahoma (0.82), South Dakota (0.72), Arkansas (0.25), and Iowa (0.16).

The size of Australian tornadoes varies from a few feet to over half a mile (Evesson 1969). One theory proposes that such a wide path could be produced by two or more tornadoes moving parallel and adjacent to each other, rather than by one with a very large diameter (Evesson 1970). However, this theory would apply to American tornadoes as well and thus would not explain the difference in size between American and the Australian storms.

Maximum tangential velocities are estimated mainly according to minimum force values that could produce the observed damage. These values do not imply any upper limits, just lower values; therefore, the real wind velocities could be much higher than that. These estimates include the following storms: 120–50 mph for the Piedmont tornado on July 1, 1861; 140–50 mph for the Nairobi tornado on October 13, 1893; and 110+ mph for the Erina tornado on December 5, 1966.

Tornadoes, as the term is usually applied, are natural phenomena not limited to the North American continent that can be observed in many other parts of the world, especially within the band within 20 and 50° of latitude. Those countries that report frequent tornado occurrences include: the United States, the former Soviet Union, many European countries, Japan, India, South Africa, Argentina, and New Zealand. Without covering the entire spectrum of tornado occurrence on such a vast geographical scale, only few specific areas are briefly investigated herein to provide a concise indication of the necessity to extend tornado-resistant design criteria to buildings outside the jurisdiction of U.S. codes.

A major storm that left vivid memories in the minds of those who visited the scene after its occurrence was a *tromba d'aria* or tornado that struck a Venetian lagoon and the territory of the province of Padova on September 11, 1970. In using both the Italian and the English terms, we note that many authors in the scientific literature have hesitated to adopt one or the other, since the conventional usage of the terms in Italy refers to the Italian term for mild storms and the English term for the much more violent phenomena that characterize typical U.S. storms. This hesitation was also evident in reports of the 1970 storm, which was exceptionally violent for the area.

Most of the data used in the present context to describe this special event derives from an accurate account published by R. Janeselli, then associated with the Meteorological Observatory of the Cavanis Institute of Venice (see Janeselli 1970). After careful consideration of the two alternatives, he eventually decided that the storm in question definitely demonstrated an intensity that would put it in the same category as North American storms, so that he used the term "tornado" rather than *tromba d'aria* to define it. In comparing it to American tornadoes, Janeselli used such parameters as the high level of electrical activity throughout the occurrence of the storm and the typical response of building structures to the pressure differential within the funnel.

The upper floor portions of multistory buildings, including roofs and walls, burst under the sudden pressure change on the outside surfaces. Similar to American vortices, this funnel moved over the lagoon and had enough suction power to lift heavy motorboats from the water, pulling them up to a considerable height and letting them drop back after short horizontal translation. As is typically observed during American tornadoes, motor vehicles were tossed and dragged for several hundred yards, reducing them to twisted masses of metal. This storm also totally uprooted trees, which is also typical of North American tornadoes. In fact, thousands of trees fell after the root system gave way.

The Hydrographic Office of the Magistrato delle Acque of Venice (a government department that originated in the Republic of Venice which is in charge of managing the hydrology of the territory) issued reports of particular interest about the characteristics of this storm. A meteorological observatory under the jurisdiction of this office located at the Lido happened to be located along the path of this storm and received substantial damage. Fortunately, a part of it which contained instruments was miraculously saved, so that measurements recorded by two instruments survived. The surviving instruments were an anemometer with a graphic recording device and a equally equipped barometer, and both were capable of recording wind gustiness and sudden pressure changes. These two instruments comprised the most useful means used to characterize this storm. The account of this storm may remind the reader of a quite similar episode that occurred in St. Louis, Missouri, during the May 27, 1896 tornado, when a barometer left intact by a passing storm recorded a pressure drop of 2.05" of Hg.

The Lido barometer measurements also indicated the pressure changes that occurred as the storm passed overhead. The original atmospheric pressure prior to storm arrival was 756 mm of mercury (29.7" of Hg), which suddenly dropped to 740 mm of Hg (29.1" of Hg) after a few seconds. The graph then indicated a sudden pressure rise, returning to the original value of 756 mm. Following the first swing, the pressure again dropped suddenly to 751 mm of Hg (29.5" of Hg) and again returned to its original value.

These instrumental readings reveal quite an interesting event that could be interpreted two different ways. One could have been assumed that the same funnel passed over the same point twicc, which would confirm the erratic nature of the path (observed quite often in other tornadoes). On the other hand, it would have been equally feasible that two nearby funnels passed over the same site one after another. The first pressure drop value, the larger of the two, was only 16 mm (0.629") of Hg, indicating a substantial but not extraordinary number, and the second value (5 mm of Hg) was modest. These measurements cannot be taken as maximum pressure drops that actually occurred within the eye of the funnel, because the instrument may not have been exactly within the eye. In fact, the actual pressure drop must have been much larger to match the severity of damage caused. It is also possible that the sensitivity of the barometer to sudden changes may not have been adequate to register lower values that occurred instantaneously. However, all this is speculation.

The gustiness recorded by the anemometer indicated a maximum velocity of only 220 km/hr, since this was the upper limit of the scale, leaving unmeasured the real value due to instrumental inadequacy.

With regard to the general state of the atmosphere in this region, the account reports the existence of most favorable conditions for tornado formation throughout September 11, which confirms a weather report describing the southeast area of the Po River Valley on that day. Data furnished by the Italian Air Force were also confirmed by those issued by the Meteorological Observatory of Padova for the same day. These conditions seem to quite closely resemble those typically found in the Gulf of Mexico during March and April, where they imply the presence of a thick layer of dry air from the Rocky Mountains superimposed over a layer of hot humid air from the Gulf. Similarly, in the Po Valley, a large influx of hot humid air from the Adriatic on the east took place in the lower levels of the atmosphere for a number of days prior to September 11 while dry hot air was flowing in from the west in the upper levels. As the temperature of these air masses rose, atmospheric instability was increasing. The day before, a cold front advanced from the Alps to the north. Quite a number of convective storms were present in the area on the morning of the 11th. The cold front finally reached the region in the afternoon, creating just the right conditions for formation of a vortex, which began precisely at 8:45 p.m. Further climatologic data for this area at the time of the storm include the following measurements from the Meteorological Observatory of Padova: 100% of R.H., a temperature of 28.7°C, and an atmospheric pressure at 7:00 p.m. of 1009.2 mb.At its first formation of 8:45 p.m., the air funnel descended from the sky in an elongated tubular probe and followed an erratic path, often rising without touching the ground and then suddenly descending, while often translating in a zigzag pattern. Its was first sighted over the hills of Colli Euganei between Teolo and Rovolon. It then continued to Abano and from there to the southern outskirts of Padova. The storm moved toward Venice at about 9:15 p.m., reaching Vigovono, and then proceeded over Sanbruson di Dolo and Dogaletto e Giare di Mira, reaching Fusina at 9:27 p.m.

Seven bungalows and many tents at a campground in Fusina were completely destroyed, leaving the occupants suddenly in the open but fortunately untouched. Amazingly, more than 2000 trees were uprooted in this area. Then lifting up for a short distance, the funnel skipped along the ground, and then descended into the waters of the lagoon, cutting down 11 of the 50 reinforced concrete transmission towers between the island of Sacca Sessola and the Hospital of Santa Maria delle Grazie. These concrete frameworks carried electric cables that exhibited peculiar behavior under the wind forces: they were cut at the base as by a sharp razor. The steel reinforcement was spared, but bent so much that utility poles were flattened horizontally though they remained connected at the base.

Then passing on the south side of the island of Giudecca, the storm reached the north side of the Hospital of Santa Maria delle Grazie at 9:32 p.m., pulverizing the baluster on the lagoon. At 9:35, the funnel reached the basin of St. Marco near St. Elena and, passing over the *S.S. Aquileia*, it blew off the entire upper structure of the ship, dragging it several hundred meters. Only a minute later, the funnel engulfed Ferryboat No. 130 en route from St. Marco to the docks of St. Elena. The ferry was lifted from the water, turned over, and plunged again into the water, as reported by the pilot, whose life was spared. The boat was found a few hours later by divers near a St. Elena dock lying at the bottom of the lagoon (about 3 m below the surface), still containing the bodies of 21 passengers.

Then passing over the island of St. Elena, great devastation was done to the urban structures in this area, despite evidence that the storm had lost some of its energy. Leaving this island, the storm progressed towards St. Nicolo del Lido, reaching it at 9:40 p.m., damaging more than 50 homes, uprooting many trees, and dragging an airplane more than 100 m. After crossing Porti di Lido, it reached Punta Sabbioni, destroying a campsite where over 300 tourists were housed in 80 tents and 56 bungalows — all totally destroyed without killing anyone. Continuing over water, the storm dissipated near Jesolo, having progressed throughout its path with an average translational speed of about 54 km/hr.

The effects left behind by this storm indicate that the average width of the damaged area was about 100 m and its general translation direction was from southwest to northeast. The observed counterclockwise rotation of the vortex reflects the typical motion of tornado vortices in the Northern Hemisphere, confirming the true nature of this storm in all its details.

Tornadoes in the United Kingdom

While tornadoes have been observed in numerous geographical areas throughout the world, there are some zones that are officially considered free from such storms. There is much doubt about the truth of such statements based on the ambiguity that exists as to the exact definition of the term "tornado," especially when the term is translated into other languages. The term "tornado" has been used in technical papers to indicate only those storms occurring in North America, differentiating those occurring in other parts of the world as less intense storms with similar dynamic mechanisms. The *tromba d'aria* occurring over the territory of Italy constitute an example of such a distinction of terms that implies the possibility of two different phenomena, without definite conclusions as to whether the phenomena in question are of the same nature as those that occur in North America.

The occurrence of tornadoes in United Kingdom is denied by some authorities. The regulatory agencies which control the design of nuclear power plants do not include tornado-generated forces in their design parameters. The Atomic Energy Authority (AEA) clearly excludes tornado forces from the design of nuclear buildings, since they deem the probability of their occurrence to be sufficiently small that the hazard can be disregarded. In accordance with the "Safety Assessment Principles for Nuclear Power Plants" prepared by the Health and Safety Executive of the AEA, critical facilities have to be designed to withstand extreme weather conditions which having a minimum probability of occurrence of once every 10,000 years. Since the period for the sites in question exceeds this threshold for tornadoes, they are not considered. While this has been the established design procedure in the UK, it is possible that tornado forces will be included in the future, as confirmed in a recent communication from AEA Technology (Standerwick 1993).

Tornadoes in Australia

Though the highest tornado frequencies occur in certain regions of the United States, they can occur with considerable magnitude in other parts of the world, as demonstrated above. Australia is certainly not free from this type of storm. Many tornadoes have not been identified as such in the past, creating the impression that tornadoes are not as frequent here as they really are. Evesson (1969) clarified this point and furnished valuable information on storm activity over New South Wales. Much of the data prove that interpretations of many severe storms were incorrect in failing to identify the presence of tornado activity. Other data also indicate that the tornado frequency in New South Wales is on the same order of magnitude as in some zones of the United States considered to be particularly tornado-prone.

The first report of a tornado in Australia was on October 20, 1826, and more methodical records start in 1887. The earliest reports include newspaper accounts by eyewitnesses and other unofficial documents whose reliability is questionable since they were not verified by qualified experts. Between the late 1800s and the early 1900s, H.C. Russell, government astronomer and meteorologist in New South Wales, contributed a great deal to the accumulated data on tornadoes.

Several types of local Australian storms have similar manifestations that can lead lay observers to erroneous conclusions. Some mistakes are due to a misunderstanding of the terminology. Confusion arises from the misuse of the following nomenclature: "cyclone," "tornadic squall," "hurricane," "cockeyed bob," "willy-willy," and "tornado" (refer to Clarke's statement above).

As mentioned above, the average number of reported tornado days per year in New South Wales according to Evesson (1969, 1970) is 15.2: 11.1 inland and 4.1 costal. To compare, the areas of the United States with the highest tornado frequency are Kansas (24.9), Oklahoma (23.7), and Nebraska (15.4). The average annual reported tornado days per population density in New South Wales is 0.16 (0.12 inland and 0.04 coastal). Equivalent U.S. data include: Kansas (1.13), Nebraska (1.10), Oklahoma (0.82), South Dakota (0.72), Arkansas (0.25), and Iowa (0.16). However, Evesson states that a direct comparison in terms of density with U.S. territory is not very meaningful because of the high population concentration in Sydney.

From the same source, we learn that the size of Australian tornadoes varies from a few feet to over half a mile, with a percentage distribution of the damage path width as depicted in Figure 3.14. Remember that one theory proposes that such a wide path could be produced by two or more tornadoes moving parallel and adjacent to each other, rather than by one with a very large diameter (Evesson 1970). However, this theory would apply to American tornadoes as well and thus would not explain the difference in size between American and the Australian storms.

Maximum tangential velocities are estimated mainly according to minimum force values that could produce the observed damage. These values do not imply any upper limits, just lower values; therefore, the real wind velocities could be much higher than that. These estimates include the following storms: 120–50 mph for the Piedmont tornado on July 1, 1861; 140–50 mph for the Nairobi tornado on October 13, 1893; and 110+ mph for the Erina tornado on December 5, 1966.

To gather new information concerning design of structures to withstand high-velocity winds, it is useful to refer to the many nuclear power stations that have built throughout the world, since these buildings are designed for the most stringent loading conditions that local climatology and geography demand. Tornado-prone areas in the most remote parts of the world can be efficiently analyzed based on criteria used to design nuclear power plants in these areas. In Australia, however, this invaluable source of information is not available because nuclear power plants have not been built there, at least up to 1993 according to the relevant literature. The use of nuclear power was considered in the 1960s but was eventually rejected as uneconomical. More current data on present and future planning for nuclear power stations can obtained from the Australian Atomic Energy Commission (45 Beach Street, Coogee, New South Wales 2034, Australia). The feasibility of future exploitation of nuclear power is enhanced by the presence of large uranium deposits (amounting to 20% of the known recoverable uranium outside of the former Communist bloc countries).

6. GOVERNMENT TORNADO RESEARCH IN THE UNITED STATES

The input of the U.S. Government into research on tornado-related topics constitutes the most active government participation in the world, not surprising since the frequency with which tornadoes strike U.S. territory makes them a much more relevant problem for this country than for any other. The most efficient and up-to-date scientific data related to tornado research can thus be found at such U.S. Government organizations as the National Severe Storms Laboratory (NSSL) of the National Oceanic and Atmospheric Administration (NOAA) in Norman, Oklahoma.

A good introduction to the roll of the NSSL is provided by the following quotation from one of its numerous publications:

> The National Severe Storms Laboratory conducts a broad program of research to develop basic understanding of severe weather phenomena, including large-scale and mesoscale environments that evolve and interact to produce intense storms. The focus is on observational and theoretical

> studies of hazardous middle-latitude weather phenomena such as tornadoes, hailstorms, lightning, windstorms, floods, and blizzards. NSSL often participates in research field programs, usually in coordination with other ERL components and external agencies. Sophisticated numerical models are frequently employed as research tools. The laboratory's research projects provide sound scientific foundations upon which improvements to NOAA's weather and climate forecasting services can be built. The Laboratory works directly with the National Weather Service to improve capabilities to detect, forecast, and warn of hazardous weather events. (NOAA 1992)

Derived from the National Severe Storm Project, the NSSL was established at its present site in 1962 for the specific purpose of ensuring "safety of flight, and probing severe storms by aircraft." The ground-based lab facilities originally included radar equipment for weather surveillance, height-finder radar, and an L=band aircraft transponder/interrogator. The NSSL developed a 3-cm pulse Doppler radar system in 1964. This system, which remained in service until 1969, was one of the first of its kind and used analog methods to record data at 10 range locations. Its facilities are the state of the art at present and its scope of activity has been greatly enlarged.

7. U.S. CODES FOR TORNADO FORCES ON STRUCTURES

Within the jurisdiction of the United States, there are several codes which directly address application of tornado-generated forces to specific buildings and structures. The major codes include the following:

- U.S. Atomic Energy Commission Regulations of 1974
- U.S. Nuclear Regulatory Commission Regulations of 1978
- U.S. Nuclear Regulatory Commission Regulations amended 1989
- American National Standards Institute of 1983
- Electric Power Research Institute Regulations: the interim report on "Full-Scale Tornado Missile Impact Tests" of April 1976, the second general report of July 1977, and "Local Response of Reinforced Concrete to Missile Impacts" of October 1979

In the last few decades, these agencies have contributed to development of a regulatory framework with guidelines that now constitute the most comprehensive source for guidance in the design of any structures subjected to tornado action. These guidelines have gradually evolved, prompted by development of new information on the dynamics of these storms and on the behavior of structures. Emerging from the meteorological sciences and structural engineering, research data continuously accumulated by various institutes have produced codified guidelines that were improvements of their predecessors. Each agency acted more or less independent of one another without official coordination, so that some overlaps occur, and one set of regulations does not necessarily make previous ones invalid or obsolete. To this effect, we include brief presentations on regulations formulated by the American National Standards Institute and by the Electric Power Research Institute (see also Appendix A), and we present the following case studies: Calvert Cliffs, the Donald C. Cook Nuclear Power Plant, the Susquehanna Nuclear Power Plant, and the Vogtle Electric Generating Plant.

8. ANSI REGULATIONS ON TORNADO LOADS

The American National Standards Institute (ANSI) regulations of 1983 on tornado loads (a review of ANSI/ANS-2.3 1983) indicate the following general specifications for inclusion of tornado forces in design procedures for buildings and other structures. If the probability of tornado occurrence on a given site is equal to 10^{-4} per year, the following parameters are recommended.

General Parameters

The Design Basis Tornado Characteristics are presented in Table 3.10 in accordance with eight individual cases that vary in terms of different size of the tornado vortex from 1090 to 370 ft in diameter. From this table, we can see that the vortex radii involved were 540, 453, 435, 355, 320, 270, 253, and 185 ft. Corresponding to each size, the table includes specific values for maximum wind speed, translational velocity, and maximum atmospheric pressure drop. Maximum wind speed was assumed to be higher for larger-sized vortices and diminished as vortex radius decreased. Maximum wind speed was assumed to be 320 mph, and minimum 100 mph. Translational wind velocity was also assumed to be higher for large-sized vortices and to diminish as vortex radius decreased. The maximum translational speed values were assumed to be 70 mph, gradually decreasing to a minimum 25 mph for the smallest storm. Similarly, pressure drop within the vortex core varied from a maximum value of 1.96 psi (for the largest vortex) to lower values as vortex size decreased, reaching 0.20 psi for the smallest storm considered.

Missiles

The spectrum of missile types carried aloft by a tornado vortex to be used in standard design under these specifications only includes a 4000-lb automobile and a wide-flange steel beam traveling at different speeds that coincide with the values previously indicated for each of the eight vortex sizes. These air speeds varied from 320 to 100 mph and included eight different cases. For each air speed, the table gives three entries indicating maximum missile horizontal velocity, maximum missile altitude, and maximum missile range. Naturally, the difference between air speed and the missile speed is due to the aerodynamic characteristics of each missile. Not all missiles are expected to acquire the speed of the air stream that propels them. For example, a 4000-lb automobile propelled by a 320 mph wind, as would be expected in a tornado with a 540 ft radius, is assumed to travel at a maximum horizontal velocity of 115 mph, at a maximum altitude of 45 ft, and to have a maximum range of 1100 ft.

9. ELECTRIC POWER RESEARCH INSTITUTE TORNADO STANDARDS

The Electric Power Research Institute (EPRI) was founded in 1973 by the international electric power utilities, who pooled their resources to concentrate their efforts in research and development into one organization. The institute was funded by voluntary contribution from the 600 member utilities. Its basic purposes included technological research for the most cost-effective and environmentally sound generation, delivery, and utilization of

Table 3.10. Design Basis Tornado Characteristics

Maximum tornado wind speed V		Translational velocity T		Radius R		Maximum atmospheric pressure drop ΔP[1]	
mph	ms^{-1}	mph	ms^{-1}	feet	m	psi	mb
320	143	70	31	540	165	1.96	135
260	116	57	25	453	138	1.46	101
250	112	55	24	435	133	1.35	93
200	89	45	20	355	108	0.85	58
180	80	40	18	320	98	0.70	48
150	67	33	16	270	82	0.47	32
140	63	32	14	253	77	0.41	28
100	45	25	11	185	56	0.20	14

[1] $AP = PV^2 = 3.546 \times 10^{-5} V_t^2$ (V_t in mph).

electric energy. Based in Palo Alto, California, the organization employs several hundred scientists and engineers who manage projects around the world. Industrial and commercial firms, universities, utility companies, and government laboratories are involved in the work of generating services and information for electric utility companies. EPRI technical reports cover more than 40 different categories, which in turn are classified into the following divisions: Customer Systems, Electrical Systems, Environment, Generation, Membership, Nuclear Power, Office of Explanatory Research, and Planning. The nuclear power division is the one most concerned with design specifications for critical structures with respect to tornado forces.

The EPRI requirements for tornado loads (chapter 1, section 4.4.C.8) state that the design should include ANSI/ANS-2.3, "Standard for Estimating Tornado and Extreme Wind Characteristics at Nuclear Power Sites," as the guidelines to be used under EPRI regulations. The probability value of 10^{-6} per year is to be used for selection of proper parameters involved in design of tornado-resistant structures.

While EPRI and ANSI specifications follow generally similar design criteria, some differences do occur:

	NRC, Region 1	EPRI
Maximum tornado wind speed, V	360 mph	260 mph
Translational velocity, T	70 mph	57 mph
Radius of maximum rotational speed	150 ft	453 ft
Maximum pressure drop, Δp	3.0 psi	1.46 psi

(Note that the territory of the United States is subdivided into three regions with respect to tornado intensity, with region 1 having the highest value. This chart only represents region 1; the reader is referred to the original source for the other regions.) We can see that the NRC values are higher than those indicated by the EPRI. We should note that the NRC values are based on an occurrence probability of 10^{-7} per year, while the EPRI uses 10^{-6} per year. In addition, the NRC requirements are based on previous data, while the EPRI values are more realistic as they are based on more recent data made available by tornado research. In fact, significant modifications were made by the NRC in their January 9, 1989 revisions based on the lower values recommended by the EPRI. This was also reflected in the new Table of Design Basis Tornado Characteristics (Recommended Modification to the RG1.76: Tornado Design Basis for the ALWR).

10. CALVERT CLIFFS: A CASE STUDY

This case study includes units 1 and 2 in Lusby, Maryland, owned by the Baltimore Gas and Electric Company. Unit 1 has a net capacity of 825 NW_e and entered into commercial operation in 1975. Unit 2 has an equal net capacity and became commercially operational in 1977. Because of the construction dates, the designs of the structures of both units reflect NRC specifications prior to 1978.

Taken from the Final Safety Analysis Report for this plant (vol. 1, chap. 2), the enclosed data include meteorological studies conducted for the site based on records from the Patuxent Naval Air Test Center-PAX (now Patuxent Naval Air Station-NHK). The records include the periods 1955–60 and 1949–64. In addition, meteorological data from Washington National Airport (DCA), Byrd Field in Richmond, Virginia (RIC), and Annapolis, Maryland (ANA), were used to evaluate the frequency of various weather parameters and certain meteorological extremes. Statistical data for severe weather parameters were derived from the Environmental Science Services Administration (ESSA) in Asheville, North Carolina.

The records of tornado occurrence include five storms between 1953 and 1962 proximal to the site (within a 1° latitude × 1° longitude quadrant). The mean annual frequency was 0.5 tornadoes, and the period of occurrence between two consecutive storms was calculated at 2700 years. These values reflect quite a high probability of occurrence in compared to other U.S. sites. Based on these data and using Thom's formula, the tornado occurrence probability for this site was calculated to be 3.75×10^{-4} per year.

The basic design parameters for tornado-generated loads include the following:

- Maximum rotational wind speed: 300 mph
- Maximum translational speed: 60 mph
- Maximum tangential wind speed (effective velocity): 300 mph
- Maximum pressure differential: 3 psi
- Maximum pressure differential rate: 1 psi/sec
- Pressure differential sequence: initial pressure drop of 3 psi in 3 seconds, followed by 2 seconds of calm and subsequent repressurization at 1 psi/sec.

It is important to note that only reactor containment structures are designed to withstand full outburst pressure in the design of structures for pressure differential. Other structures designed to resist tornado loads are considered adequately vented to avoid any pressure differential load.

Missile Impacts

Within the spectrum of tornado-generated missiles, only two types are considered: a 4000-lb automobile and a wood plank. An automobile traveling at 50 mph was assumed to move below 25 ft above ground and to impact the structure with a contact area of 20 ft^2. A 4" × 12" × 12′ wooden plank weighing 108 lb traveling end-on at 300 mph was assumed to be a probable missile to strike the structure at any height.

The following tornado-induced loads — i.e., maximum wind pressure, maximum pressure differential, and missile impact — were assumed to act simultaneously within the structure. The allowable stress in design of structures under tornado loads were assumed to be 90% of the yield stress of the reinforcing bars and 85% of the ultimate strength of the concrete, except for the effect of local crushing due to missile impacts. We should point out that, when load combinations including wind forces were used, tornado loads were always considered without any reduction factor.

11. THE DONALD C. COOK NUCLEAR POWER PLANT: A CASE STUDY

This case refers to the Donald C. Cook Nuclear Power Plant, owned and operated by the American Electric Power Service Corporation. The pertinent data presented herein were taken from documentation contained in the Updated Final Safety Analysis Report (UFSAR) (these records are kept by the Indiana and Michigan Electric Company and the Westinghouse Electric Corporation). This plant is located in Bridgman, Michigan, on the shore of Lake Michigan. It consists of two units: unit 1 went into operation in 1975, and unit 2 went was operational in 1978.

General criteria for application of wind forces to critical structures including reactor containment structures were based on ASCE Paper 3269: "Wind Forces on Structures." The fastest-mile wind speed for this site was assumed to be 90 mph, adjusted by correction factors for coastal areas. The basic data for tornado-generated loads include the following:

- Tornado translational velocity: 60 mph
- Tangential velocity: 300 mph
- Maximum tangential velocity on one side of the funnel: 300 + 60 = 360 mph
- Tangential velocity on the other side of the funnel, opposite the location of the previous value: 300 – 60 = 240 mph
- Size of funnel assumed for design includes to different cases:
    - a larger-sized tornado funnel with a diameter not less than 130 ft, so that the reactor containment structure could be totally engulfed within the maximum wind speed zone, and
    - a small-sized tornado funnel that would generate asymmetrical wind pressure.

Quoting directly from the UFSAR, the following tornado loading conditions were adopted:

The containment structure was analyzed considering:

a. A large diameter "funnel" tornado with a distribution of maximum tangential velocity in a band at least 130 ft in width and 180 ft in height such that the full effects of the band are felt by the structure. The maximum tangential velocity is considered additive to forward progression, so that the maximum wind velocity is considered to be 360 mph. This results in a tornado velocity pressure (at one atmosphere) of 330 psf, which is evaluated with an internal positive pressure of 3 psi.

b. To determine the effects of a nonuniform pressure distribution, a narrow-diameter "funnel" tornado is assumed to pass directly over the structure, such that the center of the "funnel" is aligned with the centerline of the containment structure. The maximum tangential velocity is considered additive to the forward progression in the right leading quadrant and subtractive in the left trailing quadrant, so that the maximum wind velocities are 360 and 240 mph, respectively. Tornado velocity pressures are evaluated with an internal positive pressure of 3 psi. This type of pressure diagram results in smaller internal forces in the containment structure than the wide-band distribution. Torsional stresses in the containment structure due to tornado loading were determined to be negligible.

The two reactor containment structures present the special aerodynamic problem of interference based on their relative location on the site. A funnel effect arises as the wind passes between the two buildings due to their proximity. A special study was made to analyze the funnel effect and its conclusion included the following:

> The nonuniform pressure distribution over the containment as related to the conditions in this case were found in the technical literature and plotted on a circular graph. The suction on the containment walls that face each other was found to be greater than that on the walls that face away from the line of symmetry of the two containments. The greatest local increase in suction due to the funneling effect was about 15%. The pressures on the windward and leeward sides were practically unchanged by the funneling effect. The 15% increase of suction was not significant enough to justify a revision of the analysis performed for the symmetrical loading.

In this case, the pressure drop within the tornado funnel was assumed to be 3 psi. Using a 3-psi pressure drop combined with the positive wind pressure generated by the 360-mph wind yields results that offset the funneling effect generated by the proximity of the two buildings.

Missile Impacts

Missile protection for critical structures in the design of this plant include a series of potential missile types within the interior of the structures and on their exterior. The thickness of the reactor containment walls and of other barriers are based on effective protection from missile penetration. To this effect, the method used in this part of design was the modified Petri's formula.

12. THE SUSQUEHANNA NUCLEAR POWER PLANT: A CASE STUDY

This case study includes units 1 and 2 at the Susquehanna Nuclear Power Plant in Berwick, Pennsylvania, operated by the Pennsylvania Power and Light Company. Unit 1 had a capacity of 1052 (net MW_e) and became operational in 1983; unit 2 had the same capacity and became operational in 1985. All the data presented for this case are derived from the Final Safety Analysis Report provided by Pennsylvania Power and Light.

The regional meteorological conditions for design and operation included in the report referred to hurricanes, tornadoes, thunderstorms, lightning, hail, extreme winds, freezing rain, duststorms, high air pollution potential, snowpack, design basis tornado, and ultimate heat sink. We neglecting most of the meteorological data, and include only those parts directly related to tornado climatology and tornado forces.

The section on tornadoes first discusses tornado incidence. The frequency for this region is quite low: 38 tornadoes between 1950 and 1973 within 50 miles of the site. Tornadoes have been noticed to happen in this area during the summer months, especially in the late afternoon or evening. Data gathered between 1953 and 1962 on tornado frequency and intensity in the region indicate frequency by month, hour, and intensity within a 3 × 3° square centered on the site. These data are presented in a graph within the report which indicates that July and September are the months of highest tornado frequency (with 11 tornadoes), followed by June (9) and August (8). With respect to time of day, the highest frequency is at 6:00 p.m. (with 9 tornadoes), followed by 5:00 and 7:00 p.m. (each with 8). The tornadoes were rated using the seven-class Fujita tornado intensity classification. The records for this area indicate that the most intense tornado were class 3, occurring with maximum frequency during September, with three such tornadoes recorded in this month. Only one class 3 tornado was observed in July (see Fig. 2.3.1 in the Final Safety Analysis Report for more specific data).

In a section entitled "Design Basis Tornado," the report provides a rationale for inclusion or exclusion of tornado forces within design parameters. To this effect, probability occurrence value, P_s, equal to 10^{-7} per year is the threshold used. This value was determined based on an amplitude of 3° latitude × 3° longitude. The equation used for calculation of tornado occurrence probability is written as follows:

$$P_s = n(a/A),$$

where P_s is the mean annual probability of a tornado striking a certain point, n is the mean number of tornadoes which occur within area A per year, and a is the mean path area determined from a log-normal distribution.

When we introduce parameters a = 0.388 mi^2, A = 32,265 mi^2, and n = 63/24 yr = 2.625/yr, we have

$$P_s = 3.157 \times 10^{-5} \text{ per year.}$$

This site had an amplitude of 3° of longitude × 3° of latitude (as stated above) and extended between 75–78° longitude and 40–43° latitude. There were 63 tornadoes observed in this areas between 1950 and 1973, only 34 of which were recorded in terms of path length and width. These general data were used to derive the above parameters. Quoting from the report (at p. 2.3-7):

> The design basis wind speed is one which satisfies the condition $P_s P_i < 10^{-7}$ per year, where P_i is the acceptable intensity probability and is determined from the plot of cumulative F-scale

intensity frequencies on log-probability paper. The F-scale is an estimate of tornadic wind speed range based upon damage inspection and has been compiled for the years of 1971 and 1972.

The highest tornadic intensity recorded in this region was F2, or 157 mph. Thus, substituting $P_s P_i = 10^{-7}$ per year, we obtain:

$$P_i = 3.168 \times 10^{-3} \text{ per year.}$$

The pressure drop rate within the tornado funnel considered for structural design was calculated based on the following formula:

$$dp/dt = pT/2r_m,$$

where p is the pressure change, t is time, T is the translational speed, and r_m is the radial rotational wind speed, equal to 150′.

However, the total pressure drop, p, is given by the following cyclostrophic wind equation:

$$\int_0^r \frac{\partial p}{\partial r} dr = \int_0^r 2\rho A \frac{V_m^2}{r_m} d_r,$$

where V_m is the maximum rotational wind speed and ρ is the air density (1×10^{-3} gm/cm³). Then the pressure drop rate and total pressure drop are determined from the above equations to be

$$\text{Pressure drop rate} = 0.9 \text{ psi/sec,}$$

and

$$\text{Total pressure drop} = 1.9 \text{ psi.}$$

Interpolating between regions I and II, the maximum translational wind speed is determined to be

$$T = 55 \text{ mph}$$

To conclude, the calculated basic design parameters for this case are as follows:

Total maximum wind speed = 260 mph,
Rotational wind speed = 205 mph,
Maximum translational speed = 55 mph,
Minimum translational speed = 5 mph,
Total pressure drop = 1.9 psi,
Rate of pressure drop = 0.9 psi/sec.

Tornado-Generated Missiles

The criteria adopted for selection of tornado-generated missiles are clearly exhibited in Tables 3.11 and 3.12. Table 3.11 refers to missiles applicable for design of all tornado-resistant structures, except for diesel generator building *E*, and describes the geometric and

Table 3.11. Tornado-Generated Missile Parameters For All Tornado-Resistant Structures Except Diesel Generator Building "E"

Missile	Weight (lbs)	Velocity (mph)
Wood plank, 4" × 12" × 12′, traveling end-on	108	300
Steel pipe, 3" diam., Schedule 40, 10′ long, traveling end-on	76	100
Automobile flying through the air no more than 25 ft above ground with 20 ft^2 contact area	4000	50
Steel rod 1" diam. × 3′ long	8	216
Utility pole 13-1/2" diam., 35′ long acting no more than 30 ft above the ground	1490	144

Vertical velocities are considered equal to 80% of horizontal velocities mentioned above.

Table 3.12. Tornado-Generated Missile Parameters for Diesel Generator Building "E"

Missile	Weight (lbs)	Impact velocity (fps)
Wood plank, 4" × 12" × 12′, traveling end-on	108	440
Steel pipe, 3" diam., Schedule 40, 10′ long, traveling end-on	72	147
Steel pipe, 6" diam., Schedule 40, 15′ long	285	170
Steel pipe, 12" diam., Schedule 40, 15′ long	750	155
Steel rod 1" diam. × 3′ long	8	317
Automobile flying through the air at no more than 25 ft above the ground with 20 ft^2 contact area	4000	195
Utility pole 13.5" diam., 35′ long	1490	211

Vertical velocities are considered equal to 80% of horizontal velocities mentioned above.

dynamic characteristics. Table 3.12 includes equivalent values for tornado-generated missiles applicable for diesel generator building *E*.

13. THE VOGTLE ELECTRIC GENERATING PLANT: A CASE STUDY

This case study refers to the Vogtle Electric Generating Plant, which is operated by the Georgia Power Company. The plant is located on the southwest bank of the Savannah River, 23 miles upstream from the intersection of the Savannah River and U.S. Highway 301 in Burke County, Georgia, near the South Carolina border. The basic data presented hereinafter are taken from the Final Safety Analysis Report submitted in support of the application by Georgia Power Company for the original operating license. The application was for two units, each with a reactor core rated at 3411 MW. The application conformed with section 103(b) of the Atomic Energy Act of 1954, as amended, and the NRC regulations set forth in Part 50 of Title 10 of the Code of Federal Regulations (10 CFR 50).

The plant was built as a two-unit plant, with each unit essentially identical. Of course, the most critical structures in the plant are the two reactor containments. The coordinates of the center of the containment for each unit are given in both latitude and longitude and in Universal Transverse Mercator (UTM) coordinates. For unit 1, 33° 08′ 30″ N lat. and 81° 45′ 44″ W long.; in UTM coordinates: Zone 17SMG(m), N 3,666,900 and E 428,900. For unit 2, 33° 08′ 30″ N lat. and 81° 45′ 48″ long.; in UTM coordinates: Zone 17SMG(m), N 3,666,900 and E 428,800. (Note the accuracy used in this report to give the location of the containment buildings.)

Construction of unit 1 was completed in January 1987 and commercial power operation was initiated in June of the same year. Construction of unit 2 was completed in February 1989 and commercial power operation initiated in June of the same year.

Regional Climatology

Regional meteorological conditions relevant plant design with respect to severe weather includes heavy precipitation, hail, snow, ice storms, hurricanes, thunderstorms, and lightning. Focusing on the hurricanes, it is pointed out that the plant is located approximately 100 miles inland from the coast of the Atlantic, so that it is assumed that the wind force during hurricanes will be considerably reduced. Records indicate that 11 hurricanes and 26 tropical storms passed through a 2.5° (longitude and latitude) quadrant containing the site between 1886 to 1969. The probability that one tropical storm or hurricane will enter this quadrant per season is thus ~0.36.

Probability of Tornadoes

Specific data on tornado probability for this site are furnished, starting with the number of tornadoes recorded in this area. Based on 40 years of observations, between one and two tornadoes per year occurred in the 2° of latitude and longitude quadrant where the plant is located. However, if we consider the 12-year observation period of 1955 to 1967, where the average number of tornadoes for the 1° quadrant including the site was about 1½ per year, this corresponds to 6 tornadoes per year for the 2° quadrant. The conflict between these two values can be explained by the fact that more tornadoes have been reported in recent years due to more sophisticated detection instruments and modern communication technology. Further comments on tornado frequency in the report assume that even more tornadoes actually occur than even the last numbers indicate, and that assuming a frequency of three tornadoes per year for the 1° quadrant is a much truer value. Quoting from the report:

> A typical tornado is about a quarter of a mile wide, and covers an area of about 2½ mi^2. The 1° square at this latitude has an area of approximately 4000 mi^2. A conservative estimate of the probability of a given point being affected by a tornado in a given year is approximately: 2½ × 3/4000, i.e., ~1/500. That is, a given point can be expected to be affected by a tornado once in 500 years, on the average.

Tornado Design Parameters

The following design parameters are specifically included in the report:

- Maximum peripheral tangential velocity: 290 mph
- Translational velocity: 70 mph maximum and 5 mph minimum
- Maximum wind velocity: 360 mph
- Radial center of tornado where maximum wind velocity occurs: 150 ft
- Atmospheric pressure drop: 3 psi
- Rate of pressure drop: 2 psi/second

Tornado Forces

The procedure specified to transform tornado wind loading and differential pressure loading into effective loads on structures followed a report entitled *Tornado and Extreme Wind Design Criteria for Nuclear Power Plants* (Betchel Power Corporation 1974). Dynamic wind pressure was applied to the structure following a conventional procedure, though without including the gust factor.

The loading combinations and load factors were as follows:

$$W_t = W_w,$$

$$W_t = W_p,$$

$$W_t = W_m,$$

$$W_t - W_w + 0.5W_p,$$

$$W_t = W_w + W_m,$$

$$W_t = W_w + 0.5W_p + W_m,$$

where W_t is the total tornado load, W_w is the total wind load, W_p is the total differential pressure load, and W_m is the total missile load. W_p is assumed to be equal to 3 psi for unvented structures; however, smaller values can be used if the structure is assumed to be partially vented.

Missile Protection

Safety-related structures are designed in accordance with 10 CFR 50 (Appendix A, GDC-2 and 4) and are protected by being located within missile-proof structures, by providing separation, or by providing missile shields or barriers.

The missile selection used in the design of this plant was based on the following sources:

A. Internally generated missiles:
   internally generated missiles outside containment
   internally generated missiles inside containment

B. Externally generated missiles:
- missiles generated by natural phenomena: mostly from tornadoes
- missiles generated by events near the site: nonexistent in this case
- aircraft hazards: nonexistent in this case
- gravity-generated missiles

Of the different types of externally generated missiles, only those indicated as missiles generated by natural phenomena are discussed. The list of missiles along with their physical characteristics and their impact velocities indicates the criteria adopted for design of this plant (see Table 3.13).

Table 3.13. Tornado Missiles Considered in the Design of the Vogtle Electric Generating Plant

Description of missile	Weight (lb)	Height limit (ft)	End-on horizontal velocity (ft/s)	End-on vertical velocity (ft/s)
Wood plank, 4" × 12" × 12′	200	216	200	160
Steel pipe, 3" diam., Schedule 40, 10′ long	78.5	212	200	160
Steel rod, 1" diam., 3′ long	8	Unlimited	317	254
Steel pipe, 6" diam., Schedule 40, 15′ long	285	101	160	128
Steel pipe, 12" diam., Schedule 40, 15′ long	744	46	150	120
Utility pole, 13-1/2" diam., 35′ long	1490	30[1]	211	169
Automobile, frontal area 20 ft^2	4000	Ground level	75	60

[1]To 30 ft above all grade levels within half a mile of facility structures.

The structures to be protected and the barriers used to protect them missile impacts generated by natural phenomena (tornadoes) for this plant are given in Table 3.14.

It is important to note that, when concrete exterior walls and roofs are used as barriers to offer missile protection, these walls have a 24" minimum thickness and the roofs are at least 21" thick. The concrete has a 28-day compressive strength of at least 4000 psi (91-day strength for concrete containing pozzolan). Where interior walls and slabs with concrete compressive strength of 5000 psi are used as missile barriers, such walls have an 18" minimum thickness and the slabs are at least 14" thick.

Table 3.14. Protected System and Component Barriers Against Externally Generated Missiles

Protected systems and components	Missile barrier	Minimum concrete thickness (inches)			Design concrete strength (psi)
		Walls	Roof	Floor	
Reactor equipment, reactor coolant system, containment piping and valves, containment electrical, instrumentation, and control systems and containment engineered safety features actuation systems, transfer tube	Containment shell & dome	45	45	–	6000
	Containment basemat	–	–	126	5000
	Internal structures:	varies, 102 min.			5000
	Primary shield wall	36	–	–	5000
	Secondary shield wall	–	–	24	5000
	Floor at 220′				
Penetrations in containment shell	Missile shield wall/ building walls	24	21	–	5000
Control room and protected electrical, instrumentation, control, and ventilation equipment in control building	Control building	24	21	–	4000
Safety injection, containment spray cooling water, ventilation, electrical, instrumentation and control equipment	Auxiliary building	24	21	–	5000
Reactor auxiliaries, e.g., chemical and volume control system, boric acid storage tanks and transfer pumps, and CCW system	Auxiliary building	24[1]	21	–	5000
Spent fuel pool, transfer tube	Fuel handling building	24	21	–	4000
	Fuel pool walls	60	–	–	4000
Diesel generators, diesel generator fuel oil system, combustion air intake	Diesel generator building	24	21	–	4000
Diesel generator combustion air exhaust	Concrete barrier	24	21	–	6000
Diesel fuel storage tank, diesel fuel transfer pumps and pump motors	Diesel fuel storage tank pumphouse	24	21	–	4000

Main steam line isolation valves	Auxiliary and control building main steam valve rooms	24	21	–	5000-AB[2] 4000-CB
Category 1 water storage tanks	Cylindrical walls and sloping roof	24	21	–	4000
Category 1 water storage pumps, valves, and piping	Enclosures adjoining tanks	24	21	–	4000
Nuclear service cooling water tower fan motors	Enclosures	24	21	–	4000
Nuclear service cooling water pumps	Nuclear service cooling tower valve houses	24	21	–	4000
Auxiliary feedwater pumps, motors, valves, and piping house	Auxiliary feedwater pump	24	21	–	4000
Category 1 piping and electrical cables	Category 1 tunnels, or buried min. 6′ backfill cover or 21" concrete cover or a total 6′ combination of backfill and concrete	24	21	–	4000
Auxiliary building HVAC intakes and exhausts	Auxiliary building	24	21	–	5000
Control building HVAC intakes and exhausts	Control building	24	21	–	4000
Fuel handling building HVAC intakes and exhausts	Fuel handling building	24	21	–	4000
Diesel generator building HVAC intakes and exhausts	Diesel generator building	24	21	–	4000
Auxiliary feedwater pumphouse HVAC intakes and exhausts	Auxiliary feedwater pumphouse	24	21	–	4000
Auxiliary feedwater piping steam valve	Control building main room entrance	24	21	–	4000

[1] Interior walls and roof slabs credited with providing missile protection each have a minimum concrete thickness of 18" and 14", respectively, and a design strength of 5000 psi.
[2] AB = auxiliary building; CB = control building.

Systems located both inside and outside of containment are considered when identifying the above potential missiles. Quoting directly from the report: "The basic approach is to ensure design adequacy against generation of missiles rather than to allow missile formation and design plant features to contain their effects." The two sources of possible missiles outside containment are rotating component failures and pressurized component failures. After careful evaluation, missiles due to pressurized component failure were found to be nonexistent.

APPENDIX A

Nuclear Power Plants in the United States and Worldwide

The following lists are included for use as a general reference to determine the existence of a nuclear power plant in a given region so that the design specifications used for the structure can be employed as guidelines in the design of other structures in the region. The significance of the design of a nuclear power plant to that of other structures with respect to tornado forces cannot be overemphasized given the critical nature of such structures. U.S. plants are listed by plant name, unit number (if more than one unit), location, and date ordered. Plants in other countries are listed by country, plant name, unit number (if more than one unit), operating organization (where available), power capacity, and order date.

Nuclear Power Plants Within the United States

Allen's Creek Nuclear Power Plant-1, Wallis, TX, 1973
Allen's Creek Nuclear Power Plant-2, Wallis, TX, 1973
Arkansas Nuclear One Steam Electric Station-1, Russellville, 1967
Arkansas Nuclear One Steam Electric Station-2, Russellville, 1970
Atlantic Nuclear Power Plant-1, Atlantic City, NJ, 1972
Atlantic Nuclear Power Plant-2, Atlantic City, NJ, 1972
Bailly Nuclear Power Plant, Baillytown, IN, 1971
Barton (Alan R.) Nuclear Power Plant-1, Clanton, AL, 1972
Barton (Alan R.) Nuclear Power Plant-2, Clanton, AL, 1974
Beaver Valley Power Station-1, Shippingport, PA, 1967
Beaver Valley Power Station-2, Shippingport, PA, 1971
Bellefonte Nuclear Power Plant-1, Hollywood, AL, 1970
Bellefonte Nuclear Power Plant-2, Hollywood, AL, 1970
Big Rock Point Nuclear Power Plant, Charlevoix, MI, 1959
Black Fox Nuclear Power Plant-1, Inola, OK, 1973
Black Fox Nuclear Power Plant-2, Inola, OK, 1973

Blue Hills Nuclear Power Plant-1, Mayflower, TX, 1973
Blue Hills Nuclear Power Plant-2, Mayflower, TX, 1974
Bodega Bay Nuclear Power Plant, Bodega Bay, CA, 1961
Braidwood Station-1, Braceville, IL, 1973
Braidwood Station-2, Braceville, IL, 1973
Brown's Ferry Nuclear Plant-1, Decatur, AL, 1966
Brown's Ferry Nuclear Plant-2, Decatur, AL, 1966
Brown's Ferry Nuclear Plant-3, Decatur, AL, 1967
Brunswick Nuclear Plant-1, Southport, NC, 1968
Brunswick Nuclear Plant-2, Southport, NC, 1968
Byron Station-1, Byron, IL, 1971
Byron Station-2, Byron, IL, 1973
Callaway Nuclear Plant, Fulton, MO, 1973
Callaway Nuclear Plant, Fulton, MO, 1973
Calvert Cliffs Nuclear Power Plant-1, Lusby, MD, 1967
Calvert Cliffs Nuclear Power Plant-2, Lusby, MD, 1967
Carolinas–Virginia Tube Reactor (CVTR), Parr, SC, 1955
Carroll County Nuclear Power Plant-1, Savanna, IL, 1970
Carroll County Nuclear Power Plant-2, Savanna, IL, 1970
Catawba Nuclear Station-1, Clover, SC, 1970
Catawba Nuclear Station-2, Clover, SC, 1970
Cherokee Nuclear Power Plant-1, Blacksburg, SC, 1974
Cherokee Nuclear Power Plant-2, Blacksburg, SC, 1974
Cherokee Nuclear Power Plant-3, Blacksburg, SC, 1974
Clinch River Breeder Reactor Project (CRBRP), Oak Ridge, TN, 1972
Clinton Nuclear Power Station, Clinton, IL, 1973
Clinton Nuclear Power Station, Clinton, IL, 1973
Comanche Peak Nuclear Power Plant, Glen Rose, TX, 1972
Comanche Peak Nuclear Power Plant, Glen Rose, TX. 1972
Cook (Donald C.) Nuclear Plant-1, Bridgman, MI, 1967
Cook (Donald C.) Nuclear Plant-2, Bridgman, MI, 1967
Cooper Nuclear Power Plant, Brownville, ME, 1966
Crystal River Energy Complex, Crystal River, FL, 1967
Davis–Besse Nuclear Power Plant-1, Oak Harbor, OH, 1968
Davis–Besse Nuclear Power Plant-2, Oak Harbor, OH, 1973
Davis–Besse Nuclear Power Plant-3, Oak Harbor, OH, 1973
Diablo Canyon Nuclear Power Plant-1, Avila Beach, CA, 1967
Diablo Canyon Nuclear Power Plant-2, Avila Beach, CA, 1968
Douglas Point Nuclear Power Plant-1, Nanjemoy, MD, 1972
Douglas Point Nuclear Power Plant-2, Nanjemoy, MD, 1972
Dresden Nuclear Station-1, Morris, IL, 1955
Dresden Nuclear Station-2, Morris, IL, 1965
Dresden Nuclear Station-3, Morris, IL, 1965
Duane Arnold Energy Center (DAEC), Palo, IA, 1968
Elk River Nuclear Power Plant, Elk River, MN, 1958
Enrico Fermi 1 Fast Breeder Reactor Project, Newport, MI, 1955
Enrico Fermi 2 Nuclear Power Plant, Newport, MI, 1968

Enrico Fermi 3 Nuclear Power Plant, Newport, MI, 1971
Farley (Joseph M.) Nuclear Power Station-1, Ashford, AL, 1969
Farley (Joseph M.) Nuclear Power Station-2, Ashford, AL, 1970
Fitzpatrick (James A.) Nuclear Power Plant, Lycoming, NY, 1966
Forked River Nuclear Power Plant, Forked River, NJ, 1969
Fort Calhoun Generating Station, Fort Calhoun, SC, 1966
Fort St. Vrain Nuclear Power Plant, Platteville, CO, 1965
Fulton Nuclear Power Plant-1, Lancaster, PA, 1971
Fulton Nuclear Power Plant-2, Lancaster, PA, 1971
Ginna (Robert Emmet) Nuclear Power Plant, Ontario, NY, 1965
Grand Gulf Nuclear Station-1, Port Gibson, MS, 1972
Grand Gulf Nuclear Station-2, Port Gibson, MS, 1973
Greene County Nuclear Power Plant, Clementon, NJ, 1974
Greenwood Nuclear Power Plant-2, Port Huron, MI, 1972
Greenwood Nuclear Power Plant-3, Port Huron, MI, 1972
Haddam Neck Nuclear Power Plant, Haddam Neck, CT, 1963
Hallam Nuclear Power Plant, Hallam, NB, 1957
Shearon Harris Nuclear Power Plant-1, New Hill, NC, 1971
Shearon Harris Nuclear Power Plant-2, New Hill, NC, 1971
Shearon Harris Nuclear Power Plant-3, New Hill, NC, 1971
Shearon Harris Nuclear Power Plant-4, New Hill, NC, 1971
Hartsville Nuclear Power Plant-A1, Hartsville, TN, 1972
Hartsville Nuclear Power Plant-A2, Hartsville, TN, 1972
Hartsville Nuclear Power Plant-B1, Hartsville, TN, 1972
Hartsville Nuclear Power Plant-B2, Hartsville, TN, 1972
Hatch (Edwin I.) Nuclear Power Plant-1, Baxley, GA 1967
Hatch (Edwin I.) Nuclear Power Plant-2, Baxley, GA 1967
Haven Nuclear Power Plant-1, Haven, WI, 1973
Haven Nuclear Power Plant-2, Haven, WI, 1974
Hope Creek Nuclear Generating Station-1, Hancocks Bridge, NJ, 1969
Hope Creek Nuclear Generating Station-2, Hancocks Bridge, NJ, 1969
Humboldt Bay Nuclear Power Plant-3, Eureka, CA, 1958
Indian Point Nuclear Generating Station-1, Buchanan, NY, 1955
Indian Point Nuclear Generating Station-2, Buchanan, NY, 1965
Indian Point Nuclear Power Plant-3, Buchanan, NY, 1967
Isolte Nuclear Power Plant-1, Arecibo, PR, 1973
Isolte Nuclear Power Plant-2, Arecibo, PR, 1973
Jamesport Nuclear Power Plant-1, Jamesport, NJ, 1973
Jamesport Nuclear Power Plant-2, Jamesport, NJ, 1973
Kewaunee Nuclear Plant, Kewaunee, WI, 1967
Lasalle County Nuclear Station-1, Marseilles, IL, 1970
Lasalle County Nuclear Station-2, Marseilles, IL, 1970
Limerick Generating Station-1, Saratoga, PA, 1969
Limerick Generating Station-2, Saratoga, PA, 1969
Maine Yankee Nuclear Power Plant, Wicasset, ME, 1967
Malibu Nuclear Power Plant, Malibu, CA, 1963
Marble Hill Nuclear Power Plant-1, Paynesville, IN, 1973

Marble Hill Nuclear Power Plant-2, Paynesville, IN, 1973
Mayport Nuclear Power Plant-1, Mayport, FL, 1969
Mayport Nuclear Power Plant-2, Mayport, FL, 1969
McGuire Nuclear Station-1, Huntersville, NC, 1969
McGuire Nuclear Station-2, Huntersville, NC, 1969
Midland Nuclear Power Plant-1, Midland, MI, 1968
Midland Nuclear Power Plant-2, Midland, MI, 1968
Millstone Nuclear Power Plant-1, Waterford, CT, 1965
Millstone Nuclear Power Plant-2, Waterford, CT, 1968
Montague Nuclear Power Plant-1, Montague, MA, 1974
Montague Nuclear Power Plant-2, Montague, MA, 1974
Monticello Nuclear Generating Plant, Monticello, MN, 1966
NEP Nuclear Power Plant-1, Charlestown, RI, 1974
NEP Nuclear Power Plant-2, Charlestown, RI, 1974
New Haven Nuclear Power Plant-1, New Haven, NJ, 1977
New Haven Nuclear Power Plant-2, New Haven, NJ, 1977
Nine Mile Point Nuclear Power Plant-1, Lycoming, NY, 1963
Nine Mile Point Nuclear Power Plant-2, Lycoming, NY, 1972
North Anna Nuclear Power Station-1, Mineral, VA, 1967
North Anna Nuclear Power Station-2, Mineral, VA, 1967
North Anna Nuclear Power Station-3, Mineral, VA, 1971
North Anna Nuclear Power Station-4, Mineral, VA, 1971
Oconee Nuclear Station-1, Seneca, SC, 1966
Oconee Nuclear Station-2, Seneca, SC, 1966
Oconee Nuclear Station-3, Seneca, SC, 1967
Oyster Creek Nuclear Generating Station, Forked River, NJ, 1963
Palisades Nuclear Plant, Covert, MI, 1966
Palo Verde Nuclear Generating Station-1, Phoenix, AZ, 1973
Palo Verde Nuclear Generating Station-2, Phoenix, AZ, 1973
Palo Verde Nuclear Generating Station-3, Phoenix, AZ, 1973
Pathfinder Nuclear Power Plant, Sioux Falls, SD, 1957
Peach Bottom Nuclear Power Plant-1, Delta, PA, 1953
Peach Bottom Nuclear Power Plant-2, Delta, PA, 1966
Peach Bottom Nuclear Power Plant-3, Delta, PA, 1967
Perkins Nuclear Power Plant-1, Mocksville, NC, 1973
Perkins Nuclear Power Plant-2, Mocksville, NC, 1973
Perkins Nuclear Power Plant-3, Mocksville, NC, 1973
Perry Nuclear Power Plant-1, Perry, OH, 1972
Perry Nuclear Power Plant-2, Perry, OH, 1974
Phipps Bend Nuclear Power Plant-1, Surgionsville, TN, 1974
Phipps Bend Nuclear Power Plant-2, Surgionsville, TN, 1974
Pilgrim Station-1, Plymouth, MA, 1967
Pilgrim Station-2, Plymouth, MA, 1972
Piqua Nuclear Power Plant, Piqua, OH, 1959
Point Beach Nuclear Plant-1, Two Rivers, WI, 1965
Point Beach Nuclear Plant-2, Two Rivers, WI, 1966
Prairie Island Nuclear Plant-1, Welch, MN, 1967

Prairie Island Nuclear Plant-2, Welch, MN, 1967
PSEG Offshore Nuclear Power Plant-1, Newark, NJ, 1973
PSEG Offshore Nuclear Power Plant-2, Newark, NJ, 1973
Quad Cities Station-1, Cordova, IL, 1966
Quad Cities Station-2, Cordova, IL, 1966
Rancho Seco Nuclear Generating Station, Herald, CA, 1966
River Bend Station-1, St. Francisville, LA, 1972
River Bend Station-2, St. Francisville, LA, 1972
Robinson (H.B.) Electric Power Plant-2, Hartsville, SC, 1966
Salem Nuclear Generating Station-1, Hancocks Bridge, NJ, 1966
Salem Nuclear Generating Station-2, Hancocks Bridge, NJ, 1966
San Onofre Nuclear Generating Station-1, San Clemente, CA, 1963
San Onofre Nuclear Generating Station-2, San Clemente, CA, 1970
San Onofre Nuclear Generating Station-3, San Clemente, CA, 1970
Santa Susana Sodium–Graphite Reactor Experiment (SRE), Santa Susana, CA, 1954
Seabrook Station-1, Seabrook, NH, 1972
Seabrook Station-2, Seabrook, NH, 1972
Sequoyah Nuclear Plant-1, Saddy-Daisy, TN, 1968
Sequoyah Nuclear Plant-2, Saddy-Daisy, TN, 1968
Shippingport Atomic Power Station, Shippingport, PA, 1953
Shoreham Nuclear Power Plant, Wading River, NY, 1967
Skagit/Hanford Nuclear Project-1, Richland, WA, 1973
Skagit/Hanford Nuclear Project-2, Richland, WA, 1973
South Dade Nuclear Power Plant-1, South Dade, FL, 1975
South Dade Nuclear Power Plant-2, South Dade, FL, 1975
South Texas Project Electric Generating Station-1, Wadsworth, TX, 1973
South Texas Project Electric Generating Station-2, Wadsworth, TX, 1973
Sterling Nuclear Power Plant, Sterling, NJ, 1973
St. Lucie Nuclear Power Station-1, Fort Pierce, FL, 1968
St. Lucie Nuclear Power Station-2, Fort Pierce, FL, 1972
St. Rosalie Nuclear Power Plant-1, Alliance, LA, 1974
St. Rosalie Nuclear Power Plant-2, Alliance, LA, 1974
Summer (Virgil C.) Nuclear Station, Jenkinsville, SC, 1971
Summit Nuclear Power Plant-1, Middletown, DE, 1971
Summit Nuclear Power Plant-2, Middletown, DE, 1971
Sundesert Nuclear Power Plant-1, Blythe, CA, 1975
Sundesert Nuclear Power Plant-2, Blythe, CA, 1975
Surry Nuclear Power Station-1, Surry, VA, 1966
Surry Nuclear Power Station-2, Surry, VA, 1966
Surry Nuclear Power Station-3, Surry, VA, 1972
Surry Nuclear Power Station-4, Surry, VA, 1972
Susquehanna Steam Electric Station-1, Berwick, PA, 1970
Susquehanna Steam Electric Station-2, Berwick, PA, 1970
Three Mile Island Nuclear Power Plant-1, Middletown, PA, 1966
Three Mile Island Nuclear Power Plant-2, Middletown, PA, 1969
Trojan Nuclear Plant, Rainier, OR, 1968
Turkey Point Nuclear Station-3, Princeton, FL, 1965

Turkey Point Nuclear Station-4, Princeton, FL, 1965
Tyrone Nuclear Power Plant-1, Durand, WI, 1973
Tyrone Nuclear Power Plant-2, Durand, WI, 1973
Vallecitos Nuclear Power Plant, Pleasonton, CA, 1952
Vandalia Nuclear Power Plant, Prairie City, IA, 1976
Vermont Yankee Nuclear Power Plant, Vernon, VT, 1966
Vidal Nuclear Power Plant-1, Vidal, CA, 1971
Vidal Nuclear Power Plant-2, Vidal, CA, 1974
Vogtle Nuclear Power Station-1, Waynesboro, GA, 1971
Vogtle Nuclear Power Station-2, Waynesboro, GA, 1974
Washington Nuclear Project-1 (WNP-1), Richland, WA, 1973
Washington Nuclear Project-2 (WNP-2), Richland, WA, 1971
Washington Nuclear Project-3 (WNP-3), Richland, WA, 1973
Washington Nuclear Project-4 (WNP-4), Richland, WA, 1974
Washington Nuclear Project-5 (WNP-5), Richland, WA, 1974
Waterford Power Station-3, Kilona, LA, 1970
Watts Bar Nuclear Plant-1, Spring City, TN, 1970
Watts Bar Nuclear Plant-2, Spring City, TN, 1970
Wolf Creek Nuclear Power Plant, Burlington, KS, 1973
Yankee Rowe Nuclear Power Plant, Rowe, MA, 1956
Yellow Creek Nuclear Power Plant-1, Corinth, MS, 1974
Yellow Creek Nuclear Power Plant-2, Corinth, MS, 1974
Zimmer (William H.) Nuclear Power Plant-1, Moscow, OH, 1969
Zimmer (William H.) Nuclear Power Plant-2, Moscow, OH, 1974
Zion Station-1, Zion, IL, 1967
Zion Station-2, Zion, IL, 1967

Nuclear Power Plants Outside the United States

Argentina

Atucha-1, Comisión Nacional de Energia Atomica, 345 MW, 1974
Atucha-2, Comisión Nacional de Energia Atomica, 560 MW, 1987
Embalse, Comisión Nacional de Energia Atomica, 600 MW, 1980

Austria

Tullnerfeld, Gemeinschaftskernkraftwerk Tullnerfeld, 692 MW (construction begun 1971; further development halted 1975; postponed indefinitely)
Zwentendorf (commercial operation begun 1971, ceased 1978)

Belgium

BR-3, Groupe Independent pour l'Exploration du BR-3, 10.5 MW, 1962
Doel-1, Societé Reunie d'Energie de Baffin de l'Escaut SA, 392.5 MW, 1975
Doel-2, Societé Reunie d'Energie de Baffin de l'Escaut SA, 392.5 MW, 1975

Doel-3, Societé Reunie d'Energie de Baffin de l'Escaut SA, 897 MW, 1981
Doel-4, Societé Reunie d'Energie de Baffin de l'Escaut SA, 1984
Tihange-1, Societé Intercommunale Belge de Gas et d'Electricité, 880 MW, 1975
Tihange-2, Societé Intercommunale Belge de Gas et d'Electricité, 902 MW, 1981
Tihange-3, Societé Intercommunale Belge de Gas et d'Electricité, 1006 MW, 1984

Brazil

Angra dos Reis-1, 630 MW, 1979
Angra dos Reis-2 and 3, 1200 MW

Bulgaria

Kozloduy-1, State Economic Corporation, 408 MW, 1974
Kozloduy-2, State Economic Corporation, 408 MW, 1975
Kozloduy-3, State Economic Union/Energetics and Coal, 408 MW, 1979
Kozloduy-4, State Economic Union/Energetics and Coal, 420, 1980

Canada

Bruce-1 and 2, Ontario Hydro, 740 MW, 1977
Bruce-3, Ontario Hydro, 740 MW, 1978
Bruce-4, Ontario Hydro, 740 MW, 1979
Bruce-5, Ontario Hydro, 750 MW, 1984
Bruce-6, Ontario Hydro, 750 MW, 1983
Bruce-7, Ontario Hydro, 750 MW, 1985
Bruce-8, Ontario Hydro, 750 MW, 1986
Darlington-1, Ontario Hydro, 854 MW, 1988
Darlington-2, Ontario Hydro, 854 MW, 1987
Darlington-3, Ontario Hydro, 854 MW, 1989
Darlington-4, Ontario Hydro, 854 MW, 1990
Douglass Point Generating Station, Ontario Hydro, 206 MW, 1968
Gentilly-1, Hydro Quebec, 250 MW, 1972
Gentilly-2, Hydro Quebec, 638 MW, 1980
NPD, Ontario Hydro, 22.5 MW, 1962
Pickering-1 and 2, Ontario Hydro, 514 MW, 1971
Pickering-3, Ontario Hydro, 514 MW, 1972
Pickering-4, Ontario Hydro, 514 MW, 1973
Pickering-5, Ontario Hydro, 516 MW, 1981
Pickering-6 and 7, Ontario Hydro, 516 MW, 1982
Pickering-8, Ontario Hydro, 516 MW, 1983
Point Lepreau, New Brunswick Electric Power Commission, 633 MW, 1981

Cuba

Cienfuegos-1, 408 MW, 1983
Cienfuegos-2, 408 MW, 1984

Czechoslovakia

A-1 Bohunice, Atomova Electraren, 110 MW, 1972
Bohunice-1, Slovak Energy Board, 380.5 MW, 1979
Bohunice-2, Slovak Energy Board, 380.5 MW, 1980
Brno, 420 MW
Dukovany-1 and 2, Czechoslovak Energy Board, 380.5 MW, 1983
Dukovany-3 and 4, Czechoslovak Energy Board, 420 MW
Levice-1 and 2 and 3 and 4, Slovak Energy Board, 420 MW

Egypt

Inchasse, 2000 kW, 1961

Finland

Loviisa-1, Imatran Voima OY, 420 MW, 1977
Loviisa-2, Imatran Voima OY, 420 MW, 1979
Loviisa-3, Imatran Voima OY, 1000 MW, 1988
TVO-1, Tellisuuden Voima Oy, 660 MW, 1978
TVO-2, Tellisuuden Voima Oy, 660 MW, 1980

France

Ardennes, Societé d'Energie Nucleaire Franco-Belge des Ardennes, 270 MW, 1967
Blayais-1 and 2, Electricité de France 925 MW, 1981
Blayais-3, Electricité de France, 925 MW, 1982
Blayais-4, Electricité de France, 925 MW, 1983
Bugey-1, Electricité de France, 540 MW, 1972
Bugey-2 and 3, Electricité de France, 925 MW, 1978
Bugey-4, Electricité de France, 905 MW, 1979
Bugey-5, Electricité de France, 905 MW, 1979
Chinon-2, Electricité de France, 200 MW, 1966
Chinon-3, Electricité de France, 320 MW, 1967
Chinon-B1 and B2, Electricité de France 905 MW, 1982
Cruas-1 and 2, Electricité de France, 905 MW, 1984
Cruas-3 and 4, Electricité de France, 905 MW, 1984
Dampierre-1, Electricité de France, 905 MW, 1979
Dampierre-2 and 3, Electricité de France, 905 MW, 1980
Dampierre-4, Electricité de France, 905 MW, 1981
EL-4, Commissariat à l'Energie Atomique and Electricité de France, 70 MW, 1968
Fessenheim-1, Electricité de France, 890 MW, 1977
Fessenheim-2, Electricité de France, 890 MW, 1978
Flamanville-2, Electricité de France, 1300 MW, 1986
Gravelines B1, Electricité de France, 925 MW, 1979
Gravelines B2, Electricité de France, 925 MW, 1980
Gravelines B3 and B4, Electricité de France, 925 MW, 1981
Paluel-1 and 2, Electricité de France, 1300 MW, 1983

Paluel-3, Electricité de France, 1300 MW, 1984
Paluel-4, Electricité de France, 1300 MW, 1985
Phenix, Commissariat à l'Energie Atomique and Electricité de France, 250 MW, 1974
St. Laurent A1, Electricité de France, 480 MW, 1969
St. Laurent A2, Electricité de France, 515 MW, 1971
St. Laurent B1 and B2, Electricité de France, 905 MW, 1981
St. Maurice-1, Electricité de France, 1300 MW, 1985
St. Maurice-2, Electricité de France, 1300 MW, 1986
Super Phenix, Groupement Centrale, Nucleaire Europeénne de Neutrons Rapides, 1200 MW, 1983
Tricastin-1 and 2, Electricité de France, 925 MW, 1979
Tricastin-3 and 4, Electricité de France, 925 MW, 1980

Germany (former East Germany)

Bruno Leuschner-1 and 2, VEB KKW Bruno Leuschner, 408 MW, 1974
Bruno Leuschner-3, VEB KKW Bruno Leschner, 408 MW, 1978
Bruno Leuschner-4, VEB KKW Bruno Leschner, 408 MW, 1980
Màgdeburg-1, 2, 3, and 4, 408 MW, 1980
Magdeburg-5, 6, 7, and 8, 408 MW
Rheinsberg AKW, VEB KKW Bruno Leuschner, 62.5 MW, 1966

Germany (former West Germany)

AVR Juelich, Arbeitsgemeinschaft Versuchs-Reaktor GmbH, 13.5 MW, 1969
Biblis-A, Rheinisch Westfaelisches Elektrizitaetswerk AG, 1146 MW, 1975
Biblis-B, Rheinisch Westfaelisches Elektrizitaetswerk AG, 1178 MW, 1977
Bilbis-C, Rheinisch-Westfaelisches Elektrizitaetswerk AG, 1166 MW, 1983
Brokdorf, Kernkraftwerk Brokdorf GmbH, 1288 MW, 1983
Emsland, KLE, 1232 MW, 1986
GKN Neckar-1, Kernkraftwerk Brunsbuettel, 770 MW, 1977
GKN Neckar-2, Kernkraftwerk Neckar, 750 MW, 1986
Grohnde, Geneinschaftskraftwerk Grohnde GmbH, 1289 MW, 1981
KKG Grafen-Rheinfeld, Bayernwerk AG, 122 MW, 1980
KKH Hamm, KLE, 1231 MW, 1985
KKI-Isar-2, Kernkraftwerk Isar, 1230 MW, 1986
KKI Philippsburg-1, Kernkraftwerk Philippsburg, 864 MW, 1979
KKK Kruemmel, Kernkraftwerk Kruemmel, 1260 MW, 1980
KKP Philippsburg-2, Kernkraftwerk Philippsburg, 1281 MW, 1983
KKS Stade, Kernkraftwerk Stade GmbH, 630 MW, 1972
KKU Unterwasser, Kernkraftwerk Unterwasser, 1230 MW, 1979
KKW Kalker, Gemeinsames Europaeisches Unternehmen, 292 MW, 1983
KNK II, Kernkraftwerk-Betriebsgesellschaft mbH, 17.8 MW, 1973
KRB Gundremmingen-A, Kernkraftwerk RWE-Bayernwerk GmbH, 237 MW, 1967
KRB Gundremmingen-B and C, Kernkraftwerk RWE-Bayernwerk GmbH, 1244 MW, 1982
KWB Borken, Preussenelektra, 1200 MW

KWL Lingen, Kernkraftwerk Lingen GmbH, 256 MW, 1968
KWO Obrigheim, Kernkraftwerk Obrigheim GmbH, 328 MW, 1969
KWS Wyhl-1, Kernkraftwerk Sued, 1284 MW, 1986
Meulheim-Kaerlich, Rheinisch-Westfaelisches Elektrizitaetswerk AG, 1215 MW, 1981
MZFR, Kernkraftwerk-Betriebsgesellschaft mbH, 52 MW, 1966
Neupotz-1 and 2, Rheinisch-Westfaelisches Elektrizitaetswerk AG, 1245 MW, 1985
Rehling, 1300
SNR-2, ESK, 1300 MW, 1989
THTR-300, Hochtemperatur-Kernkraftwerk GmbH, 300 MW, 1981
Vahnum-1 and 2, Rheinisch-Westfaelisches Elektrizitaetswerk AG, 1232 MW
VAK Kahl, Versuchsatomkraftwerk GmbH, 15 MW, 1962
Weurgassen, Pruessische Elektrizitaets AG, 640 MW, 1975

Hungary

Paks-1, Hungarian Electric Works, 408 MW, 1980
Paks-2, Hungarian Electric Works, 408 MW, 1983
Paks-3, Hungarian Electric Works, 408 MW, 1984
Paks-5, Hungarian Electric Works, 408 MW, 1985

India

Kalpakkam-1, Department of Atomic Energy, 220 MW, 1981
Kalpakkam-2, Department of Atomic Energy, 220 MW, 1983
Narora-1, Department of Atomic Energy, 220 MW, 1984
Narora-2, Department of Atomic Energy, 220 MW, 1985
Rajasthan-2, Department of Atomic Energy, 207 MW, 1980
Tarapur-1 and 2, Department of Atomic Energy, 198 MW, 1969

Iran

Iran-1, Atomic Energy Organization of Iran, 1200 MW, 1980
Iran-2, Atomic Energy Organization of Iran, 1200 MW, 1981
Iran-3, Atomic Energy Organization of Iran, 891 MW, 1983
Iran-4, Atomic Energy Organization of Iran, 991 MW, 1984
Iran-5 and 6, 1200 MW
Iran-7 and 8, 1200 MW

Israel

Israel-1, 600 MW, 1985

Italy

Caorso, ENEL, 840 MW, 1978
Cirene, ENEL, 36 MW, 1982
ENEL-6, ENEL, 980 MW, 1983

ENEL-8, ENEL, 980 MW, 1984
ENEL-5 and 7, ENEL, 1984
Garigliano, ENEL, 150 MW, 1964
Latina, ENEL, 150 MW, 1964
Trino Vercellese, ENEL, 242 MW, 1965

Japan

Fugen ATR, Power Reactor and Nuclear Fuel Development Corporation, 150 MW, 1979
Fukushima I-2, Tokyo Electric Power Company, 760 MW, 1974
Fukushima I-3, Tokyo Electric Power Company, 760 MW, 1976
Fukushima I-4 and I-5, Tokyo Electric Power Company, 760 MW, 1978
Fukushima I-6, Tokyo Electric Power Company, 1067 MW, 1979
Fukushima II-1, Tokyo Electric Power Company, 1067 MW, 1982
Fukushima II-2, Tokyo Electric Power Company, 1067 MW, 1983
Fukushima II-3, Tokyo Electric Power Company, 1067 MW, 1984
Fukushima II-4, Tokyo Electric Power Company, 1067 MW, 1985
Genkai-1, Kyushu Electric Power Company, 529 MW, 1975
Genkai-2, Kyushu Electric Power Company, 529 MW, 1981
Hamaoka-1, Chubu Electric Power Company, 515 MW, 1976
Hamaoka-2, Chubu Electric Power Company, 814 MW, 1978
Hamaoka-3, Chubu Electric Power Company, 1050 MW, 1985
Ikata-1, Shikoku Electric Power Company, 538 MW, 1977
Ikata-2, Shikoku Electric Power Company, 538 MW, 1982
JPDR-II, Japan Atomic Energy Research Institute, 10.3 MW, 1963
Kashiwazaki-1, Tokyo Electric Power Company, 1067 MW, 1984
Mihama-1, Kansai Electric Power Company, 320 MW, 1970
Mihama-2, Kansai Electric Power Company, 470 MW, 1972
Mihama-3, Kansai Electric Power Company, 780 MW, 1976
Monju, Power Reactor and Nuclear Fuel Development Corporation, 250 MW, 1987
Ohi-1 and 2, Kansai Electric Power Company, 1120 MW, 1979
Onagawa-1, Tohoku Electric Power Company, 500 MW, 1983
Sendai-1, Kyushu Electric Power Company, 846 MW, 1984
Sendai-2, Kyushu Electric Power Company, 1985
Shimane, Chugoku Electric Power Company, 439 MW, 1974
Takahama-1, Kansai Electric Power Company, 780 MW, 1974
Takahama-2A, Kansai Electric Power Company, 780 MW, 1974
Takahama-2B, Kansai Electric Power Company, 780 MW, 1975
Takahama-3, Kansai Electric Power Company, 830 MW, 1983
Takahama-4, Kansai Electric Power Company, 830 MW, 1984
Tokai-1, Japan Atomic Power Company, 158.5 MW, 1966
Tokai-2, Japan Atomic Power Company, 1056 MW, 1978
Tsuruga-1, Japan Atomic Power Company, 340 MW, 1970
Tsuruga-2, Japan Atomic Power Company, 1115 MW, 1986

Korea (Republic of South Korea)

Ko-Ri-1, Korea Electric Company, 564 MW, 1978
Ko-Ri-2, Korea Electric Company, 605 MW, 1983
Ko-Ri-3, Korea Electric Company, 930 MW, 1982
Ko-Ri-4, Korea Electric Company, 930 MW, 1986
Wolsung-1, Korea Electric Company, 628.6 MW, 1982

Mexico

Laguna Verde-1, Comisión Federal de Electricidad, 654 MW, 1982
Laguna Verde-2, Comisión Federal de Electricidad, 654 MW, 1983

Netherlands

Borssele, Provinciale Zeeuwse Energie Maatschappij, 447 MW, 1973
Dodewaard, Kernkraftwerk Neckar, 51.5 MW, 1969

Pakistan

Kanupp, PAEC, 125 MW, 1972
Chasnupp, PAEC, 600 MW, 1986

Philippines

PNPP-1, Philippine National Power Company, 620.8 MW, 1982

Poland

Zarnowiec-1, North of Poland Power Authority, 408 MW, 1985
Zarnowiec-2, North of Poland Power Authority, 408 MW, 1986

Republic of South Africa

Koeberg, 921 MW, 1982 or 1983

Romania

Olt, 408 MW, 1983

Spain

Almaraz-1 and 2, Compania Sevillana de Electricidad Hidro-electrica Espanola SA, and Union Elecrtica SA, 900 MW, 1980
Asco-1 and 2, Fuerzas Electricas de Cataluna SA, 881.5 MW, 1980
Cofrentes, Hidroelectrica Espanola SA, 935 MW, 1981
Jose Cabera-1, Union Electrica SA, 153.2 MW, 1969
Lemoniz-1, Iberduero SA, 902 MW, 1981

Lemoniz-2, Iberduero SA, 902 MW, 1983
Regodola, Fuerzas Electricas de Cataluna SA, 900 MW, 1986
Santa Maria de Garona, Centrales Nucleares del Norte SA, 440 MW, 1971
Sayago, Iberduero SA, 1036 MW, 1984
Trillo-1, Union Electrica SA, 990 MW, 1987
Trillo-2, Union Electrica SA, 1000 MW, 1988
Valde-Caballeros-1, Hidroelectrica Espanola SA and Compania Sevillana de Electricidad, 938.8 MW, 1984
Valde-Caballeros-2, Hidroelectrica Espanola SA and Compania Sevillana de Electricidad, 938.8 MW, 1985
Vandellos-1, Sociedad Hispano-Francesa de Energia Nuclear SA, 480 MW, 1972
Vandellos-2, ANV, 930 MW, 1984
Vandellos-3, Fuerzas Electricas de Cataluna SA, 1000 MW, 1987

Sweden

Barsebaeck-1, Sydsvenska Kraftakltiebolaget, 570 MW, 1975
Barsebaeck-2, Sydsvenska Kraftakltiebolaget, 570 MW, 1977
Barsebaeck-3, Sydsvenska Kraftakltiebolaget, 900 MW, 1987
Forsmark-1, Swedish State Power Board, 899 MW, 1979
Forsmark-2, Swedish State Power Board, 899 MW, 1980
Forsmark-3, Swedish State Power Board, 1060 MW, 1984
Oskarshamn-1, Oskarshamnsverkets Kraftgrupp AB, 440 MW, 1972
Oskarshamn-2, Oskarshamnsverkets Kraftgrupp AB, 570 MW, 1974
Oskarshamn-3, Oskarshamnsverkets Kraftgrupp AB, 1000 MW, 1987
Ringhals-1, Swedish State Power Board, 750 MW, 1976
Ringhals-2, Swedish State Power Board, 800 MW, 1975
Ringhals-3, Swedish State Power Board, 912 MW 1979
Ringhals-4, Swedish State Power Board, 912 MW, 1980

Turkey

Akkuyu, Turkish Power Authority, 620 MW, 1985

Union of Soviet Socialist Republics (former)

APS-1 Obninsk, Institute of Planning and Energetics, 5 MW, 1954
Armenia-1, 400 MW, 1976
Armenia-2, 400 MW, 1979
Atash-1, 1000 MW
Beloyarsk-1, Ministry of Power Stations of the USSR, 102 MW, 1964
Beloyarsk-2, Ministry of Power Stations of the USSR, 175 MW, 1967
Bilbin, 4 × 11 MW, 1974-1976
BN-350, State Committee for the Use of Atomic Energy in the USSR, 135 MW, 1973
BN-600, 600 MW, 1979
BN-1600, 1600 MW
Bor-60, 11 MW, 1969

Chernobyl-1 and 2, 1000 MW, 1978
Chernobyl-3, 1000 MW, 1980
Chernobyl-4, 1000 MW, 1982
Ignalino-1, 1500 MW, 1984
Ignalino-2, 1500 MW, 1985
Kalinin-1, 1000 MW, 1981
Kalinin-2, 1000 MW, 1982
Kalinin-3, 1000 MW, 1987
Kalinin-4, 1000 MW, 1988
Kola-1, 370 MW, 1973
Kola-2, 370 MW, 1974
Kola-3, 420 MW, 1981
Kola-4, 420 MW, 1982
Kursk-1, 1000 MW, 1976
Kursk-2, 1000 MW, 1979
Kursk-3, 12000 MW, 1980
Kursk-4, 1000 MW
Leningrad-1, 1000 MW, 1974
Leningrad-2, 1000 MW, 1975
Leningrad-3, 1000 MW, 1979
Leningrad-4, 1000 MW, 1980
Nikolajev-1, 1000 MW, 1980
Nikolajev-2 and 3 and 4, 1000 MW each
Novo Voronezh-1, State Committee for the Use of Atomic Energy in the USSR, 196 MW, 1964
Novo Voronezh-2, State Committee for the Use of Atomic Energy in the USSR, 338 MW, 1970
Novo Voronezh-3, State Committee for the Use of Atomic Energy in the USSR, 410 MW, 1972
Novo Voronezh-4, State Committee for the Use of Atomic Energy in the USSR, 410 MW, 1973
Novo Voronezh-5, 1100 MW, 1979
Rovno-1, 420 MW, 1979
Rovno-2 and 3, 420 MW, 1981
Smolensk-1, 1000 MW, 1980
Smolensk-2, 1000 MW, 1981
VK-50, Scientific Research Institute for Atomic Reactors, 50 MW, 1966
West Ukrainian-3 and 4, 1000 MW each
Zaporozhe-1, 1000 MW

United Kingdom

Berkeley, Central Electricity Generating Board, 2 × 143 MW, 1962
Bradwell, Central Electricity Generating Board, 2 × 125 MW, 1962
Calder Hall, British Nuclear Fuels Ltd., 4 × 50 MW, 1956-1959
Chapelcross, British Nuclear Fuels Ltd., 4 × 50 MW, 1959-1960
CDFR, 1250 MW, 1986

Dungeness-A, Central Electricity Generating Board, 2 × 205 MW, 1965
Dungeness-B, Central Electricity Generating Board, 2 × 607 MW, 1979
Hartpool, Central Electricity Generating Board, 2 × 265 MW, 1980
Heysham-A, Central Electricity Generating Board, 2 × 625 MW, 1981
Heysham-B, Central Electricity Generating Board, 2 × 611 MW, 1986
Hinkley Point-A, Central Electricity Generating Board, 2 × 230 MW, 1965
Hinkley Point-B, Central Electricity Generating Board, 2 × 616 MW, 1977
Hunterston-A, South of Scotland Electricity Board, 2 × 150 MW, 1964
Hunterston-B, South of Scotland Electricity Board, 2 × 616 MW, 1977
Oldsbury-A, Central Electricity Generating Board, 2 × 205 MW, 1968
PFR Dounreay, UKAEA, 230 MW, 1977
Sizewell-A, Central Electricity Generating Board, 2 × 210 MW, 1966
Torness, South of Scotland Electricity Board, 2 × 611, 1986
Trawsfynydd, Central Electricity Generating Board, 2 × 195 MW, 1965
Winfrith, UKAEA, 92 MW, 1968
Windscale, UKAEA, 32 MW, 1963
Wylfa, Central Electricity Generating Board, 2 × 240 MW, 1971

Yugoslavia

Krsko, Nuklearna Elektrana Krsko, 632 MW, 1980

APPENDIX B

Roof Sheathing Fastening Schedules for Wind Uplift (Published by the American Plywood Association)

INTRODUCTION

This data file provides recommended nailing schedules for wood structural panel roof sheathing — plywood, COM-PLYO, and oriented strand board (OSB). These schedules were established to provide resistance to wind uplift pressure, with particular emphasis on high-wind exposures.

Recommendations were developed through computer analysis and verified by full-scale laboratory testing. Wet, as well as dry specimens of plywood and OSB panels were tested for full panel withdrawal under uniform pressure. The results of this testing were compared with wind loads calculated in accordance with the design provisions of ASCE 7-88, Minimum Design Loads for Buildings and Other Structures (American Society of Civil Engineers 1990).

RECOMMENDED FASTENING SCHEDULES

The fastening schedules presented in Table B1 reflect the differences in wind uplift pressures that may be anticipated over various portions of roof systems. Higher pressures at eaves, corners, edges and gable-ends require more restrictive schedules than at interior portions of the roof system. For this reason, fastening schedules may have different requirements for each of the three roof fastening zones illustrated in Figure B1.

Three fastening schedules are provided in Table B1 for roof applications with framing spaced at 24 inches on center or less. These schedules assume the use of wood structural panels 5/8 inch thick or less and are appropriate for buildings with a mean roof height of up

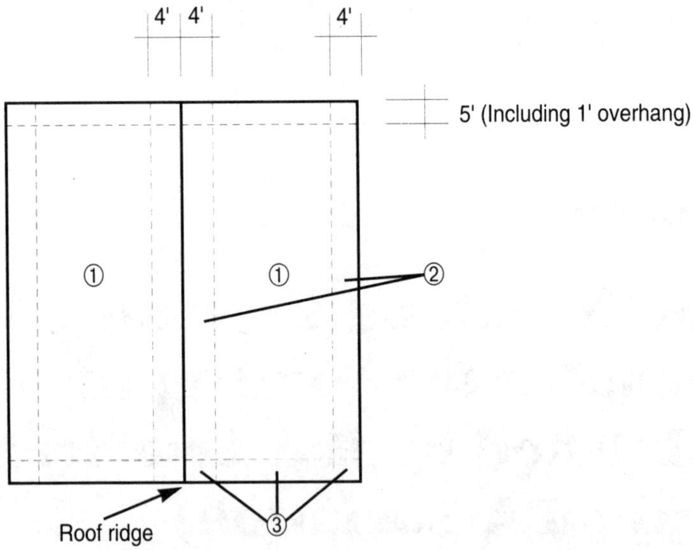

Figure B1. Roof fastening zones for wind uplift.

Table B1. Roof Sheathing Fastening Schedule

Region	Nails	Panel location	Roof fastening zone 1	2	3
			Fastening schedule (inches on center)		
High-wind uplift	8d	Panel edges[1]	6	6	4[2]
	common	Panel field	6	6	6[2]
Intermediate uplift	8d	Panel edges[1]	6	6	4
	common	Panel field	12	6	6
Basic uplift	8d	Panel edges[1]	6	6	6
	common	Panel field	12	12	12

[1]Edge spacing also applies over gable-end walls or trusses. [2]Use 8d ring-shank nails in this zone if mean roof height is greater than 25′.

to 35 feet. All fasteners listed in the tables are minimum 8d common nails with smooth or ring shanks, depending on the basic wind speed and fastener location. All recommendations are based on the use of full-length nails.

The three schedules provided give nailing recommendations for **basic uplift, intermediate uplift**, and **high-wind uplift** conditions as follows:

Basic uplift — The basic uplift fastening schedule is appropriate for buildings located in areas where the basic wind speed, as determined by your local building department, is 80 mph or less. These areas are normally included under the prescriptive sections of building codes. As such, the nailing schedule is the familiar 6 inches on center at supported panel edges, including gable-end walls, and 12 inches on center over intermediate panel supports. Note, however, that *minimum 8d nails are recommended* for all panels 5/8 inch thick or less. Former APA minimum fastening recommendations included the use of 6d nails for panels 1/2 inch thick or less.

Intermediate uplift — The intermediate uplift fastening schedule is appropriate in inland areas with a basic wind speed above 80 mph and below basic wind speeds for which the high-wind uplift schedule is recommended.

High-wind uplift — The schedule for high-wind uplift is appropriate for *all hurricane oceanline regions* (Atlantic and Gulf of Mexico coastal areas). In addition, this schedule should be considered for the *transition zone between hurricane oceanline and inland regions*. The section on High-Wind Uplift Fastening Schedule for Inland Regions is provided to assist in determining at which basic wind speed (for inland regions) the high-wind uplift schedule is recommended. Contact your local building department for basic wind speed used for design in your area.

For conditions that are not addressed by these general guidelines, such as the "special wind regions" identified in ASCE 7-88, engineered design is recommended.

HIGH-WIND UPLIFT FASTENING SCHEDULE FOR INLAND REGIONS

The high-wind uplift fastening schedule is recommended at hurricane oceanlines and transition regions. To determine the basic wind speed at which the high-wind uplift fastening schedule is recommended for a specific structure in an inland region, consider the following:

1. The ability of a roof sheathing panel to resist high winds is directly related to how well it is secured to the roof framing. The type and number of fasteners required for a specific application is obviously an important consideration. Another important consideration is the wood species of the roof framing members into which the sheathing fasteners are driven. Wood of more dense species such as Douglas-fir and southern pine provide greater nail withdrawal resistance and significantly improve the performance of the sheathing nailing. As shown in Table B2, if less dense species such as hem-fir or spruce-pine-fir are specified and used, the high-wind uplift schedule is recommended at lower basic wind speeds than if the denser species are used.

2. Another consideration relates to the condition of the building envelope during the high wind event. If the building envelope remains intact during the storm the destructive forces of the wind are considerably less than experienced if a large window, sliding glass door, or garage door are breached, or if there are permanent openings. Generally speaking, well designed and installed shutter systems are intended to keep the building envelope intact during high wind conditions. In addition to maintaining the building envelope intact and lowering the wind forces on the structure, shutters also serve to protect the interior of the building from water damage caused by failed doors and glass. As can be seen from Table B2, the high-wind uplift schedule is recommended at lower basic wind speeds when there is a possibility that the envelope may be breached by breakage or by permanent openings.

Table B2. Basic Wind Speeds[1] for Which the High-Wind Uplift Schedule Is Recommended for Inland Regions

Wood species of roof framing	Building envelope intact	Building envelope breached
Hemlock, eastern spruce, hem-fir, white pine, northern pine or spruce-pine-fir (0.42 ≥ s.g. ≥ 0.49)	≥100 mph	≥90 mph
Douglas-fir or Southern Pine, (0.50 ≥ s.g. ≥ 0.55)	≥110 mph	≥100 mph

[1]Contact your local building department for basic wind speed used for design in your area.

Table B3. Depths of Nail Penetration in Withdrawal Specimens

Nail type	Nail length (in)	Relief (in)	Penetration (in)
6d Common	2.0	0.498	1.502
8d Common	2.5	0.498	2.002
8d Ring-Shank	2.5	0.498	2.002

Table B4. Nail Withdrawal Test Results

	6d common		8d common		8d ring-shank	
	As tested (lbf)	Normal-ized (lbf/in)	As tested (lbf)	Normal-ized (lbf/in)	As tested (lbf)	Normal-ized (lbf/in)
Average[1]	185.6	123.5	544.8	272.1	697.5	348.4
COV (%)	20.6	20.6	20.1	20.1	13.4	13.4
Minimum	115.3	76.8	379.0	189.3	474.0	236.8
Maximum	256.8	171.0	738.0	368.6	886.0	442.6

[1]40 replications for each nail type.

NAIL WITHDRAWAL TEST RESULTS

Results are presented in two formats. The first format is simply direct presentation of test results, or results "as tested." The second format represents test results normalized to a "per inch of penetration" basis. Nail withdrawal design values are provided in normalized format. Test results are summarized in Table B4.

APPENDIX C

Engineering Tables

Reprinted from the *Manual of Steel Construction*, with permission from the American Institute of Steel Construction.

PROPERTIES OF GEOMETRIC SECTIONS

SQUARE

Axis of moments through center

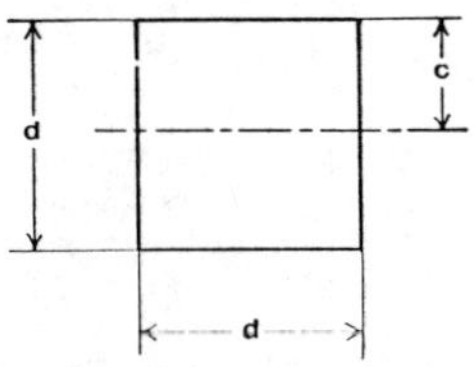

$$A = d^2$$

$$c = \frac{d}{2}$$

$$I = \frac{d^4}{12}$$

$$S = \frac{d^3}{6}$$

$$r = \frac{d}{\sqrt{12}} = .288675\ d$$

$$Z = \frac{d^3}{4}$$

SQUARE

Axis of moments on base

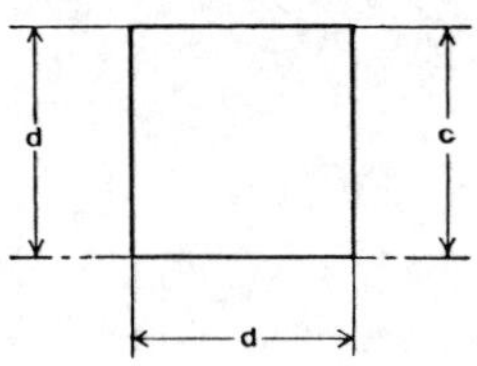

$$A = d^2$$

$$c = d$$

$$I = \frac{d^4}{3}$$

$$S = \frac{d^3}{3}$$

$$r = \frac{d}{\sqrt{3}} = .577350\ d$$

SQUARE

Axis of moments on diagonal

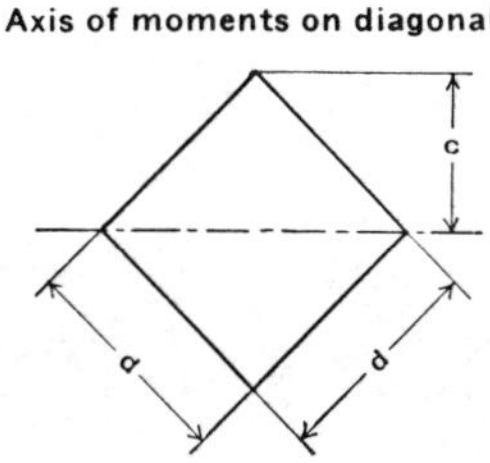

$$A = d^2$$

$$c = \frac{d}{\sqrt{2}} = .707107\ d$$

$$I = \frac{d^4}{12}$$

$$S = \frac{d^3}{6\sqrt{2}} = .117851\ d^3$$

$$r = \frac{d}{\sqrt{12}} = .288675\ d$$

$$Z = \frac{2c^3}{3} = \frac{d^3}{3\sqrt{2}} = .235702d^3$$

RECTANGLE

Axis of moments through center

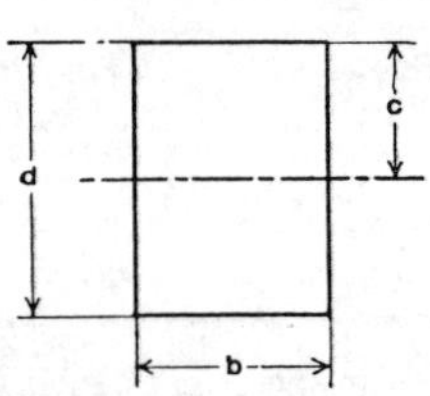

$$A = bd$$

$$c = \frac{d}{2}$$

$$I = \frac{bd^3}{12}$$

$$S = \frac{bd^2}{6}$$

$$r = \frac{d}{\sqrt{12}} = .288675\ d$$

$$Z = \frac{bd^2}{4}$$

PROPERTIES OF GEOMETRIC SECTIONS

RECTANGLE

Axis of moments on base

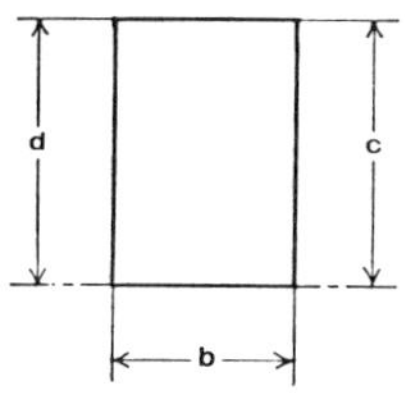

$$A = bd$$

$$c = d$$

$$I = \frac{bd^3}{3}$$

$$S = \frac{bd^2}{3}$$

$$r = \frac{d}{\sqrt{3}} = .577350\ d$$

RECTANGLE

Axis of moments on diagonal

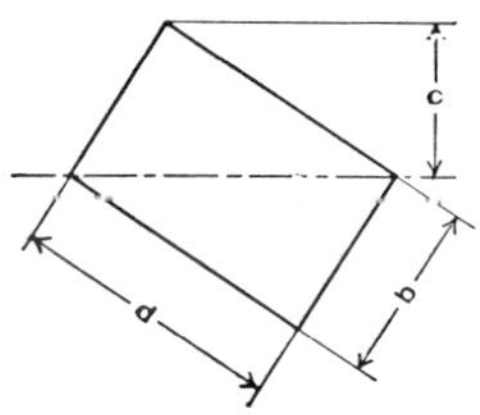

$$A = bd$$

$$c = \frac{bd}{\sqrt{b^2 + d^2}}$$

$$I = \frac{b^3d^3}{6\,(b^2 + d^2)}$$

$$S = \frac{b^2d^2}{6\sqrt{b^2 + d^2}}$$

$$= \frac{bd}{\sqrt{6\,(b^2 + d^2)}}$$

RECTANGLE

Axis of moments any line through center of gravity

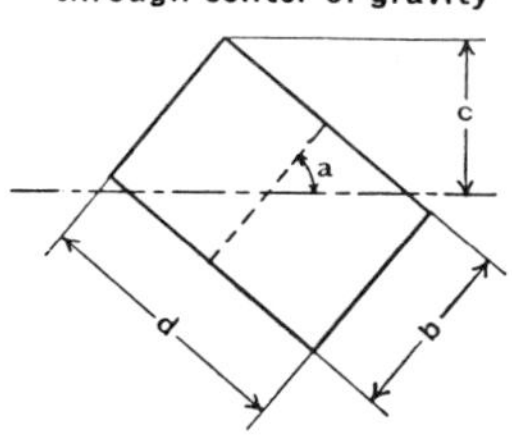

$$A = bd$$

$$c = \frac{b \sin a + d \cos a}{2}$$

$$I = \frac{bd\,(b^2 \sin^2 a + d^2 \cos^2 a)}{12}$$

$$S = \frac{bd\,(b^2 \sin^2 a + d^2 \cos^2 a)}{6\,(b \sin a + d \cos a)}$$

$$r = \sqrt{\frac{b^2 \sin^2 a + d^2 \cos^2 a}{12}}$$

HOLLOW RECTANGLE

Axis of moments through center

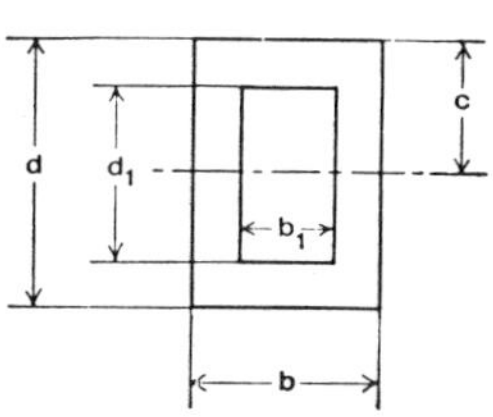

$$A = bd - b_1d_1$$

$$c = \frac{d}{2}$$

$$I = \frac{bd^3 - b_1d_1^3}{12}$$

$$S = \frac{bd^3 - b_1d_1^3}{6d}$$

$$r = \sqrt{\frac{bd^3 - b_1d_1^3}{12\,A}}$$

$$Z = \frac{bd^2}{4} - \frac{b_1d_1^2}{4}$$

PROPERTIES OF GEOMETRIC SECTIONS

EQUAL RECTANGLES

Axis of moments through center of gravity

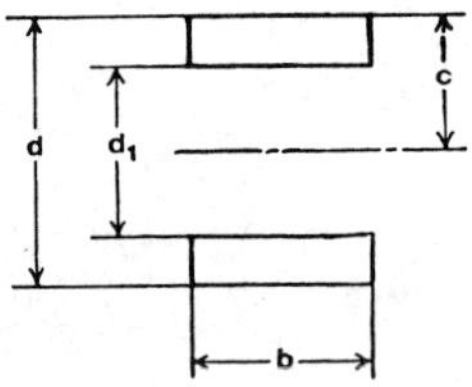

$A = b(d - d_1)$

$c = \dfrac{d}{2}$

$I = \dfrac{b(d^3 - d_1^3)}{12}$

$S = \dfrac{b(d^3 - d_1^3)}{6d}$

$r = \sqrt{\dfrac{d^3 - d_1^3}{12(d - d_1)}}$

$Z = \dfrac{b}{4}(d^2 - d_1^2)$

UNEQUAL RECTANGLES

Axis of moments through center of gravity

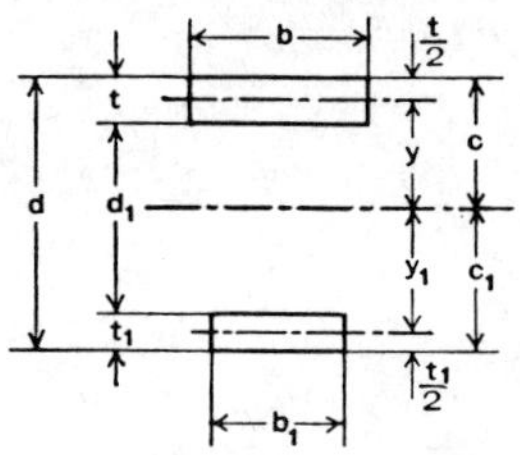

$A = bt + b_1 t_1$

$c = \dfrac{\frac{1}{2}bt^2 + b_1 t_1 (d - \frac{1}{2} t_1)}{A}$

$I = \dfrac{bt^3}{12} + bty^2 + \dfrac{b_1 t_1^3}{12} + b_1 t_1 y_1^2$

$S = \dfrac{I}{c} \qquad S_1 = \dfrac{I}{c_1}$

$r = \sqrt{\dfrac{I}{A}}$

$Z = \dfrac{A}{2}\left[d - \left(\dfrac{t + t_1}{2}\right)\right]$

TRIANGLE

Axis of moments through center of gravity

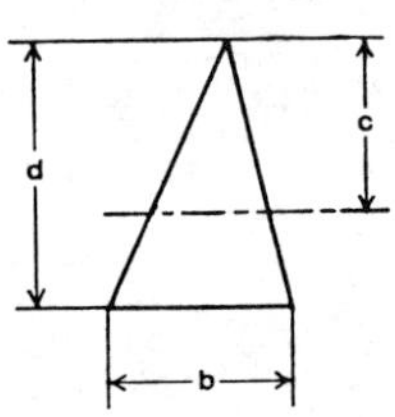

$A = \dfrac{bd}{2}$

$c = \dfrac{2d}{3}$

$I = \dfrac{bd^3}{36}$

$S = \dfrac{bd^2}{24}$

$r = \dfrac{d}{\sqrt{18}} = .235702\ d$

TRIANGLE

Axis of moments on base

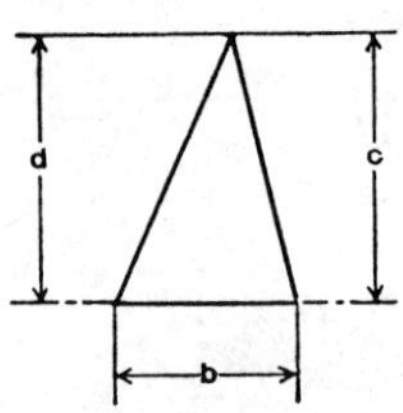

$A = \dfrac{bd}{2}$

$c = d$

$I = \dfrac{bd^3}{12}$

$S = \dfrac{bd^2}{12}$

$r = \dfrac{d}{\sqrt{6}} = .408248\ d$

PROPERTIES OF GEOMETRIC SECTIONS

TRAPEZOID

Axis of moments through center of gravity

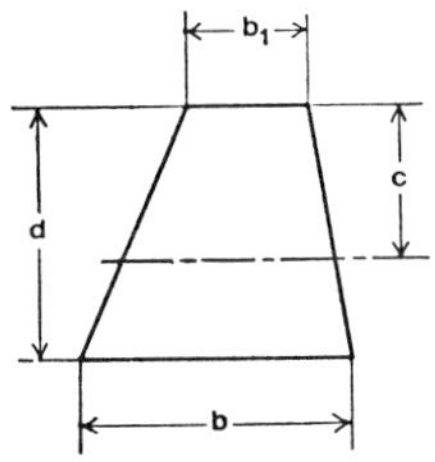

$$A = \frac{d(b + b_1)}{2}$$

$$c = \frac{d(2b + b_1)}{3(b + b_1)}$$

$$I = \frac{d^3 (b^2 + 4\,bb_1 + b_1^2)}{36\,(b + b_1)}$$

$$S = \frac{d^2 (b^2 + 4\,bb_1 + b_1^2)}{12\,(2b + b_1)}$$

$$r = \frac{d}{6(b + b_1)} \sqrt{2\,(b^2 + 4\,bb_1 + b_1^2)}$$

CIRCLE

Axis of moments through center

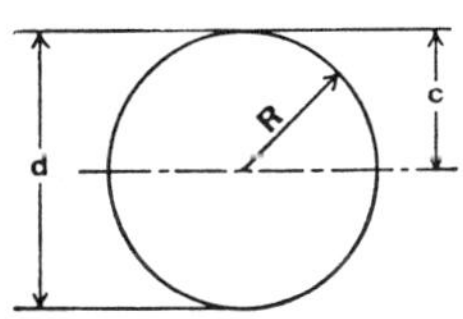

$$A = \frac{\pi d^2}{4} = \pi R^2 = .785398\,d^2 = 3.141593\,R^2$$

$$c = \frac{d}{2} = R$$

$$I = \frac{\pi d^4}{64} = \frac{\pi R^4}{4} = .049087\,d^4 = .785398\,R^4$$

$$S = \frac{\pi d^3}{32} = \frac{\pi R^3}{4} = .098175\,d^3 = .785398\,R^3$$

$$r = \frac{d}{4} = \frac{R}{2}$$

$$Z = \frac{d^3}{6}$$

HOLLOW CIRCLE

Axis of moments through center

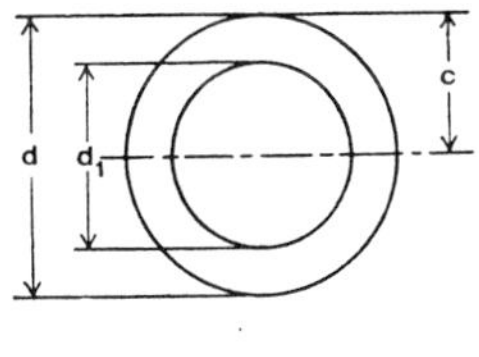

$$A = \frac{\pi(d^2 - d_1^2)}{4} = .785398\,(d^2 - d_1^2)$$

$$c = \frac{d}{2}$$

$$I = \frac{\pi(d^4 - d_1^4)}{64} = .049087\,(d^4 - d_1^4)$$

$$S = \frac{\pi(d^4 - d_1^4)}{32d} = .098175\,\frac{d^4 - d_1^4}{d}$$

$$r = \frac{\sqrt{d^2 + d_1^2}}{4}$$

$$Z = \frac{d^3}{6} - \frac{d_1^3}{6}$$

HALF CIRCLE

Axis of moments through center of gravity

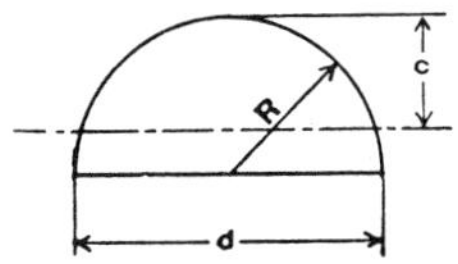

$$A = \frac{\pi R^2}{2} = 1.570796\,R^2$$

$$c = R\left(1 - \frac{4}{3\pi}\right) = .575587\,R$$

$$I = R^4\left(\frac{\pi}{8} - \frac{8}{9\pi}\right) = .109757\,R^4$$

$$S = \frac{R^3}{24}\,\frac{(9\pi^2 - 64)}{(3\pi - 4)} = .190687\,R^3$$

$$r = R\,\frac{\sqrt{9\pi^2 - 64}}{6\pi} = .264336\,R$$

PROPERTIES OF GEOMETRIC SECTIONS

PARABOLA

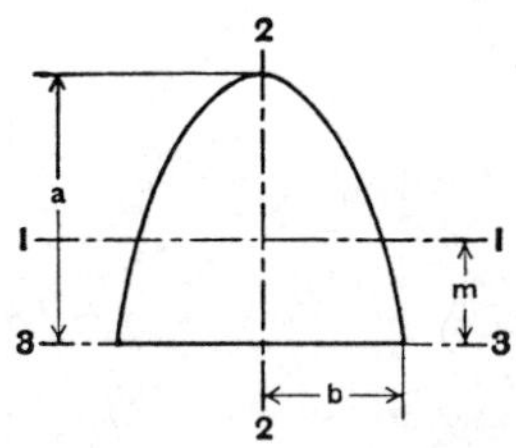

$$A = \frac{4}{3}ab$$

$$m = \frac{2}{5}a$$

$$I_1 = \frac{16}{175}a^3b$$

$$I_2 = \frac{4}{15}ab^3$$

$$I_3 = \frac{32}{105}a^3b$$

HALF PARABOLA

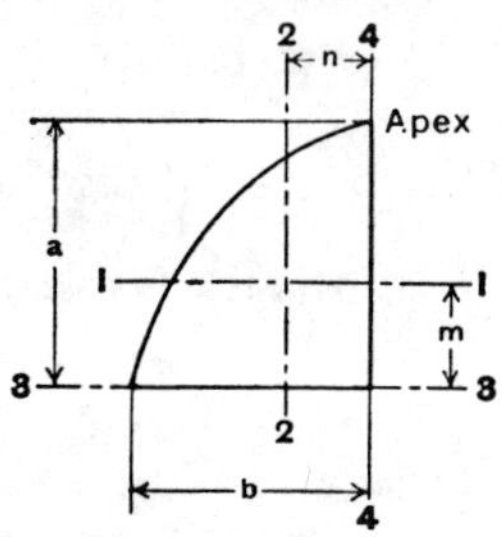

$$A = \frac{2}{3}ab$$

$$m = \frac{2}{5}a$$

$$n = \frac{3}{8}b$$

$$I_1 = \frac{8}{175}a^3b$$

$$I_2 = \frac{19}{480}ab^3$$

$$I_3 = \frac{16}{105}a^3b$$

$$I_4 = \frac{2}{15}ab^3$$

COMPLEMENT OF HALF PARABOLA

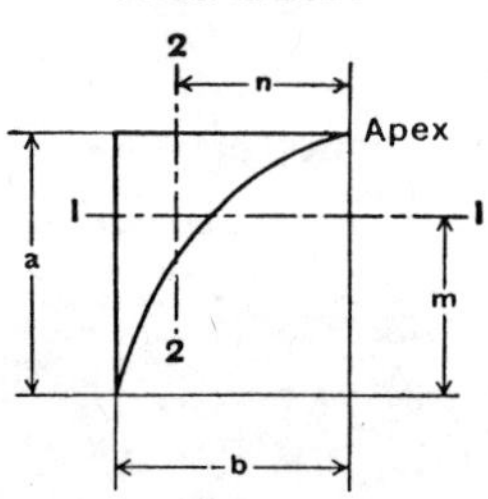

$$A = \frac{1}{3}ab$$

$$m = \frac{7}{10}a$$

$$n = \frac{3}{4}b$$

$$I_1 = \frac{37}{2100}a^3b$$

$$I_2 = \frac{1}{80}ab^3$$

PARABOLIC FILLET IN RIGHT ANGLE

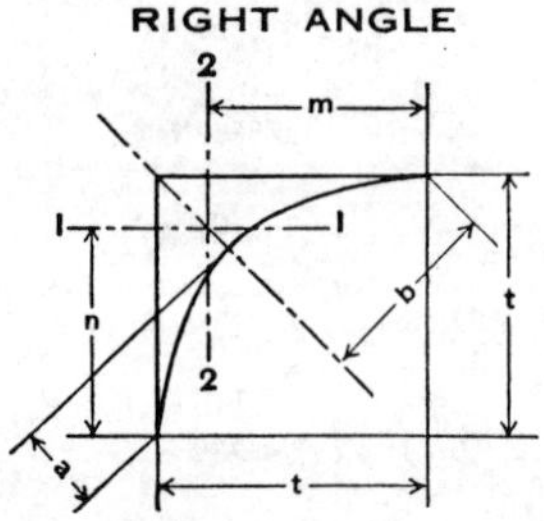

$$a = \frac{t}{2\sqrt{2}}$$

$$b = \frac{t}{\sqrt{2}}$$

$$A = \frac{1}{6}t^2$$

$$m = n = \frac{4}{5}t$$

$$I_1 = I_2 = \frac{11}{2100}t^4$$

PROPERTIES OF GEOMETRIC SECTIONS

* HALF ELLIPSE

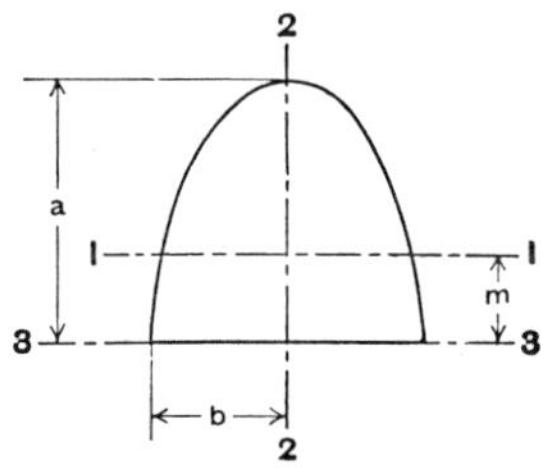

$$A = \frac{1}{2}\pi ab$$

$$m = \frac{4a}{3\pi}$$

$$I_1 = a^3b\left(\frac{\pi}{8} - \frac{8}{9\pi}\right)$$

$$I_2 = \frac{1}{8}\pi ab^3$$

$$I_3 = \frac{1}{8}\pi a^3b$$

* QUARTER ELLIPSE

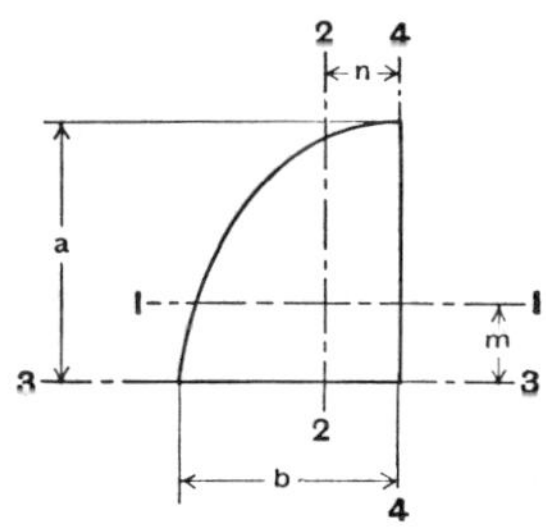

$$A = \frac{1}{4}\pi ab$$

$$m = \frac{4a}{3\pi}$$

$$n = \frac{4b}{3\pi}$$

$$I_1 = a^3b\left(\frac{\pi}{16} - \frac{4}{9\pi}\right)$$

$$I_2 = ab^3\left(\frac{\pi}{16} - \frac{4}{9\pi}\right)$$

$$I_3 = \frac{1}{16}\pi a^3b$$

$$I_4 = \frac{1}{16}\pi ab^3$$

* ELLIPTIC COMPLEMENT

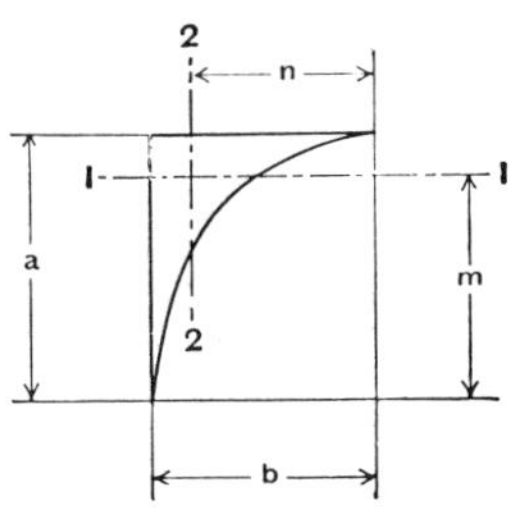

$$A = ab\left(1 - \frac{\pi}{4}\right)$$

$$m = \frac{a}{6\left(1 - \frac{\pi}{4}\right)}$$

$$n = \frac{b}{6\left(1 - \frac{\pi}{4}\right)}$$

$$I_1 = a^3b\left(\frac{1}{3} - \frac{\pi}{16} - \frac{1}{36\left(1 - \frac{\pi}{4}\right)}\right)$$

$$I_2 = ab^3\left(\frac{1}{3} - \frac{\pi}{16} - \frac{1}{36\left(1 - \frac{\pi}{4}\right)}\right)$$

* To obtain properties of half circle, quarter circle and circular complement substitute a = b = R.

PROPERTIES OF GEOMETRIC SECTIONS AND STRUCTURAL SHAPES

REGULAR POLYGON

Axis of moments through center

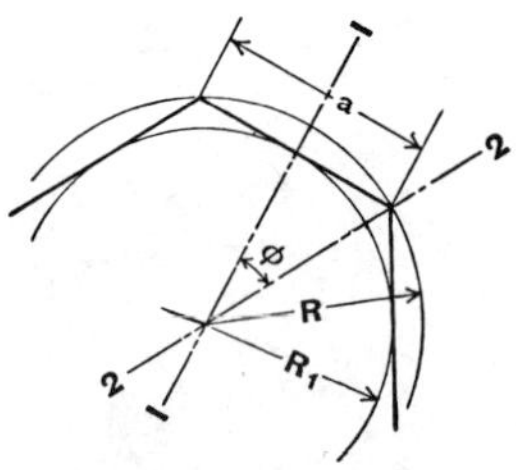

$$n = \text{Number of sides}$$

$$\phi = \frac{180^\circ}{n}$$

$$a = 2\sqrt{R^2 - R_1^2}$$

$$R = \frac{a}{2\sin\phi}$$

$$R_1 = \frac{a}{2\tan\phi}$$

$$A = \frac{1}{4}na^2\cot\phi = \frac{1}{2}nR^2\sin 2\phi = nR_1^2\tan\phi$$

$$I_1 = I_2 = \frac{A(6R^2 - a^2)}{24} = \frac{A(12R_1^2 + a^2)}{48}$$

$$r_1 = r_2 = \sqrt{\frac{6R^2 - a^2}{24}} = \sqrt{\frac{12R_1^2 + a^2}{48}}$$

ANGLE

Axis of moments through center of gravity

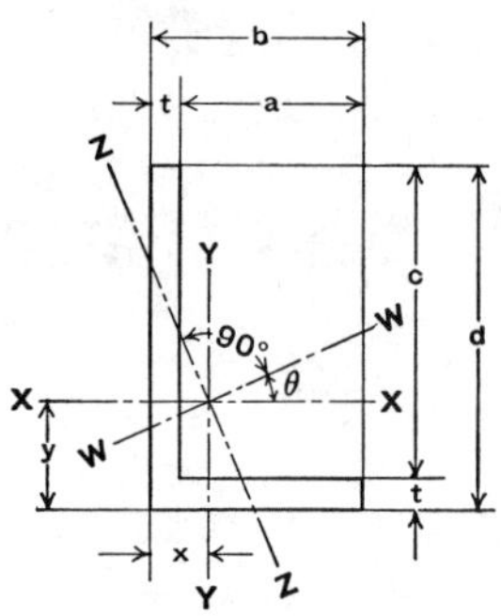

Z-Z is axis of minimum I

$$\tan 2\theta = \frac{2K}{I_Y - I_X}$$

$$A = t(b + c) \quad x = \frac{b^2 + ct}{2(b + c)} \quad y = \frac{d^2 + at}{2(b + c)}$$

K = Product of Inertia about X-X & Y-Y

$$= \mp \frac{abcdt}{4(b + c)}$$

$$I_X = \frac{1}{3}\left(t(d - y)^3 + by^3 - a(y - t)^3\right)$$

$$I_Y = \frac{1}{3}\left(t(b - x)^3 + dx^3 - c(x - t)^3\right)$$

$$I_Z = I_X\sin^2\theta + I_Y\cos^2\theta + K\sin 2\theta$$

$$I_W = I_X\cos^2\theta + I_Y\sin^2\theta - K\sin 2\theta$$

K is negative when heel of angle, with respect to c. g., is in 1st or 3rd quadrant, positive when in 2nd or 4th quadrant.

BEAMS AND CHANNELS

Transverse force oblique through center of gravity

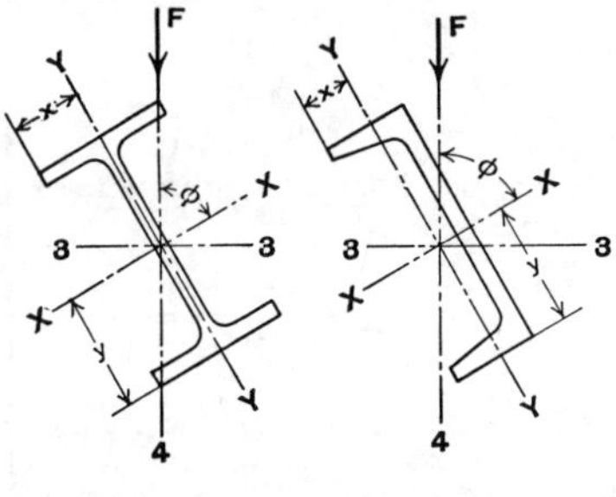

$$I_3 = I_X\sin^2\phi + I_Y\cos^2\phi$$

$$I_4 = I_X\cos^2\phi + I_Y\sin^2\phi$$

$$f_b = M\left(\frac{y}{I_X}\sin\phi + \frac{x}{I_Y}\cos\phi\right)$$

where M is bending moment due to force F.

TRIGONOMETRIC FORMULAS

TRIGONOMETRIC FUNCTIONS

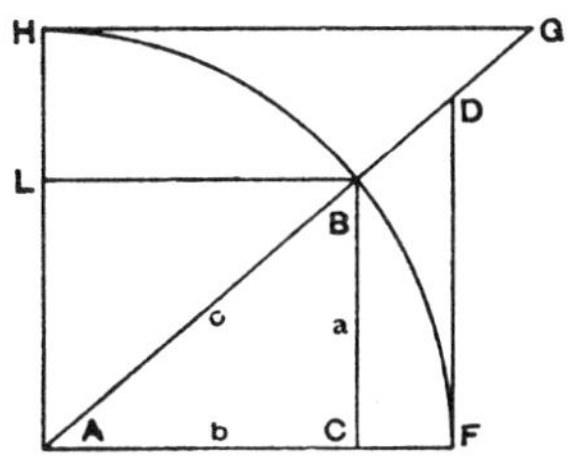

Radius AF $= 1$
$= \sin^2 A + \cos^2 A = \sin A \operatorname{cosec} A$
$= \cos A \sec A = \tan A \cot A$

Sine A $= \dfrac{\cos A}{\cot A} = \dfrac{1}{\operatorname{cosec} A} = \cos A \tan A = \sqrt{1-\cos^2 A} = BC$

Cosine A $= \dfrac{\sin A}{\tan A} = \dfrac{1}{\sec A} = \sin A \cot A = \sqrt{1-\sin^2 A} = AC$

Tangent A $= \dfrac{\sin A}{\cos A} = \dfrac{1}{\cot A} = \sin A \sec A = FD$

Cotangent A $= \dfrac{\cos A}{\sin A} = \dfrac{1}{\tan A} = \cos A \operatorname{cosec} A = HG$

Secant A $= \dfrac{\tan A}{\sin A} = \dfrac{1}{\cos A} = AD$

Cosecant A $= \dfrac{\cot A}{\cos A} = \dfrac{1}{\sin A} = AG$

RIGHT ANGLED TRIANGLES

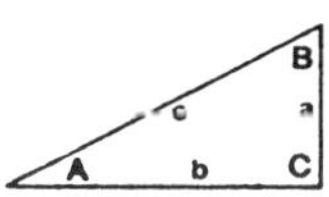

$$a^2 = c^2 - b^2$$
$$b^2 = c^2 - a^2$$
$$c^2 = a^2 + b^2$$

Known	Required: A	B	a	b	c	Area
a, b	$\tan A = \frac{a}{b}$	$\tan B = \frac{b}{a}$			$\sqrt{a^2 + b^2}$	$\frac{ab}{2}$
a, c	$\sin A = \frac{a}{c}$	$\cos B = \frac{a}{c}$		$\sqrt{c^2 - a^2}$		$\frac{a\sqrt{c^2-a^2}}{2}$
A, a		$90° - A$		$a \cot A$	$\frac{a}{\sin A}$	$\frac{a^2 \cot A}{2}$
A, b		$90° - A$	$b \tan A$		$\frac{b}{\cos A}$	$\frac{b^2 \tan A}{2}$
A, c		$90° - A$	$c \sin A$	$c \cos A$		$\frac{c^2 \sin 2A}{4}$

OBLIQUE ANGLED TRIANGLES

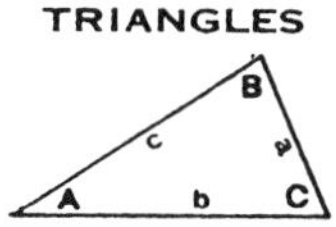

$$s = \frac{a+b+c}{2}$$

$$K = \sqrt{\frac{(s-a)(s-b)(s-c)}{s}}$$

$$a^2 = b^2 + c^2 - 2bc \cos A$$
$$b^2 = a^2 + c^2 - 2ac \cos B$$
$$c^2 = a^2 + b^2 - 2ab \cos C$$

Known	Required: A	B	C	b	c	Area
a, b, c	$\tan \frac{1}{2} A = \frac{K}{s-a}$	$\tan \frac{1}{2} B = \frac{K}{s-b}$	$\tan \frac{1}{2} C = \frac{K}{s-c}$			$\sqrt{s(s-a)(s-b)(s-c)}$
a, A, B			$180° - (A+B)$	$\frac{a \sin B}{\sin A}$	$\frac{a \sin C}{\sin A}$	
a, b, A		$\sin B = \frac{b \sin A}{a}$			$\frac{b \sin C}{\sin B}$	
a, b, C	$\tan A = \frac{a \sin C}{b - a \cos C}$				$\sqrt{a^2+b^2-2ab\cos C}$	$\frac{ab \sin C}{2}$

APPENDIX D

Modulus of Elasticity of Building Materials

Concrete:

$$E_c = 33W^{1.5}\sqrt{f_c},$$

where E_c is the modulus of elasticity of concrete (psi), f_c is the ultimate strength of concrete at 28 days (psi), and W is the unit weight of concrete (pcf).

Steel:

$$E_s = 29 \text{ million psi (for all grades).}$$

Aluminum:

$$E = 10{,}600{,}000 \text{ psi}$$

for 2014-T6 structural alloy, and

$$E = 10{,}600{,}000 \text{ psi}$$

for 6061-T6 structural alloy.

Iron:

$$E = 28{,}000{,}000 \text{ psi}$$

for cast gray wrought iron.

Clay Brick:

$$E = 2{,}000{,}000 \text{ psi.}$$

Bluestone:

$$E = 7{,}000{,}000 \text{ psi.}$$

Granite/Gneiss:

$$E = 7{,}000{,}000 \text{ psi.}$$

Limestone/Marble:

$$E = 7{,}000{,}000 \text{ psi.}$$

Sandstone:

$$E = 3{,}000{,}000 \text{ psi.}$$

Slate:

$$E = 14{,}000{,}000 \text{ psi.}$$

APPENDIX E

Modulus of Elasticity for Design of Wood Members

Species and commercial grade	Size classification	E
Aspen (surfaced dry or surfaced green. Used at 19% max. m.c.)		
Select structural		1,100,000
No. 1	2" to 4" thick	1,100,000
No. 2	2" to 4" wide	1,000,000
No. 3		900,000
Appearance		1,100,000
Stud		900,000
Construction	2" to 4" thick	900,000
Standard	4" wide	900,000
Utility		900,000
Select structural		1,100,000
No. 1	2" to 4" thick	1,100,000
No. 2	5" and wider	1,000,000
No. 3		900,000
Appearance		1,100,000
Stud		900,000
Balsam Fir (Surfaced dry or surfaced green. Used at 19% max. m.c.)		
Select structural	2" to 4" thick	1,500,000
No. 1	2" to 4" wide	1,500,000
No. 2		1,300,000
No. 3		1,200,000
Appearance		1,500,000
Stud		1,200,000
Construction	2" to 4" thick	1,200,000
Standard	4" wide	1,200,000
Utility		1,200,000

Balsam Fir, cont'd		
Select structural	2" to 4" thick	1,500,000
No. 1	5" and wider	1,500,000
No. 2		1,300,000
No. 3		1,200,000
Appearance		1,500,000
Stud		1,200,000
Select structural	Beams and stringers	1,400,000
No. 1		1,400,000
Select structural	Posts and timbers	1,400,000
No. 1		1,400,000
Select decking	Decking	1,500,000
Commercial decking		1,300,000
Black Cottonwood (Surfaced dry or surfaced green. Used at 19% max. m.c.)		
Select structural		1,200,000
No. 1	2" to 3" thick	1,200,000
No. 2	2" to 4" wide	1,100,000
No. 3		900,000
Appearance		1,200,000
Stud		900,000
Construction	2" to 4" thick	900,000
Standard	4" wide	900,000
Utility		900,000
Select structural		1,200,000
No. 1	2" to 4" thick	1,200,000
No. 2	5" and wider	1,100,000
No. 3		900,000
Appearance		1,200,000
Stud		900,000
California Redwood (Surfaced dry or surfaced green. Used at 19% max. m.c.)		
Clear heart structural	≤4" thick	1,400,000
Clear structural	any width	1,400,000
Select structural		1,400,000
Select structural, open grain		1,100,000
No. 1	2" to 4" thick	1,400,000
No. 1, open grain		1,100,000
No. 2	2" to 4" wide	1,250,000
No. 2, open grain		1,000,000
No. 3		1,100,000
No. 3, open grain		900,000
Stud		900,000
Construction	2" to 4" thick	900,000
Standard		900,000
Utility	4" wide	900,000

Select structural		1,400,000
Select structural, open grain		1,100,000
No. 1	2" to 4" thick	1,400,000
No. 1, open grain		1,100,000
No. 2	5" and wider	1,250,000
No. 2, open grain		1,000,000
No. 3		1,100,000
No. 3, open grain		900,000
Stud		1,100,000
Clear heart structural or Clear structural		1,300,000
Select structural	5" by 5" and larger	1,300,000
No. 1		1,300,000
No. 2		1,100,000
No. 3		1,000,000
Select decking, close grain	Decking	1,400,000
Select decking	2" thick	1,100,000
Commercial decking	6" and wider	1,000,000

Coast Sitka Spruce (Surfaced dry or surfaced green. Used at 19% max. m.c.)

Select structural		1,700,000
No. 1	2" to 3" thick	1,700,000
No. 2	2" to 4" wide	1,500,000
No. 3		1,300,000
Appearance		1,700,000
Stud		1,300,000
Construction	2" to 4" thick	1,300,000
Standard	4" wide	1,300,000
Utility		1,300,000
Select structural	2" to 4" thick	1,700,000
No. 1	5" and wider	1,700,000
No. 2		1,500,000
No. 3		1,300,000
Appearance		1,700,000
Stud		1,300,000
Select structural	Beams and stringers	1,500,000
No. 1		1,500,000
Select structural	Posts and timbers	1,500,000
No. 1		1,500,000
Select decking	Decking	1,700,000
Commercial decking		1,500,000

Coast Species (Surfaced dry or surfaced green. Used at 19% max. m.c.)

Select structural		1,500,000
No. 1	2" to 3" thick	1,500,000
No. 2	2" to 4" wide	1,400,000
No. 3		1,200,000
Appearance		1,500,000
Stud		1,200,000

Coast Species, cont'd		
Construction	2" to 4" thick	1,200,000
Standard	4" wide	1,200,000
Utility		1,200,000
Select structural		1,500,000
No. 1	2" to 4" thick	1,500,000
No. 2	5" and wider	1,400,000
No. 3		1,200,000
Appearance		1,500,000
Stud		1,200,000
Select decking	Decking	1,500,000
Commercial decking		1,400,000
Cottonwood (Surfaced dry or surfaced green. Used at 19% max. m.c.)		
Stud	2" to 3" thick, 2" to 4" wide	1,000,000
Construction	2" to 4" thick	1,000,000
Standard	4" wide	1,000,000
Utility		1,000,000
Douglas Fir-Larch (Surfaced dry or surfaced green. Used at 19% max. m.c.)		
Dense select structural		1,900,000
Select structural		1,800,000
Dense No. 1		1,900,000
No. 1	2" to 4" thick	1,800,000
Dense No. 2	2" to 4" wide	1,700,000
No. 2		1,700,000
No. 3		1,500,000
Appearance		1,800,000
Stud		1,500,000
Construction	2" to 4" thick	1,500,000
Standard	4" wide	1,500,000
Utility		1,500,000
Dense select structural		1,900,000
Select structural		1,800,000
Dense No. 1	2" to 4" thick	1,900,000
No. 1	5" and wider	1,800,000
Dense No. 2		1,700,000
No. 2		1,700,000
No. 3		1,500,000
Appearance		1,800,000
Stud		1,500,000
Dense select structural		1,700,000
Select structural	Beams and stringers	1,600,000
Dense No. 1		1,700,000
No. 1		1,600,000
Dense select structural		1,700,000
Select structural	Posts and timbers	1,600,000
Dense No. 1		1,700,000
No. 1		1,600,000

Select Dex	Decking	1,800,000
Commercial Dex		1,700,000
Dense select structural		1,700,000
Select structural	Beams and stringers	1,600,000
Dense No. 1		1,700,000
No. 1		1,600,000
Dense select structural		1,700,000
Select structural	Posts and timbers	1,600,000
Dense No. 1		1,700,000
No. 1		1,600,000
Selected decking	Decking	1,800,000
Commercial decking		1,700,000
Selected decking	Decking	1,900,000
Commercial decking		1,700,000

Douglas Fir-Larch (North) (Surfaced dry or surfaced green. Used at 19% max. m.c.)

Select structural	2" to 3" thick	1,800,000
No. 1	2" to 4" wide	1,800,000
No. 2		1,700,000
No. 3		1,500,000
Appearance		1,800,000
Stud		1,500,000
Construction	2" to 4" thick	1,500,000
Standard	4" wide	1,500,000
Utility		1,500,000
Select structural		1,800,000
No. 1	2" to 4" thick	1,800,000
No. 2	5" and wider	1,700,000
No. 3		1,500,000
Appearance		1,800,000
Stud		1,500,000
Select structural	Beams and stringers	1,600,000
No. 1		1,600,000
Select structural	Posts and timbers	1,600,000
No. 1		1,600,000
Select decking	Decking	1,800,000
Commercial decking		1,700,000

Douglas Fir-Larch (South) (Surfaced dry or surfaced green. Used at 19% max. m.c.)

Select structural	2" to 4" thick	1,400,000
No. 1	2" to 4" wide	1,400,000
No. 2		1,300,000
No. 3		1,100,000
Appearance		1,400,000
Stud		1,100,000

Douglas Fir-Larch (South), cont'd		
Construction	2" to 4" thick	1,100,000
Standard	4" wide	1,100,000
Utility		1,100,000
Select structural		1,400,000
No. 1	2" to 4" thick	1,400,000
No. 2	5" and wider	1,300,000
No. 3		1,100,000
Appearance		1,400,000
Stud		1,100,000
Select structural	Beams and stringers	1,200,000
No. 1		1,200,000
Select structural	Posts and timbers	1,200,000
No. 1		1,200,000
Selected decking	Decking	1,400,000
Commercial decking		1,300,000
Selected decking	Decking	1,500,000
Commercial decking		1,300,000
Eastern Hemlock (Surfaced dry or surfaced green. Used at 19% max. m.c.)		
Select structural	2" to 4" thick	1,200,000
No. 1	2" to 4" wide	1,200,000
No. 2		1,100,000
No. 3		1,000,000
Appearance		1,200,000
Stud		1,000,000
Construction	2" to 4" thick	1,000,000
Standard	4" wide	1,000,000
Utility		1,000,000
Select structural		1,200,000
No. 1	2" to 4" thick	1,200,000
No. 2	5" and wider	1,100,000
No. 3		1,000,000
Appearance		1,200,000
Stud		1,000,000
Select structural	Beams and stringers	1,200,000
No. 1		1,200,000
Select structural	Posts and timbers	1,200,000
No. 1		1,200,000
Eastern Hemlock-Tamarack (Surfaced dry or surfaced green. Used at 19% max. m.c.)		
Select structural	2" to 4" thick	1,300,000
No. 1	2" to 4" wide	1,300,000
No. 2		1,100,000
No. 3		1,000,000
Appearance		1,300,000
Stud		1,000,000

Construction	2" to 4" thick	1,000,000
Standard	4" wide	1,000,000
Utility		1,000,000
Select structural		1,300,000
No. 1	2" to 4" thick	1,300,000
No. 2	5" and wider	1,100,000
No. 3		1,000,000
Appearance		1,300,000
Stud		1,000,000
Select structural	Beams and stringers	1,200,000
No. 1		1,200,000
Select structural	Posts and timbers	1,200,000
No. 1		1,200,000
Select decking	Decking	1,300,000
Commercial decking		1,100,000

Eastern Hemlock-Tamarack (North) (Surfaced dry or surfaced green. Used at 19% max. m.c.)

Select structural	2" to 3" thick	1,300,000
No. 1	2" to 4" wide	1,300,000
No. 2		1,100,000
No. 3		1,000,000
Appearance		1,300,000
Stud		1,000,000
Construction	2" to 4" thick	1,000,000
Standard	4" wide	1,000,000
Utility		1,000,000
Select structural		1,300,000
No. 1	2" to 4" thick	1,300,000
No. 2	5" and wider	1,100,000
No. 3		1,000,000
Appearance		1,300,000
Stud		1,000,000
Select structural	Beams and stringers	1,300,000
No. 1		1,300,000
Select structural	Posts and timbers	1,300,000
No. 1		1,300,000
Select decking	Decking	1,300,000
Commercial decking		1,100,000

Eastern Softwoods (Surfaced dry or surfaced green. Used at 19% max. m.c.)

Select structural	2" to 4" thick	1,200,000
No. 1	2" to 4" wide	1,200,000
No. 2		1,100,000
No. 3		1,000,000

Eastern Softwoods, cont'd		
Stud	2" to 4" thick, 2" to 4" wide	1,000,000
Construction	2" to 4" thick	1,000,000
Standard	4" wide	1,000,000
Utility		1,000,000
Select structural		1,200,000
No. 1	2" to 4" thick	1,200,000
No. 2	5" and wider	1,100,000
No. 3		1,000,000
Appearance		1,200,000
Stud	2" to 4" thick, 5" and wider	1,000,000
Eastern Spruce (Surfaced dry or surfaced green. Used at 19% max. m.c.)		
Select structural	2" to 4" thick	1,500,000
No. 1	2" to 4" wide	1,500,000
No. 2		1,400,000
No. 3		1,200,000
Appearance		1,500,000
Stud		1,200,000
Construction	2" to 4" thick	1,200,000
Standard	4" wide	1,200,000
Utility		1,200,000
Select structural		1,500,000
No. 1	2" to 4" thick	1,500,000
No. 2	5" and wider	1,400,000
No. 3		1,200,000
Appearance		1,500,000
Stud		1,200,000
Select structural	Beams and stringers	1,400,000
No. 1		1,400,000
Select structural	Posts and timbers	1,400,000
No. 1		1,400,000
Select decking	Decking	1,500,000
Commercial decking		1,400,000
Eastern White Pine (Surfaced dry or surfaced green. Used at 19% max. m.c.)		
Select structural	2" to 4" thick	1,200,000
No. 1	2" to 4" wide	1,200,000
No. 2		1,100,000
No. 3		1,000,000
Appearance		1,200,000
Stud		1,000,000
Construction	2" to 4" thick	1,000,000
Standard	4" wide	1,000,000
Utility		1,000,000

Select structural		1,200,000
No. 1	2" to 4" thick	1,200,000
No. 2	5" and wider	1,100,000
No. 3		1,000,000
Appearance		1,200,000
Stud		1,000,000
Select structural	Beams and stringers	1,100,000
No. 1		1,100,000
Select structural	Posts and timbers	1,100,000
No. 1		1,100,000
Select decking	Decking	1,200,000
Commercial decking		1,100,000

Eastern White Pine (North) (Surfaced dry or surfaced green. Used at 19% max. m.c.)

Select structural		1,200,000
No. 1	2" to 3" thick	1,200,000
No. 2	2" to 4" wide	1,100,000
No. 3		1,000,000
Appearance		1,200,000
Stud		1,000,000
Construction	2" to 4" thick	1,000,000
Standard	4" wide	1,000,000
Utility		1,000,000
Select structural		1,200,000
No. 1	2" to 4" thick	1,200,000
No. 2	5" and wider	1,100,000
No. 3		1,000,000
Appearance		1,200,000
Stud		1,000,000
Select decking	Decking	1,200,000
Commercial decking		1,100,000

Eastern Woods (Surfaced dry or surfaced green. Used at 19% max. m.c.)

Select structural		1,100,000
No. 1	2" to 4" thick	1,100,000
No. 2	2" to 4" wide	1,000,000
No. 3		900,000
Appearance		1,100,000
Stud		900,000
Construction	2" to 4" thick	900,000
Standard	4" wide	900,000
Utility		900,000
Select structural		1,100,000
No. 1	2" to 4" thick	1,100,000
No. 2	2" to 4" wider	1,000,000
No. 3		900,000
Appearance		1,100,000
Stud		900,000

Engelmann Spruce-Alpine Fir (Engelmann Spruce-Lodgepole Pine)		
Select structural		1,300,000
No. 1	2" to 4" thick	1,300,000
No. 2	2" to 4" wide	1,100,000
No. 3		1,000,000
Appearance		1,300,000
Stud		1,000,000
Construction	2" to 4" thick	1,000,000
Standard	4" wide	1,000,000
Utility		1,000,000
Select structural		1,300,000
No. 1	2" to 4" thick	1,300,000
No. 2	5" and wider	1,100,000
No. 3		1,000,000
Appearance		1,300,000
Stud		1,000,000
Select structural	Beams and stringers	1,100,000
No. 1		1,100,000
Select structural	Posts and timbers	1,100,000
No. 1		1,100,000
Select decking	Decking	1,300,000
Commercial decking		1,100,000
Selected decking	Decking	1,300,000
Commercial decking		1,200,000
Hem-Fir (Surfaced dry or surfaced green. Used at 19% max. m.c.)		
Select structural		1,500,000
No. 1	2" to 4" thick	1,500,000
No. 2	2" to 4" wide	1,400,000
No. 3		1,200,000
Appearance		1,500,000
Stud		1,200,000
Construction	2" to 4" thick	1,200,000
Standard	4" wide	1,200,000
Utility		1,200,000
Select structural		1,500,000
No. 1	2" to 4" thick	1,500,000
No. 2	5" and wider	1,400,000
No. 3		1,200,000
Appearance		1,500,000
Stud		1,200,000
Select structural	Beams and stringers	1,300,000
No. 1		1,300,000
Select structural	Posts and timbers	1,300,000
No. 1		1,300,000

Select Dex	Decking	1,500,000
Commercial Dex		1,400,000
Select structural	Beams and stringers	1,300,000
No. 1		1,300,000
Select structural	Posts and timbers	1,300,000
No. 1		1,300,000
Selected decking	Decking	1,500,000
Commercial decking		1,400,000
Selected decking	Decking	1,600,000
Commercial decking		1,400,000
Hem-Fir (North) (Surfaced dry or surfaced green. Used at 19% max. m.c.)		
Select structural		1,600,000
Commercial decking	2" to 3" thick	1,500,000
No. 1	2" to 4" wide	1,500,000
No. 2		1,400,000
No. 3		1,200,000
Appearance		1,500,000
Stud		1,200,000
Construction	2" to 4" thick	1,200,000
Standard	4" wide	1,200,000
Utility		1,200,000
Select structural		1,500,000
No. 1	2" to 4" thick	1,500,000
No. 2	5" and wider	1,400,000
No. 3		1,200,000
Appearance		1,500,000
Stud		1,200,000
Select structural	Beams and stringers	1,300,000
No. 1		1,300,000
Select structural	Posts and timbers	1,300,000
No. 1		1,300,000
Select decking	Decking	1,500,000
Commercial decking		1,400,000
Idaho White Pine (Surfaced dry or surfaced green. Used at 19% max. m.c.)		
Select structural	2" to 4" thick	1,400,000
No. 1	2" to 4" wide	1,400,000
No. 2		1,300,000
No. 3		1,200,000
Appearance		1,400,000
Stud		1,200,000
Construction	2" to 4" thick	1,200,000
Standard	4" wide	1,200,000
Utility		1,200,000

Idaho White Pine, cont'd		
Select structural		1,400,000
No. 1	2" to 4" thick	1,400,000
No. 2	5" and wider	1,300,000
No. 3		1,200,000
Appearance		1,400,000
Stud		1,200,000
Select structural	Beams and stringers	1,300,000
No. 1		1,300,000
Select structural	Posts and timbers	1,300,000
No. 1		1,300,000
Selected decking	Decking	1,400,000
Commercial decking		1,300,000
Selected decking	Decking	1,500,000
Commercial decking		1,400,000
Lodgepole Pine (Surfaced dry or surfaced green. Used at 19% max. m.c.)		
Select structural	2" to 4" thick	1,300,000
No. 1	2" to 4" wide	1,300,000
No. 2		1,200,000
No. 3		1,000,000
Appearance		1,300,000
Stud		1,000,000
Construction	2" to 4" thick	1,000,000
Standard	4" wide	1,000,000
Utility		1,000,000
Select structural		1,300,000
No. 1	2" to 4" thick	1,300,000
No. 2	5" and wider	1,200,000
No. 3		1,000,000
Appearance		1,300,000
Stud		1,000,000
Select structural	Beams and stringers	1,100,000
No. 1		1,100,000
Select structural	Posts and timbers	1,100,000
No. 1		1,100,000
Selected decking	Decking	1,300,000
Commercial decking		1,200,000
Selected decking	Decking	1,400,000
Commercial decking		1,200,000
Mountain Hemlock (Surfaced dry or surfaced green. Used at 19% max. m.c.)		
Select structural	2" to 4" thick	1,300,000
No. 1	2" to 4" wide	1,300,000
No. 2		1,100,000
No. 3		1,000,000
Appearance		1,300,000
Stud		1,000,000

Construction	2" to 4" thick	1,000,000
Standard	4" wide	1,000,000
Utility		1,000,000
Select structural		1,300,000
No. 1	2" to 4" thick	1,300,000
No. 2	5" and wider	1,100,000
No. 3		1,000,000
Appearance		1,300,000
Stud		1,000,000
Select structural	Beams and stringers	1,100,000
No. 1		1,100,000
Select structural	Posts and timbers	1,100,000
No. 1		1,100,000
Select Dex	Decking	1,300,000
Commercial Dex		1,100,000
Select structural	Beams and stringers	1,100,000
No. 1		1,100,000
Select structural	Posts and timbers	1,100,000
No. 1		1,100,000
Selected decking	Decking	1,300,000
Commercial decking		1,100,000
Selected decking	Decking	1,300,000
Commercial decking		1,200,000

Mountain Hemlock-Hem-Fir (Surfaced dry or surfaced green. Used at 19% max. m.c.)

Select structural	2" to 4" thick	1,300,000
No. 1	2" to 4" wide	1,300,000
No. 2		1,100,000
No. 3		1,000,000
Appearance		1,300,000
Stud		1,000,000
Construction	2" to 4" thick	1,000,000
Standard	4" wide	1,000,000
Utility		1,000,000
Select structural		1,300,000
No. 1	2" to 4" thick	1,300,000
No. 2	5" and wider	1,100,000
No. 3		1,000,000
Appearance		1,300,000
Stud		1,000,000
Select structural	Beams and stringers	1,100,000
No. 1		1,100,000
Select structural	Posts and timbers	1,100,000
No. 1		1,100,000

Mountain Hemlock-Hem-Fir, cont'd		
Selected decking	Decking	1,300,000
Commercial decking		1,100,000
Selected decking	Decking	1,300,000
Commercial decking		1,200,000
Northern Aspen (Surfaced dry or surfaced green. Used at 19% max. m.c.)		
Select structural	2" to 3" thick	1,400,000
No. 1	2" to 4" wide	1,400,000
No. 2		1,200,000
No. 3		1,100,000
Appearance		1,400,000
Stud		1,100,000
Construction	2" to 4" thick	1,100,000
Standard	4" wide	1,100,000
Utility		1,100,000
Select structural		1,400,000
No. 1	2" to 4" thick	1,400,000
No. 2	5" and wider	1,200,000
No. 3		1,100,000
Appearance		1,400,000
Stud		1,100,000
Northern Pine (Surfaced dry or surfaced green. Used at 19% max. m.c.)		
Select structural	2" to 4" thick	1,400,000
No. 1	2" to 4" wide	1,400,000
No. 2		1,300,000
No. 3		1,100,000
Appearance		1,400,000
Stud		1,100,000
Construction	2" to 4" thick	1,100,000
Standard	4" wide	1,100,000
Utility		1,100,000
Select structural		1,400,000
No. 1	2" to 4" thick	1,400,000
No. 2	5" and wider	1,300,000
No. 3		,100,000
Appearance		1,400,000
Stud		1,100,000
Select structural	Beams and stringers	1,300,000
No. 1		1,300,000
Select structural	Posts and timbers	1,300,000
No. 1		1,300,000
Select decking	Decking	1,400,000
Commercial decking		1,300,000

Northern Species (Surfaced dry or surfaced green. Used at 19% max. m.c.)

Select structural	2" to 3" thick	1,100,000
No. 1	2" to 4" wide	1,100,000
No. 2		1,000,000
No. 3		900,000
Appearance		1,100,000
Stud		900,000
Construction	2" to 4" thick	900,000
Standard	4" wide	900,000
Utility		900,000
Select structural		1,100,000
No. 1	2" to 4" thick	1,100,000
No. 2	5" and wider	1,000,000
No. 3		900,000
Appearance		1,100,000
Stud		900,000
Select decking	Decking	1,100,000
Commercial decking		1,000,000

Northern White Cedar (Surfaced dry or surfaced green. Used at 19% max. m.c.)

Select structural	2" to 4" thick	800,000
No. 1	2" to 4" wide	800,000
No. 2		700,000
No. 3		600,000
Appearance		800,000
Stud		600,000
Construction	2" to 4" thick	600,000
Standard	4" wide	600,000
Utility		600,000
Select structural		800,000
No. 1	2" to 4" thick	800,000
No. 2	5" and wider	700,000
No. 3		600,000
Appearance		800,000
Stud		600,000
Select structural	Beams and stringers	700,000
No. 1		700,000
Select structural	Posts and timbers	700,000
No. 1		700,000
Select decking	Decking	800,000
Commercial decking		700,000

Ponderosa Pine (Surfaced dry or surfaced green. Used at 19% max. m.c.)

Select structural	2" to 3" thick	1,200,000
No. 1	2" to 4" wide	1,200,000
No. 2		1,100,000
No. 3		1,000,000
Appearance		1,200,000
Stud		1,000,000

Ponderosa Pine, cont'd		
Construction	2" to 4" thick	1,000,000
Standard	4" wide	1,000,000
Utility		1,000,000
Select structural		1,200,000
No. 1	2" to 4" thick	1,200,000
No. 2	5" and wider	1,100,000
No. 3		1,000,000
Appearance		1,200,000
Stud		1,000,000
Select structural	Beams and stringers	1,100,000
No. 1		1,100,000
Select structural	Posts and timbers	1,100,000
No. 1		1,100,000
Select decking	Decking	1,300,000
Commercial decking		1,100,000
Ponderosa Pine-Sugar Pine (Ponderosa Pine-Lodgepole Pine) (Surfaced dry or surfaced green. Used at 19% max. m.c.)		
Select structural	2" to 4" thick	1,200,000
No. 1	2" to 4" wide	1,200,000
No. 2		1,100,000
No. 3		1,000,000
Appearance		1,200,000
Stud		1,000,000
Construction	2" to 4" thick	1,000,000
Standard	4" wide	1,000,000
Utility		1,000,000
Select structural		1,200,000
No. 1	2" to 4" thick	1,200,000
No. 2	5" and wider	1,100,000
No. 3		1,000,000
Appearance		1,200,000
Stud		1,000,000
Select structural	Beams and stringers	1,100,000
No. 1		1,100,000
Select structural	Posts and timbers	1,100,000
No. 1		1,100,000
Selected decking	Decking	1,100,000
Commercial decking		1,100,000
Selected decking	Decking	1,300,000
Commercial decking		1,100,000

Red Pine (Surfaced dry or surfaced green. Used at 19% max. m.c.)

Select structural	2" to 3" thick	1,300,000
No. 1	2" to 4" wide	1,300,000
No. 2		1,200,000
No. 3		1,000,000
Appearance		1,300,000
Stud		1,000,000
Construction	2" to 4" thick	1,000,000
Standard	4" wide	1,000,000
Utility		1,000,000
Select structural		1,300,000
No. 1	2" to 4" thick	1,300,000
No. 2	5" and wider	1,200,000
No. 3		1,000,000
Appearance		1,300,000
Stud		1,000,000
Select structural	Beams and stringers	1,100,000
No. 1		1,100,000
Select structural	Posts and timbers	1,100,000
No. 1		1,100,000
Select decking	Decking	1,300,000
Commercial decking		1,200,000

Sitka Spruce (Surfaced dry or surfaced green. Used at 19% max. m.c.)

Select structural	1,500,000	
No. 1	2" to 4" thick	1,500,000
No. 2	2" to 4" wide	1,300,000
No. 3		1,200,000
Appearance		1,500,000
Stud		1,200,000
Construction	2" to 4" thick	1,200,000
Standard	4" wide	1,200,000
Utility		1,200,000
Select structural		1,500,000
No. 1	2" to 4" thick	1,500,000
No. 2	5" and wider	1,300,000
No. 3		1,200,000
Appearance		1,500,000
Stud		1,200,000
Select structural	Beams and stringers	1,300,000
No. 1		1,300,000
Select structural	Posts and timbers	1,300,000
No. 1		1,300,000
Select Dex	Decking	1,500,000
Commercial Dex		1,300,000

Southern Pine (Surfaced at 15% maximum moisture content, K.D. Used at 15% max.)		
Select structural		1,800,000
Dense select structural		1,900,000
No. 1		1,800,000
No. 1 Dense	2" to 4" thick	1,900,000
No. 2	2" to 4" wide	1,600,000
No. 2 Dense		1,700,000
No. 3		1,500,000
No. 3 Dense		1,500,000
Stud		1,500,000
Construction	2" to 4" thick	1,500,000
Standard	4" wide	1,500,000
Utility		1,500,000
Select structural		1,800,000
Dense select structural		1,900,000
No. 1		1,800,000
No. 1 Dense	2" to 4" thick	1,900,000
No. 2	5" and wider	1,600,000
No. 2 Dense		1,700,000
No. 3		1,600,000
No. 3 Dense		1,500,000
Stud		1,500,000
Dense standard decking	2" to 4" thick	1,900,000
Select decking	2" and wider	1,600,000
Dense select decking		1,700,000
Commercial decking		1,600,000
Dense commercial decking	Decking	1,700,000
Dense structural 86	2" to 4" thick	1,900,000
Dense structural 72		1,900,000
Dense structural 65		1,900,000
Southern Pine (Surfaced dry. Used at 19% max. m.c.)		
Select structural		1,700,000
Dense select structural		1,800,000
No. 1		1,700,000
No. 1 Dense	2" to 4" thick	1,800,000
No. 2	2" to 4" wide	1,600,000
No. 2 Dense		1,600,000
No. 3		1,400,000
No. 3 Dense		1,500,000
Stud		1,400,000
Construction	2" to 4" thick	1,400,000
Standard	4" wide	1,400,000
Utility		1,400,000
Select structural		1,700,000
Dense select structural		1,800,000
No. 1		1,700,000
No. 1 Dense	2" to 4" thick	1,800,000
No. 2	5" and wider	1,600,000

No. 2 Dense		1,600,000
No. 3		1,400,000
No. 3 Dense		1,500,000
Stud		1,400,000
Dense standard decking	2" to 4" thick	1,800,000
Select decking	2" and wider	1,600,000
Dense select decking		1,600,000
Commercial decking		1,600,000
Dense commercial decking	Decking	1,600,000
Dense structural 86	2" to 4" thick	1,800,000
Dense structural 72		1,800,000
Dense structural 65		1,800,000
Southern Pine (Surfaced green. Used any condition.)		
Select structural		1,500,000
Dense select structural		1,600,000
No. 1		1,500,000
No. 1 Dense	2-1/2" to 4" thick	1,600,000
No. 2	2-1/2" to 4" wide	1,400,000
No. 2 Dense		1,400,000
No. 3		1,200,000
No. 3 Dense		1,300,000
Stud		1,200,000
Construction	2-1/2" to 4" thick	1,200,000
Standard	4" wide	1,200,000
Utility		1,200,000
Select structural		1,500,000
Dense select structural		1,600,000
No. 1		1,500,000
No. 1 Dense	2-1/2" to 4" thick	1,600,000
No. 2	5" and wider	1,400,000
No. 2 Dense		1,400,000
No. 3		1,200,000
No. 3 Dense		1,300,000
Stud		1,200,000
Dense standard decking	2-1/2" to 4" thick	1,600,000
Select decking	2" and wider	1,400,000
Dense select decking		1,400,000
Commercial decking		1,400,000
Dense commercial decking	Decking	1,400,000
No. 1 SR		1,500,000
No. 1 Dense SR	5" and thicker	1,600,000
No. 2 SR		1,400,000
No. 2 Dense SR		1,400,000
Dense structural 86	2-1/2" and thick	1,600,000
Dense structural 72		1,600,000
Dense structural 65		1,600,000

Spruce-Pine-Fir (Surfaced dry or surfaced green. Used at 19% max. m.c.)

Select structural	2" to 3" thick	1,500,000
No. 1	2" to 4" wide	1,500,000
No. 2		1,300,000
No. 3		1,200,000
Appearance		1,500,000
Stud		1,200,000
Construction	2" to 4" thick	1,200,000
Standard	4" wide	1,200,000
Utility		1,200,000
Select structural		1,500,000
No. 1	2" to 4" thick	1,500,000
No. 2	5" and wider	1,300,000
No. 3		,200,000
Appearance		1,500,000
Stud		1,200,000
Select structural	Beams and stringers	1,300,000
No. 1		1,300,000
Select structural	Posts and timbers	1,300,000
No. 1		1,300,000
Select decking	Decking	1,500,000
Commercial decking		1,300,000

Western Cedars (Surfaced dry or surfaced green. Used at 19% max. m.c.)

Select structural	2" to 4" thick	1,100,000
No. 1	2" to 4" wide	1,100,000
No. 2		1,000,000
No. 3		900,000
Appearance		1,100,000
Stud		900,000
Construction	2" to 4" thick	900,000
Standard	4" wide	900,000
Utility		900,000
Select structural		1,100,000
No. 1	2" to 4" thick	1,100,000
No. 2	5" and wider	1,000,000
No. 3		900,000
Appearance		1,100,000
Stud		900,000
Select structural	Beams and stringers	1,000,000
No. 1		1,000,000
Select structural	Posts and timbers	1,000,000
No. 1		1,000,000
Select Dex	Decking	1,100,000
Commercial Dex		1,000,000

Select structural	Beams and stringers	1,000,000
No. 1		1,000,000
Select structural	Posts and timbers	1,000,000
No. 1		1,000,000
Selected decking	Decking	1,100,000
Commercial decking		1,000,000
Selected decking	Decking	1,100,000
Commercial decking		1,000,000

Western Cedars (North) (Surfaced dry or surfaced green. Used at 19% max. m.c.)

Select structural	2" to 3" thick	1,100,000
No. 1	2" to 4" wide	1,100,000
No. 2		1,000,000
No. 3		900,000
Appearance		1,100,000
Stud		900,000
Construction	2" to 4" thick	900,000
Standard	4" wide	900,000
Utility		900,000
Select structural		1,100,000
No. 1	2" to 4" thick	1,100,000
No. 2	5" and wider	1,000,000
No. 3		900,000
Appearance		1,100,000
Stud		900,000
Select structural	Beams and stringers	1,000,000
No. 1		1,000,000
Select structural	Posts and timbers	1,000,000
No. 1		1,000,000
Select decking	Decking	1,100,000
Commercial decking		1,000,000

Western Hemlock (Surfaced dry or surfaced green. Used at 19% max. m.c.)

Select structural	2" to 4" thick	1,600,000
No. 1	2" to 4" wide	1,600,000
No. 2		1,400,000
No. 3		1,300,000
Appearance		1,600,000
Stud		1,300,000
Construction	2" to 4" thick	1,300,000
Standard	4" wide	1,300,000
Utility		1,300,000
Select structural		1,600,000
No. 1	2" to 4" thick	1,600,000
No. 2	5" and wider	1,400,000
No. 3		1,300,000

Western Hemlock, cont'd		
Appearance		1,600,000
Stud		1,300,000
Select structural	Beams and stringers	1,400,000
No. 1		1,400,000
Select structural	Posts and timbers	1,400,000
No. 1		1,400,000
Select Dex	Decking	1,600,000
Commercial Dex		1,400,000
Select structural	Beams and stringers	1,400,000
No. 1		1,400,000
Select structural	Posts and timbers	1,400,000
No. 1		1,400,000
Selected decking	Decking	1,600,000
Commercial decking		1,400,000
Selected decking	Decking	1,700,000
Commercial decking		1,500,000
Western Hemlock (North) (Surfaced dry or surfaced green. Used at 19% max. m.c.)		
Select structural	2" to 3" thick	1,600,000
No. 1	2" to 4" wide	1,600,000
No. 2		1,400,000
No. 3		1,300,000
Appearance		1,600,000
Stud		1,300,000
Construction	2" to 4" thick	1,300,000
Standard	4" wide	1,300,000
Utility		1,300,000
Select structural		1,600,000
No. 1	2" to 4" thick	1,600,000
No. 2	5" and wider	1,400,000
No. 3		1,300,000
Appearance		1,600,000
Stud		1,300,000
Select structural	Beams and stringers	1,400,000
No. 1		1,400,000
Select structural	Posts and timbers	1,400,000
No. 1		1,400,000
Select decking	Decking	1,600,000
Commercial decking		1,400,000

Western White Pine (Surfaced dry or surfaced green. Used at 19% max. m.c.)

Select structural	2" to 3" thick	1,400,000
No. 1	2" to 4" wide	1,400,000
No. 2		1,300,000
No. 3		1,200,000
Appearance		1,400,000
Stud		1,200,000
Construction	2" to 4" thick	1,200,000
Standard	4" wide	1,200,000
Utility		1,200,000
Select structural		1,400,000
No. 1	2" to 4" thick	1,400,000
No. 2	5" and wider	1,300,000
No. 3		1,200,000
Appearance		1,400,000
Stud		1,200,000
Select structural	Beams and stringers	1,300,000
No. 1		1,300,000
Select structural	Posts and timbers	1,300,000
No. 1		1,300,000
Select decking	Decking	1,400,000
Commercial decking		1,300,000

White Woods (Western Woods) (Surfaced dry or surfaced green. Used at 19% max. m.c.)

Select structural	2" to 4" thick	1,100,000
No. 1	2" to 4" wide	1,100,000
No. 2		1,000,000
No. 3		900,000
Appearance		1,100,000
Stud		900,000
Construction	2" to 4" thick	900,000
Standard	4" wide	900,000
Utility		900,000
Select structural		1,100,000
No. 1	2" to 4" thick	1,100,000
No. 2	5" and wider	1,000,000
No. 3		900,000
Appearance		1,100,000
Stud		900,000
Select structural	Beams and stringers	1,000,000
No. 1		1,000,000
Select structural	Posts and timbers	1,000,000
No. 1		1,000,000
Selected decking	Decking	1,100,000
Commercial decking		1,000,000

White Woods (Western Woods), cont'd		
Selected decking	Decking	1,100,000
Commercial decking		1,000,000
Yellow-Poplar (Surfaced dry or surfaced green. Used at 19% max. m.c.)		
Select structural	2" to 3" thick	1,500,000
No. 1	2" to 4" wide	1,500,000
No. 2		1,300,000
No. 3		1,200,000
Appearance		1,200,000
Stud		1,200,000
Construction	2" to 4" thick	1,200,000
Standard	4" wide	1,200,000
Utility		1,200,000
Select structural		1,500,000
No. 1	2" to 4" thick	1,500,000
No. 2	5" and wider	1,300,000
No. 3		1,200,000
Appearance		1,500,000
Stud		1,200,000

APPENDIX F

Weight of Building Materials

Type of substance	psf
Roof and Wall Coverings	
Wood shingles	3
Asphalt shingles	2
Cement asbestos shingles	4
Cement tile	16
Clay tile (for mortar add 10 lb)	
2" book tile	12
3" book tile	20
Roman	12
Spanish	19
Ludowici	10
Sheet-metal roofing	2
Corrugated roofing (no. 20)	2
Corrugated asbestos	3–4
Copper or tin	1
Corrugated iron	2
Gypsum sheathing 1/2"	2
Wood sheathing	3
Slag roofing	5
Composition roofing	
3-ply ready roofing	1
3-ply felt tar and gravel roofing	4
4-ply felt tar and gravel roofing	5-1/2
5-ply felt tar and gravel roofing	6
Slate roofing laid in place with 3" double lap, 3/16" thickness	7
Slate roofing laid in place with 3" double lap, 1/4" thickness	10
Slate roofing laid in place with 3" double lap, 3/8" thickness	14-1/2
Slate roofing laid in place with 3" double lap, 1/2" thickness	19 1/2
Skylight, metal frame, 3/8" wire glass	8
Plate glass per inch of thickness	14
Fiberboard 1/2"	3/4

Type of substance	psf
Ceilings	
Plaster on tile or concrete	5
Suspended metal lath and gypsum plaster	10
Suspended metal lath and cement plaster	15
Plaster on wood lath	8
Floors: Floor Fill and Finishes, Floor Slabs, and Waterproofing	
1" terrazzo + 2" stone concrete	38
1-1/2" terrazzo	18
Tile and setting bed	15–23
Marble and setting bed	25–30
1-1/2" asphalt mastic flooring	18
2" asphalt block + 1/2" mortar	30
3" wood block + 1/2" mortar	16
Linoleum or asphalt finish	2
Hardwood flooring, 7/8" thick	4
Softwood subflooring, 7/8" thick	3
Oak and yellow pine	48*
Spruce, fir, hemlock, white pine	30*
Concrete, reinforced stone, per inch of thickness	12-1/2
Concrete, reinforced, cinder, per inch of thickness	9-1/4
Concrete, reinforced, lightweight, per inch of thickness	9
Concrete, plain, stone, per inch of thickness	12
Concrete, plain, cinder, per inch of thickness	9
Concrete, plain, lightweight, per inch of thickness	7
Waterproofing	
Five-ply membrane, 1/2" thick	5
Five-ply membrane, mortar, stone concrete, 5" thick	55
2" split tile + 3" stone concrete, 5" thick	45
Patented Structural Systems	
2" Gypsteel plank	12
2-1/2" sheetrock — Pyrofil	12
2-1/2" lightweight nailing concrete	19
2-3/4" lightweight channel slabs	10
2-1/2" porete roof slabs (with 1/2" nailing fin)	15
3-1/2" porete channel slabs	12
2" cinder plank	15
Aerocrete lightweight concrete	50–80*
Nalecode	75*

*pounds per cubic foot.

Type of substance	psf	pcf
Masonry Walls and Partitions: Dead Load Without Plaster		
Solid brickwork:		
Clay brick		120
Sand lime		105
Concrete (heavy aggregate)		130
Concrete (light aggregate)		98
Concrete, stone		144
Granite, blue stone, marble		165
Limestone		156
Sandstone		144
4" brick + 4" load-bearing structural clay tile backing	60	
4" brick + 6" hollow block backing	75	
4" brick + 8" hollow block backing		80
Concrete block:		
Heavy aggregate		85
Light aggregate		55
Multiply by 1.40 for concrete block (gravel or stone aggregate)		
2" furring tile	12	
2" T.C. or cinder block (nonbearing)	16	
3" T.C. or cinder block (nonbearing)	17	
4" T.C. or cinder block (nonbearing)	18	
6" T.C. or cinder block (nonbearing)	28	
8" T.C. or cinder block (nonbearing)	34	
4" load-bearing T.C. or cinder block	22	
6" load-bearing T.C. or cinder block	32	
8" load-bearing T.C. or cinder block	39	
12" load-bearing T.C. or cinder block	56	
2" solid gypsum	10	
3" solid gypsum	13	
3" hollow gypsum	12	
4" hollow gypsum	13	
6" hollow gypsum	19	
2" solid plaster	20	
Wood studs 2 + 4 (unplastered)	4	
Plaster add each side	8	
4" glass block masonry	18	

Adapted from ACI Handbook, CRSI Handbook, and American Standards Association.

Depth in inches (Rib depth + slab thickness)	Width of rib					Add for tapered ends
	4	5	6	7	8	
Rib Slabs						
12-inch clay-tile fillers						
4 + 2	49	51	52	54	–	–
6 + 2	60	63	65	67	–	–
8 + 2-1/2	79	82	5	87	–	–
10 + 3	96	100	103	106	–	–
12 + 3	108	112	16	120	–	–
20-inch metal fillers						
6 + 2	41	43	45	47	–	4
8 + 2-1/2	51	54	57	60	–	5
10 + 3	63	67	70	74	–	5
12 + 3	69	74	78	2	86	5
14 + 3	75	81	82	87	91	5
30-inch metal fillers						
6 + 2-1/2	41	43	45	47	–	3
8 + 2-1/2	45	48	50	54	–	4
10 + 3	56	59	61	65	–	4
12 + 3	–	63	7	70	73	4
14 + 3	–	69	72	76	80	4
2-way clay tile fillers (12 × 12)						
4 + 2	1	62	64	–	–	–
6 + 2	87	87	90	–	–	–
8 + 2-1/2	100	103	107	–	–	–
10 + 3	121	126	131	–	–	–
12 + 3	136	141	146	–	–	–
2-way metal fillers (16 × 16)						
4 + 2	44	7	50	–	–	–
6 + 2	55	60	63	–	–	–
8 + 2-1/2	72	78	83	–	–	–
10 + 3	91	96	103	–	–	–
12 + 3	103	111	118	–	–	–
14 + 3	116	25	133	–	–	–
2-way metal fillers (20 × 20)						
4 + 2	42	4	46	–	–	–
6 + 2	50	54	58	–	–	–
8 + 2-1/2	66	71	76	–	–	–
10 + 3	83	88	94	–	–	–
12 + 3	93	100	107	–	–	–
14 + 3	105	113	120	–	–	–

APPENDIX G

Table for Conversion of Units

Conversion from one unit to another is frequently necessary in scientific and technological fields of work. The customary way of doing this, establishing the correct mathematical proportion and calculating the result, is time consuming and unproductive compared to using a conversion factor. In the latter case, one simply multiplies the unit by a proper factor to convert the original unit into the desired one. The following table provides an alphabetical listing of conversion factors for the major units used in science and engineering.

To convert from	To	Multiply by
Abamperes	Amperes	10
	E.M. cgs. units of current	1
	E.S. cgs. units	2.997930×10^{10}
	Faradays (chem.)/sec.	1.036377×10^{-4}
	Faradays (phys.)/sec.	1.036086×10^{-4}
	Statamperes	2.997930×10^{10}

To convert from	To	Multiply by
Abamperes/cm.	E.M. cgs. units of surface charge density	1
	E.S. cgs. units	2.997930×10^{10}
Abamperes/sq. cm.	Amperes/circ. mil	5.0670748×10^{-5}
	Amperes/sq. cm.	10
	Amperes/sq. inch	64.516
Abampere-turns	Ampere-turns	10
Abampere-turns/cm.	Ampere-turns/cm.	10
Abcoulombs	Ampere-hours	0.0027777
	Coulombs	10
	Electronic charges	6.24196×10^{19}
	E.M. cgs. units of charge	1
	E.S. cgs. units	2.997930×10^{10}
	Faradays (chem.)	1.036377×10^{-4}
	Faradays (phys.)	1.036086×10^{-4}
	Statcoulombs	2.997930×10^{10}
Abfarads	E.M. cgs. units of capacitance	1
	E.S. cgs. units	8.987584×10^{20}
	Farads	1×10^{9}
	Microfarads	1×10^{15}
	Statfarads	8.987584×10^{20}
Abhenries	E.M. cgs. units of induction	1
	E.S. cgs. units	1.112646×10^{-21}
	Henries	1×10^{-9}
Abmhos	E.M. cgs. units of conductance	1
	E.S. cgs. units	8.987584×10^{20}
	Megamhos	1000
	Mhos	1×10^{9}
	Statmhos	8.987584×10^{20}
Abohms	E.M. cgs. units of resistance	1
	Megohms	1×10^{-15}
	Microhms	0.001
	Ohms	1×10^{-9}
	Statohms	1.112646×10^{-21}
Abohm-cm.	Circ. mil-ohms/ft.	0.0060153049
	E.M. cgs. units of resistivity	1
	Microhm-inches	0.00039370079
	Ohm-cm.	1×10^{-9}

To convert from	To	Multiply by
Abvolts	Microvolts	0.01
	Millivolts	1×10^{-5}
	Volts	1×10^{-8}
	Volts (Int.)	9.99670×10^{-9}
Abvolts/cm.	E.M. cgs. units of electric field intensity	1
	E.S. cgs. units	3.335635×10^{-11}
	Volts/cm.	1×10^{-8}
	Volts/inch	2.54×10^{-8}
	Volts/meter	1×10^{-6}
Acres	Sq. cm.	40468564
	Sq. ft.	43560
	Sq. ft. (U.S. Survey)	43559.826
	Sq. inches	6272640
	Sq. kilometers	0.0040468564
	Sq. links (Gunter's)	1×10^{5}
	Sq. meters	4046.8564
	Sq. miles (statute)	0.0015625
	Sq. perches	160
	Sq. rods	160
	Sq. yards	4840
Acre-feet	Cu. feet	43,560
	Cu. meters	1233.4818
	Cu. yards	1613.333
Acre-inches	Cu. feet	3630
	Cu. meters	102.79033
	Gallons (U.S.)	27,154.286
Amperes	Abamperes	0.1
	Amperes (Int.)	1.000165
	Cgs. units of current	1
	Mks. units of current	1
	Coulombs/sec.	1
	Coulombs (Int.)/sec.	1.000165
	Faradays (chem.)/sec.	1.036377×10^{-5}
	Faradays (phys.)/sec.	1.036086×10^{-5}
	Statamperes	2.997930×10^{9}
Amperes (Int.)	Amperes	0.999835
	Coulombs/sec.	0.999835
	Coulombs (Int.)/sec.	1

To convert from	To	Multiply by
	Faradays (chem.)/sec.	1.03623×10^{-5}
	Faradays (phys.)/sec.	1.03592×10^{-5}
Amperes/meter	Cgs. units of surface current density	0.01
	E.M. cgs. units	0.001
	E.S. cgs. units	2.997930×10^{7}
	Mks. units	1
Amperes/sq. meter	Cgs. units of volume current density	0.0001
	E.M. cgs. units	1×10^{-5}
	E.S. cgs. units	299,793.0
	Mks. units	1
Amperes/sq. mil	Abamperes/sq. cm.	15,500.031
	Amperes/sq. cm.	1.5500031×10^{5}
Ampere-hours	Abcoulombs	360
	Coulombs	3600
	Faradays (chem.)	0.373096
	Faradays (phys.)	0.372991
Ampere-turns	Cgs. units of magnetomotive force	1.2566371
	E.M. cgs. units	1.2566371
	E.S. cgs. units	3.767310×10^{10}
	Gilberts	1.2566371
Ampere-turns/weber	Cgs. units of reluctance	1.256637×10^{-8}
	E.M. cgs. units	1.256637×10^{-8}
	E.S. cgs. units	1.129413×10^{13}
	Gilberts/maxwell	1.256637×10^{-8}
Ångström units	Centimeters	1×10^{-8}
	Inches	3.9370079×10^{-9}
	Microns	0.0001
	Millimicrons	0.1
	Wavelength of orange-red line of krypton 86	0.000165076373
	Wavelength of red line of cadmium	0.000155316413
Ares	Acres	0.024710538
	Sq. dekameters	1
	Sq. feet	1076.3910
	Sq. ft. (U.S. Survey)	1076.3867

To convert from	To	Multiply by
	Sq. meters	100
	Sq. miles	3.8610216×10^{-5}
Atmospheres	Bars	1.01325
	Cm. of Hg (0°C.)	76
	Cm. of H_2O (4°C.)	1033.26
	Dynes/sq. cm.	1.01325×10^{6}
	Ft. of H_2O (39.2°F.)	33.8995
	Grams/sq. cm.	1033.23
	In. of Hg (32°F.)	29.9213
	Kg./sq. cm.	1.00323
	Mm. of Hg (0°C.)	760
	Pascals (N/sq. meter)	1.01325×10^{5}
	Pounds/sq. inch	14.6960
	Tons (short)/sq. ft.	1.05811
	Torrs	760
Atomic mass units (chem.)	Electron volts	9.31395×10^{8}
	Grams	1.66024×10^{-24}
Atomic mass units (phys.)	Electron volts	9.31141×10^{8}
	Grams	1.65979×10^{-24}
Bags (Brit.)	Bushels (Brit.)	3
Barns	Sq. cm.	1×10^{-24}
Barrels (Brit.)	Bags (Brit.)	1.5
	Barrels (U.S., dry)	1.415404
	Barrels (U.S., liq.)	1.372513
	Bushels (Brit.)	4.5
	Bushels (U.S.)	4.644253
	Cu. feet	5.779568
	Cu. meters	0.1636591
	Gallons (Brit.)	36
	Liters	163.6546
Barrels (petroleum, U.S.)	Cu. feet	5.614583
	Gallons (U.S.)	42
	Liters	158.98284
Barrels (U.S., dry)	Barrels (U.S. liq.)	0.969696
	Bushels (U.S.)	3.2812195
	Cu. feet	4.083333

To convert from	To	Multiply by
	Cu. inches	7056
	Cu. meters	0.11562712
	Quarts (U.S., dry)	105
Barrels (U.S., liq.)	Barrels (U.S., dry)	1.03125
	Barrels (wine)	1
	Cu. feet	4.2109375
	Cu. inches	7276.5
	Cu. meters	0.11924047
	Gallons (Brit.)	26.22925
	Gallons (U.S., liq.)	31.5
	Liters	119.23713
Bars	Atmospheres	0.986923
	Baryes	1×10^6
	Cm. of Hg (0°C.)	75.0062
	Dynes/sq. cm.	1×10^6
	Ft. of H_2O (60°F.)	33.4883
	Grams/sq. cm.	1019.716
	In. of Hg (32°F.)	29.5300
	Kg./sq. cm.	1.019716
	Millibars	1000
	Pounds/sq. inch	14.5038
Baryes	Atmospheres	9.86923×10^{-7}
	Bars	1×10^{-6}
	Dynes/sq. cm.	1
	Grams/sq. cm.	0.001019716
	Millibars	0.001
Bels	Decibels	10
Board feet	Cu. cm.	2359.7372
	Cu. feet	0.833333
	Cu. inches	144
Bolts of cloth	Linear feet	120
	Meters	36.576
Bougie decimales	Candies (Int.)	1.00
B.t.u.	B.t.u. (IST.)	0.999346
	B.t.u. (mean)	0.998563
	B.t.u. (39°F.)	0.994982
	B.t.u. (60°F.)	0.999689
	Calorie	251.99576
	Calorie (IST.)	251.831
	Calorie (mean)	251.634

To convert from	To	Multiply by
	Calorie (20°C.)	252.122
	Cu. cm.-atm.	10,405.6
	Ergs	1.05435×10^{10}
	Foot-poundals	25020.1
	Foot-pounds	777.649
	Gram-cm.	1.07514×10^{7}
	Hp.-hours	0.000392752
	Hp.-years	4.48347×10^{-8}
	Joules	1054.35
	Joules (Int.)	1054.18
	Kg.-meters	107.514
	Kw.-hours	0.000292875
	Kw.-hours (Int.)	0.000292827
	Liter-atm.	10.4053
	Tons of refrig. (U.S. std.)	3.46995×10^{-6}
	Watt-seconds	1054.35
	Watt-seconds (Int.)	1054.18
B.t.u. (IST.)	B.t.u.	1.00065
B.t.u. (mean)	B.t.u.	1.00144
	B.t.u. (IST.)	1.00078
	B.t.u. (39°F.)	0.996415
	B.t.u. (60°F.)	1.00113
	Hp.-hours	0.000393317
	Joules	1055.87
	Kg.-meters	107.669
	Kw.-hours	0.000293297
	Kw.-hours (Int.)	0000293248
	Liter-atm.	10.4203
	Watt-hours	0.293297
	Watt-hours (Int.)	0.293248
B.t.u. (39°F.)	B.t.u.	1.00504
	B.t.u. (IST.)	1.00439
	B.t.u. (mean)	1.00360
	B.t.u. (60°F.)	1.00473
	Joules	1059.67
B.t.u. (60°F.)	B.t.u.	1.00031
	B.t.u. (IST.)	0.999657
	B.t.u. (mean)	0.998873
	B.t.u. (39°F.)	0.995291

To convert from	To	Multiply by
B.t.u./hr.	Kilocalorie/hr.	0.251996
	Ergs/sec.	2.928751×10^6
	Foot-pounds/hr.	777.649
	Horsepower	0.000392752
	Horsepower (boiler)	2.98563×10^{-5}
	Horsepower (electric)	0.000392594
	Horsepower (metric)	0.000398199
	Kilowatts	0.000292875
	Lb. ice melted/hr.	0.0069714
	Tons of refrig. (U.S. comm.)	8.32789×10^{-5}
	Watts	0.292875
B.t.u./min.	Kilocalorie/min.	0.251996
	Ergs/sec.	1.75725×10^8
	Foot-pounds/min.	777.649
	Horsepower	0.0235651
	Horsepower (boiler)	0.00179138
	Horsepower (electric)	0.0235556
	Horsepower (metric)	0.0238920
	Joules/sec.	17.5725
	Kg.-meters/min.	107.514
	Kilowatts	0.0175725
	Lb. ice melted/hr.	0.41828
	Tons of refrig. (U.S. comm.)	0.00499673
	Watts	17.5725
B.t.u. (mean)/min.	B.t.u. (mean)/hr.	60
	Kilocalorie (mean)/hr.	15.1197
	Kilocalorie (mean)/min.	0.251996
	Ergs/sec.	1.75978×10^8
	Foot-pounds/min.	778.768
	Horsepower	0.0235990
	Horsepower (boiler)	0.00179396
	Horsepower (electric)	0.0235895
	Horsepower (metric)	0.0239264
	Joules/sec.	17.5978
	Kg.-meters/min.	107.669
	Kilowatts	0.0175978
	Lb. ice-melted/hr.	0.41888

To convert from	To	Multiply by
B.t.u./lb.	Calorie/gram	0.555555
	Cu. cm.-atm./gram	22.9405
	Cu. ft.-atm./lb.	0.367471
	Cu. ft.-(lb./sq. in.)/lb.	5.40034
	Foot-pounds/lb.	777.649
	Hp.-hr./lb.	0.000392752
	Joules/gram	2.32444
B.t.u. (mean)/lb.	Calorie (mean)/gram	0.555555
	Cu. cm.-atm./gram	22.9735
	Foot-pounds/lb.	778.768
	Hp.-hr./lb.	0.000393317
	Joules/gram	2.32779
B.t.u./sec.	B.t.u./hr.	3600
	B.t.u./min.	60
	Kilocalorie/hr.	907.185
	Kilocalorie/min.	15.1197
	Cheval-vapcur	1.43352
	Ergs/sec.	1.05435×10^{10}
	Foot-pounds/sec.	777.649
	Horsepower	1.41391
	Horsepower (boiler)	0.107483
	Horsepower (electric)	1.41334
	Horsepower (metric)	1.43352
	Kg.-meters/sec.	107.514
	Kilowatts	1.05435
	Kilowatts (Int.)	1.05418
	Watts	1054.35
	Watts (Int.)	1054.18
B.t.u. (mean)/sec.	Ergs/sec.	1.05587×10^{10}
	Foot-pounds/sec.	778.768
	Horsepower	1.41594
	Horsepower (boiler)	0.107637
	Horsepower (electric)	1.41537
	Horsepower (metric)	1.43558
	Watts	1055.87
B.t.u./sq. ft.	Calorie/sq. cm.	0.271246
B.t.u./sq. ft. × min.)	Hp./sq. ft.	0.0235651
	Kw./sq. ft.	0.0175725
	Watts/sq. in.	0.122031

To convert from	To	Multiply by
Buckets (Brit.)	Cu. cm.	18,184.35
	Gallons (Brit.)	4
Bushels (Brit.)	Bags (Brit.)	0.333333
	Bushels (U.S.)	1.032056
	Cu. cm.	36368.70
	Cu. feet	1.284348
	Cu. inches	2219.354
	Dekaliters	3.636768
	Gallons (Brit.)	8
	Hectoliters	0.3636768
	Liters	36.36768
Bushels (U.S.)	Barrels (U.S.), dry	0.3047647
	Bushels (Brit.)	0.9689395
	Cu. cm.	35,239.07
	Cu. feet	1.244456
	Cu. inches	2150.42
	Cu. meters	0.03523907
	Cu. yards	0.04609096
	Gallons (U.S., dry)	8
	Gallons (U.S., liq.)	9.309177
	Liters	35.23808
	Ounces (U.S., fluid)	1191.575
	Pecks (U.S.)	4
	Pints (U.S., dry)	64
	Quarts (U.S., dry)	32
	Quarts (U.S., liq.)	37.23671
Butts (Brit.)	Bushels (U.S.)	13.53503
	Cu. feet	16.84375
	Cu. meters	0.4769619
	Gallons (U.S.)	126
Cable lengths	Fathoms	120
	Feet	720
	Meters	219.456
Caliber	Inch	0.01
	Millimeter	0.254
Calories	B.t.u.	0.0039683207
	B.t.u. (IST.)	0.00396573
	B.t.u. (mean)	0.00396262
	B.t.u. (39°F.)	0.00394841

To convert from	To	Multiply by
	B.t.u. (60°F.)	0.00396709
	Cal. (IST.)	0.999346
	Cal. (mean)	0.998563
	Cal. (15°C.)	0.999570
	Cal. (20°C.)	1.00050
	Kilocal.	0.001
	Kilocal. (IST.)	0.000999346
	Kilocal. (mean)	0.000998563
	Kilocal. (15°C.)	0.000999570
	Kilocal. (20°C.)	0.00100050
	Cu. cm.-atm.	41.2929
	Cu. ft.-atm.	0.00145824
	Ergs	4.184×10^{7}
	Foot-poundals	99.2878
	Foot-pounds	3.08596
	Gram-cm.	42,664.9
	Hp.-hours	1.55857×10^{-6}
	Joules	4.184
	Joules (Int.)	4.18331
	Kg.-meters	0.426649
	Kw.-hours	1.162222×10^{-6}
	Liter-atm.	0.0412917
	Watt-hours	0.001162222
	Watt-hours (Int.)	0.00116203
	Watt-seconds	4.184
Calories (mean)	B.t.u.	0.00397403
	Cal.	1.00144
	Cal. (IST.)	1.00078
	Cal. (20°C.)	1.00194
	Kilocal. (mean)	0.001
	Cu. cm.-atm.	41.3523
	Cu. ft.-atm.	0.00146034
	Ergs	4.19002×10^{7}
	Foot-poundals	99.4308
	Foot-pounds	3.09040
	Hp.-hours	1.56081×10^{-6}
	Joules	4.19002
	Joules (Int.)	4.18933
	Kg.-meters	0.427263
	Kw.-hours	1.16390×10^{-6}

To convert from	To	Multiply by
	Liter-atm.	0.0413511
	Watt-seconds	4.19002
Calories (15°C.)	B.t.u.	0.00397003
	Cal.	1.00043
	Cal. (IST.)	0.999776
	Cal. (mean)	0.998992
	Cal. (20°C.)	1.00093
	Joules	4.18580
	Joules (Int.)	4.18511
Calories (20°C.)	B.t.u.	0.00396633
	Cal.	0.999498
	Cal. (IST.)	0.998845
	Cal. (mean)	0.998061
	Cal. (15°C.)	0.999068
	Joules	4.18190
	Joules (Int.)	4.18121
Cal./°C	B.t.u./°F.	0.00220462
	Joules/°F.	2.324444
	Joules (Int.)/°F.	2.32406
Cal./gram	B.t.u./lb.	1.8
	Foot-pounds/lb.	1399.77
	Joules/gram	4.184
	Watt-hours/gram	0.001162222
Cal./(gram × °C.)	B.t.u./(lb. × °C.)	1.8
	B.t.u./(lb. × °F.)	1
	Kilocal./(kg. × °C.)	1
	Joules/(gram × °C.)	4.184
	Joules/(lb. × °F.)	1054.35
Cal./hr.	B.t.u./hr.	0.0039683207
	Ergs/sec.	11,622.222
	Watts	0.001162222
Cal. (mean)/hr.	B.t.u. (mean)/hr.	0.0039683207
	Ergs/sec.	11,639.0
	Watts	0.00116390
Cal./min.	B.t.u./min.	0.0039683207
	Ergs/sec.	697,333.3
	Watts	0.069733
Cal. (mean)/min.	B.t.u. (mean)/min.	0.0039683207
	Ergs/sec.	698,337

To convert from	To	Multiply by
Cal./sec.	Joules/sec.	0.0698337
	Watts	0.0698337
	B.t.u./sec.	0.0039683207
	Ergs/sec.	4.184×10^7
	Foot-pounds/sec.	3.08596
	Horsepower	0.00561084
	Watts	4.184
Cal. (mean)/sec.	Ergs/sec.	4.19002×10^7
	Watts	4.19002
Cal./(sec. × sq. cm.)	B.t.u./(hr. × sq. ft.)	13,272.1
	Cal./(hr. × sq. cm.)	3600
	Watts/sq. cm.	4.184
Cal./(sec. × sq. cm. × °C.)	B.t.u./(hr. × sq. ft. × °F.)	7373.38
Cal./sq. cm.	B.t.u./sq. ft.	3.68669
$\frac{\text{Cal.-cm.}}{(\text{hr.} \times \text{sq. cm.} \times {}^\circ\text{C.})}$	$\frac{\text{B.t.u.-ft.}}{(\text{hr.} \times \text{sq. ft.} \times {}^\circ\text{F.})}$	0.0671969
	$\frac{\text{B.t.u.-inch}}{(\text{hr.} \times \text{sq. ft.} \times {}^\circ\text{F.})}$	0.806363
Cal.-cm./sq. cm.	B.t.u.-inch/sq. ft.	1.4514530
Cal.-sec.	Planck's constant	6.31531×10^{22}
Cal.-sec./Avog. No. (chem.)	Planck's constant	1.04849×10^{10}
Cal.-sec./Avog. No. (phys.)	Planck's constant	1.04821×10^{10}
*Candles (English)	Candles (Int.)	1.04
	Hefner units	1.16
Candles (German)	Candles (English)	1.01
	Candles (Int.)	1.05
	Hefner units	1.17
Candles (Int.)	Candles (English)	0.96
	Candles (German)	0.95
	Candles (pentane)	1.00
	Hefner units	1.11
	Lumens (Int.)/steradian	1

*Candle is equivalent to candela (SI unit of luminous intensity)

To convert from	To	Multiply by
Candles (pentane)	Candles (Int.)	1.00
Candles/sq. cm.	Candles/sq. inch	6.4516
	Candles/sq. meter	10000
	Foot-lamberts	2918.6351
	Lamberts	3.1415927
Candles/sq. ft.	Candles/sq. inch	0.0069444
	Candles/sq. meter	10.763910
	Foot-lamberts	3.1415927
	Lamberts	0.0033815822
Candles/sq. inch	Candles/sq. cm.	0.15500031
	Candles/sq. foot	144
	Foot-lamberts	452.38934
	Lamberts	0.48694784
Candle power (spher.)	Lumens	12.566370
Carats (parts of gold per 24 of mixture)	Milligrams/gram	41.6666
Carats (1877)	Grains	3.168
	Milligrams	205.3
Carats (metric)	Grains	3.08647
	Grams	0.2
	Milligrams	200
Carcel units	Candles (Int.)	9.61
Centals	Kilograms	45.359237
	Pounds	100
Centares	Ares	0.01
	Sq. feet	10.763910
	Sq. inches	1550.0031
	Sq. meters	1
	Sq. yards	1.1959900
Centigrams	Grains	0.15432358
	Grams	0.01
Centiliters	Cu. cm.	10.00028
	Cu. inches	0.6102545
	Liters	0.01
	Ounces (U.S., fluid)	0.3381497
Centimeters	Ångström units	1×10^8
	Feet	0.032808399
	Feet (U.S. Survey)	0.032808333
	Hands	0.098425197

To convert from	To	Multiply by
	Inches	0.39370079
	Links (Gunter's)	0.049709695
	Links (Ramden's)	0.032808399
	Meters	0.01
	Microns	10,000
	Miles (naut., Int.)	5.3995680×10^{-6}
	Miles (statute)	6.2137119×10^{-6}
	Millimeters	10
	Millimicrons	1×10^{7}
	Mils	393.70079
	Picas (printer's)	2.3710630
	Points (printer's)	28.452756
	Rods	0.0019883878
	Wavelength of orange-red line of krypton 86	16,507.6373
	Wavelength of red line of cadmium	15,531.6413
	Yards	0.010936133
Cm. of Hg (0°C.)	Atmospheres	0.013157895
	Bars	0.0133322
	Dynes/sq. cm.	13,332.2
	Ft. of H_2O (4°C.)	0.446050
	Ft. of H_2O (60°F.)	0.446474
	In. of Hg (0°C.)	0.39370079
	Kg./sq. meter	135.951
	Pounds/sq. ft.	27.8450
	Pounds/sq. inch	0.193368
	Torrs	10
Cm. of H_2O (4°C.)	Atmospheres	0.000967814
	Dynes/sq. cm.	980.638
	Pounds/sq. inch	0.0142229
Centimeters/sec.	Feet/min.	1.9685039
	Feet/sec.	0.032808399
	Kilometers/hr.	0.036
	Kilometers/min.	0.0006
	Knots (Int.)	0.019438445
	Meters/min.	0.6
	Miles/hr.	0.022369363
	Miles/min.	0.00037282272

To convert from	To	Multiply by
Cm./(sec. × sec.)	Kilometers/(hr. × sec.)	0.036
	Miles/(hr. × sec.)	0.022369363
Centimeters/year	Inches/year	0.39370079
Centipoises	Grams/(cm. × sec.)	0.01
	Poises	0.01
	Pound/(ft. × hr.)	2.4190883
	Pounds/(ft. × sec.)	0.00067196898
Centistokes	Stokes	0.01
Chains (Gunter's)	Centimeters	2011.68
	Chains (Ramden's)	0.66
	Feet	66
	Feet (U.S. Survey)	65.999868
	Furlongs	0.1
	Inches	792
	Links (Gunter's)	100
	Links (Ramden's)	66
	Meters	20.1168
	Miles (statute)	0.0125
	Rods	4
	Yards	22
Chains (Ramden's)	Centimeters	3048
	Chains (Gunter's)	1.515151
	Feet	100
	Feet (U.S. Survey)	99.999800
Cheval-vapeur	Horsepower (metric)	1
Cheval-vapeur-heures	Joules	2,647,795
Circles	Degrees	360
	Grades	400
	Minutes	21,600
	Radians	6.2831853
	Signs	12
Circular inches	Circular mm.	645.16
	Sq. cm.	5.0670748
	Sq. inches	0.78539816
Circular mm.	Sq. cm.	0.0078539816
	Sq. inches	0.0012173696
	Sq. mm.	0.78539816

To convert from	To	Multiply by
Circular mils	Circular inches	1×10^{-6}
	Sq. cm.	5.0670748×10^{-6}
	Sq. inches	$7.8539816 + 10^{-7}$
	Sq. mm.	0.00050670748
	Sq. mils	0.78539816
Circumferences	Degrees	360
	Grades	400
	Minutes	21,600
	Radians	6.2831853
	Seconds	1,296,000
Cords	Cord-feet	8
	Cu. feet	128
	Cu. meters	3.6245734
Cord-feet	Cords	0.125
	Cu. feet	16
Coulombs	Abcoulombs	1
	Ampere-hours	0.0002777
	Ampere-seconds	1
	Coulombs (Int.)	1.000165
	Electronic charge	6.24196×10^{18}
	E.M. cgs. units of electric charge	0.1
	E.S. cgs. units of electric charge	2.997930×10^{9}
	Faradays (chem.)	1.036377×10^{-5}
	Faradays (phys.)	1.036086×10^{-5}
	Mks. units of electric charge	1
	Statcoulombs	2.997930×10^{9}
Coulombs/cu. meter	E.M. cgs. units of volume charge density	1×10^{-7}
	E.S. cgs. units	2997.930
Coulombs/sq. cm.	Abcoulombs/sq. cm.	0.1
	Cgs. units of polarization, and surface charge density	1
Cubic centimeters	Board feet	0.00042377600
	Bushels (Brit.)	2.749617×10^{-5}
	Bushels (U.S.)	2.837759×10^{-5}
	Cu. feet	3.5314667×10^{-5}

To convert from	To	Multiply by
	Cu. inches	0.061023744
	Cu. meters	1×10^{-6}
	Cu. yards	1.3079506×10^{-6}
	Drachms (Brit., fluid)	0.28156080
	Drams (U.S., fluid)	0.27051218
	Gallons (Brit.)	0.0002199694
	Gallons (U.S., dry)	0.00022702075
	Gallons (U.S., liq.)	0.00026417205
	Gills (Brit.)	0.007039020
	Gills (U.S.)	0.0084535058
	Liters	0.000999972
	Ounces (Brit., fluid)	0.03519510
	Ounces (U.S., fluid)	0.033814023
	Pints (U.S., dry)	0.0018161660
	Pints (U.S., liq.)	0.0021133764
	Quarts (Brit.)	0.0008798775
	Quarts (U.S., dry)	0.00090808298
	Quarts (U.S., liq.)	0.0010566882
Cu. cm./gram	Cu. ft./lb.	0.016018463
Cu. cm./sec.	Cu. ft./min.	0.0021188800
	Cal. (U.S.)/min.	0.015850323
	Gal. (U.S.)/sec.	0.00026417205
Cu. cm.-atm.	B.t.u.	9.61019×10^{-5}
	B.t.u. (mean)	9.59637×10^{-5}
	Cal.	0.0242173
	Cal. (mean)	0.0241824
	Cu.-ft.-atm.	3.5314667×10^{-5}
	Joules	0.101325
	Watt-hours	2.81458×10^{-5}
Cu. cm.-atm./gram.	B.t.u./lb.	0.0435911
	Cal./gram	0.0242173
	Cu. ft.-(lb./sq. in.)/lb.	0.235406
	Ft.-lb./lb.	33.8985
	Joules/gram	0.101325
	Kg.-meters/gram	0.0103323
	Kw.-hr./gram	2.81458×10^{-8}
Cubic decimeters	Cu. cm.	1000
	Cu. feet	0.035316667
	Cu. inches	61.023744

To convert from	To	Multiply by
	Cu. meters	0.001
	Cu. yards	0.0013079506
	Liters	0.999972
Cubic dekameters	Cu. decimeters	1×10^6
	Cu. feet	35,314.667
	Cu. inches	6.1023744×10^7
	Cu. meters	1000
	Liters	999,972
Cubic feet	Acre-feet	2.2956841×10^{-5}
	Board feet	12
	Bushels (Brit.)	0.7786049
	Bushels (U.S.)	0.80356395
	Cords (wood)	0.0078125
	Cord-feet	0.0625
	Cu. centimeters	28,316.847
	Cu. meters	0.028316847
	Gallons (U.S., dry)	6.4285116
	Gallons (U.S., liq.)	7.4805195
	Liters	28.31605
	Ounces (Brit., fluid)	996.6143
	Ounces (U.S., fluid)	957.50649
	Pints (U.S., liq.)	59.844156
	Quarts (U.S., dry)	25.714047
	Quarts (U.S., liq.)	29.922078
Cu. ft. of H_2O (39.2°F.)	Pounds of H_2O	62.4262
Cu. ft. of H_2O (60°F.)	Pounds of H_2O	63.3663
Cu. ft./hr.	Acre-feet/hr.	2.2956841×10^{-5}
	Cu. cm./sec.	7.8657907
	Cu. ft./day	24
	Gal. (U.S.)/hr.	7.4805195
	Liters/hr.	28.31605
Cu. ft./min.	Acre-feet/hr.	0.0013774105
	Acre-feet/min.	2.2956841×10^{-5}
	Cu. cm./sec.	471.94744
	Cu. ft./hr.	60
	Gal. (U.S.)/min.	7.4805195
	Liters/sec.	0.4719342

To convert from	To	Multiply by
Cu. ft./lb.	Cu. cm./gram	62.427961
	Millimeters/gram	62.42621
Cu. ft./sec.	Acre-inches/hr.	0.99173553
	Cu. cm./sec.	28,316.847
	Cu. yards/min.	2.222222
	Gal. (U.S.)/min.	448.83117
	Liters/min.	1698.963
	Liters/sec.	28.31605
Cu. ft. of H_2O (60°F.)/sec.	Lb. of H_2O/min.	3741.98
Cu. ft.-atm.	B.t.u.	2.72130
	Cal.	685.756
	Cu. cm.-atm.	28,316.847
	Cu. ft.-(lb/sq. in.)	14.6960
	Foot-pounds	2116.22
	Hp.-hours	0.00106880
	Joules	2869.20
	Kg.-meters	292.577
	Kw.-hours	0.000797001
Cubic inches	Barrels (Brit.)	0.0001001292
	Barrels (U.S., dry)	0.00014172336
	Board feet	0.0069444
	Bushels (Brit.)	0.0004505815
	Bushels (U.S.)	0.00046502544
	Cu. cm.	16.387064
	Cu. feet	0.00057870370
	Cu. meters	1.6387064×10^{-5}
	Cu. yards	2.1433470×10^{-5}
	Drams (U.S., fluid)	4.4329004
	Gallons (Brit.)	0.003604652
	Gallons (U.S., dry)	0.0037202035
	Gallons (U.S., liq.)	0.0043290043
	Liters	0.01638661
	Milliliters	16.38661
	Ounces (Brit., fluid)	0.5767444
	Ounces (U.S., fluid)	0.55411255
	Pecks (U.S.)	0.0018601017
	Pints (U.S., dry)	0.029761628
	Pints (U.S., liq.)	0.034632035

To convert from	To	Multiply by
	Quarts (U.S., dry)	0.014880814
	Quarts (U.S., liq.)	0.017316017
Cu. in. of H_2O (4°C.)	Pounds of H_2O	0.0361263
Cu. in. of H_2O (60°F.)	Pounds of H_2O	0.0360916
Cubic meters	Acre-feet	0.00081071319
	Barrels (Brit.)	6.110261
	Barrels (U.S., dry)	8.648490
	Barrels (U.S., liq.)	8.3864145
	Bushels (Brit.)	27.49617
	Bushels (U.S.)	28.377593
	Cu. cm.	1×10^{6}
	Cu. feet	35.314667
	Cu. inches	61,023.74
	Cu. yards	1.3079506
	Gallons (Brit.)	219.9694
	Gallons (U.S., liq.)	264.17205
	Hogshead	4.1932072
	Liters	999.972
	Pints (U.S., liq.)	2113.3764
	Quarts (U.S., liq.)	1056.6882
	Steres	1
Cu. meters/min.	Gal. (Brit.)/min.	219.9694
	Gal. (U.S.)/min.	264.1721
	Liters/min.	999.972
Cu. millimeters	Cu. cm.	0.001
	Cu. inches	6.1023744×10^{-1}
	Cu. meters	1×10^{-9}
	Minims (Brit.)	0.01689365
	Minims (U.S.)	0.016230731
Cu. yards	Bushels (Brit.)	21.02233
	Bushels (U.S.)	21.696227
	Cu. cm.	764,554.86
	Cu. feet	27
	Cu. inches	46.656
	Cu. meters	0.76455486
	Gallons (Brit.)	168.1787
	Gallons (U.S., dry)	173.56981
	Gallons (U.S., liq.)	201.97403

To convert from	To	Multiply by
	Liters	764.5335
	Quarts (Brit.)	672.7146
	Quarts (U.S., dry)	694.27926
	Quarts (U.S., liq.)	807.89610
Cu. yd./min.	Cu. ft./sec.	0.45
	Gal. (U.S.)/sec.	3.3662338
	Liters/sec.	12.74222
Cubits	Centimeters	45.72
	Feet	1.5
	Inches	18
Daltons (chem.)	Grams	1.66024×10^{-24}
Daltons (phys.)	Grams	1.65979×10^{-24}
Days (mean solar)	Days (sidereal)	1.00273791
	Hours (mean solar)	24
	Hours (sidereal)	24.065710
	Years (calendar)	0.0027397260
	Years (sidereal)	0.0027378031
	Years (tropical)	0.0027379093
Days (sidereal)	Days (mean solar)	0.99726957
	Hours (mean solar)	23.934470
	Hours (sidereal)	24
	Minutes (mean solar)	1436.0682
	Minute (sidereal)	1440
	Second (sidereal)	86,400
	Years (calendar)	0.0027322454
	Years (sidereal)	0.0027303277
	Years (tropical)	0.0027304336
Decibels	Bels	0.1
Decimeters	Centimeters	0.1
	Feet	0.32808399
	Feet (U.S. Survey)	0.328083333
	Inches	3.9370079
	Meters	0.1
Decisteres	Cu. meters	0.1
Degrees	Circles	0.0027777
	Minutes	60
	Quadrants	0.0111111
	Radians	0.017453293
	Seconds	3600

To convert from	To	Multiply by
Degrees/cm.	Radians/cm.	0.017453293
Degrees/foot	Radians/cm.	0.00057261458
Degrees/inch	Radian/cm.	0.0068713750
Degrees/min.	Degrees/sec.	0.0166666
	Radians/sec.	0.00029088821
	Revolutions/sec.	4.629629×10^{-5}
Degrees/sec.	Radians/sec.	0.017453293
	Revolutions/min.	0.166666
	Revolutions/sec.	0.0027777
Dekaliters	Pecks (U.S.)	1.135136
	Pints (U.S., dry)	18.16217
Dekameters	Centimeters	1000
	Feet	32.808399
	Feet (U.S. Survey)	32.808333
	Inches	393.70079
	Kilometers	0.01
	Meters	10
	Yards	10.93613
Demals	Gram-equiv./cu. decimeter	1
Drachms (Brit. fluid)	Cu. cm.	3.551631
	Cu. inches	0.2167338
	Drams (U.S., fluid)	0.9607594
	Milliliters	3.551531
Drams (apoth. or troy)	Drams (avdp.)	2.1942857
	Grains	60
	Grams	3887.9346
	Ounces (apoth. *or* troy)	0.125
	Ounces (avdp.)	0.13714286
	Scruples (apoth.)	3
Drams (avdp.)	Drams (apoth. *or* troy)	0.455729166
	Grains	27.34375
	Grams	1.7718452
	Ounces (apoth. *or* troy)	0.056966146
	Ounces (avdp.)	0.0625
	Pennyweights	1.1393229
	Pounds (apoth. *or* troy)	0.0047471788
	Pounds (avdp.)	0.00390625
	Scruples (apoth.)	1.3671875

To convert from	To	Multiply by
Drams (U.S., fluid)	Cu. cm.	3.6967162
	Cu. inches	0.22558594
	Drachms (Brit., fluid)	1.040843
	Gills (U.S.)	0.03125
	Milliliters	3.696588
	Minims (U.S.)	60
	Ounces (U.S., fluid)	0.125
	Pints (U.S., liq.)	0.0078125
Dynes	Grains	0.01573663
	Grams	0.001019716
	Newtons	0.00001
	Poundals	7.2330138×10^{-5}
	Pounds	2.248089×10^{-6}
Dynes/cm.	Ergs/sq. cm.	1
	Ergs/sq. mm.	0.01
	Grams/cm.	0.001019716
	Poundals/inch	0.00018371855
Dynes/cu. cm.	Grams/cu. cm.	0.001019716
	Poundals/cu. inch	0.0011852786
Dynes/sq. cm.	Atmospheres	9.86923×10^{-7}
	Bars	1×10^{-6}
	Baryes	1
	Cm. of Hg (0°C.)	7.50062×10^{-5}
	Cm. of H_2O (4°C.)	0.001019745
	Grams/sq. cm.	0.001019716
	In. of Hg (32°F.)	2.95300×10^{-5}
	In. of H_2O (4°C.)	0.000401474
	Kg./sq. meter	0.01019716
	Pascals (N/sq. meter)	0.1
	Poundals/sq. in.	0.00046664510
	Pounds/sq. in.	1.450377×10^{-5}
Dyne-centimeters	Ergs	1
	Foot-poundals	2.3730360×10^{-6}
	Foot-pounds	7.37562×10^{-8}
	Gram-cm.	0.001019716
	Inch-pounds	8.85075×10^{-7}
	Kg.-meters	1.019716×10^{-8}
	Newton-meters	1×10^{-7}

To convert from	To	Multiply by
Electron volts	Ergs	1.60209×10^{-12}
	Grams	1.78253×10^{-33}
Electronic charges	Abcoulombs	1.60209×10^{-20}
	Coulombs	1.60209×10^{-19}
	Statcoulombs	4.80296×10^{-10}
Electronic charges/kg.	Statcoulombs/dyne	4.89766×10^{-16}
E.S. cgs. units of induction flux	E.M. cgs. units	2.997930×10^{10}
E.S. cgs. units of magnetic charge	E.M. cgs. units	2.997930×10^{10}
E.S. cgs. units of magnetic field intensity	E.M. cgs. units	3.335635×10^{-11}
Ells	Centimeters	114.3
	Inches	45
Ergs	B.t.u.	9.48451×10^{-11}
	Cal.	2.39006×10^{-8}
	Kilocal.	2.39006×10^{-11}
	Kilocal. (20°C.)	2.39126×10^{-11}
	Cu. cm.-atm.	9.86923×10^{-7}
	Cu. ft.-atm.	3.48529×10^{-11}
	Cu. ft.-(lb./sq. in.)	5.12196×10^{-10}
	Dyne-cm.	1
	Electron volts	6.24196×10^{11}
	Foot-poundals	2.3730360×10^{-6}
	Foot-pounds	7.37562×10^{-8}
	Gram-cm.	0.001019716
	Joules	1×10^{-7}
	Joules (Int.)	9.99835×10^{-8}
	Kw.-hours	2.777777×10^{-14}
	Kg.-meters	1.019716×10^{-8}
	Liter-atm.	9.86895×10^{-10}
	Watt-sec.	1×10^{-7}
Ergs/(gram-mol. × °C.)	Foot-pounds/(lb.-mol. × °F.)	1.85863×10^{-5}
Ergs/sec.	B.t.u./min.	5.69071×10^{-9}
	Cal./min.	1.43403×10^{-6}
	Dyne-cm./sec.	1

To convert from	To	Multiply by
	Foot-pounds/min.	4.42537×10^{-6}
	Gram-cm./sec.	0.001019716
	Horsepower	1.34102×10^{-10}
	Joules/sec.	1×10^{-7}
	Kilowatts	1×10^{-10}
	Watts	1×10^{-7}
Ergs/sq. cm.	Dynes/cm.	1
	Ergs/sq. mm.	0.01
Ergs/sq. mm.	Dynes/cm.	100
	Ergs/sq. cm.	100
Erg-sec.	Planck's constant	1.50932×10^{26}
Farads	Abfarads	1×10^{-9}
	E.M. cgs. units	1×10^{-9}
	E.S. cgs. units	8.987584×10^{11}
	Farads (Int.)	1.000495
	Microfarads	1×10^{6}
	Statfarads	8.98758×10^{11}
Farads (Int.)	Farads	0.999505
Fathoms	Centimeters	182.88
	Feet	6
	Inches	72
	Meters	1.8288
	Miles (naut., Int.)	0.00098747300
	Miles (statute)	0.001136363
	Yards	2
Feet	Centimeters	30.48
	Chains (Gunter's)	0.01515151
	Fathoms	0.166666
	Feet (U.S. Survey)	0.99999800
	Furlongs	0.00151515
	Inches	12
	Meters	0.3048
	Microns	304,800
	Miles (naut., Int.)	0.00016457883
	Miles (statute)	0.000189393
	Rods	0.060606
	Ropes (Brit.)	0.05
	Yards	0.333333

To convert from	To	Multiply by
Feet (U.S. Survey)	Centimeters	30.480061
	Chains (Gunter's)	0.015151545
	Chains (Ramden's)	0.010000020
	Feet	1.0000020
	Inches	12.000024
	Links (Gunter's)	1.5151545
	Links (Ramden's)	1.0000020
	Meters	0.30480061
	Miles (statute)	0.00018939432
	Rods	0.060606182
	Yards	0.33333400
Feet of air (1 atm., 60°F.)	Atmospheres	3.6083×10^{-5}
	Ft. of Hg (32°F.)	0.00089970
	Ft. of H_2O (60°F.)	0.0012244
	In. of Hg (32°F.)	0.0010796
	Pounds/sq. inch	0.00053027
Feet of Hg (32°F.)	Cm. of Hg (0°C.)	30.48
	Ft. of H_2O (60°F.)	13.6085
	In. of H_2O (60°F.)	163.302
	Ounces/sq. inch	94.3016
	Pounds/sq. inch	5.80385
Feet of H_2O (4°C.)	Atmospheres	0.0294990
	Cm. of Hg (0°C.)	2.24192
	Dynes/sq. cm.	29889.8
	Grams/sq. cm.	30.4791
	In. of Hg (32°F.)	0.882646
	Kg./sq. meter	304.791
	Pascals (N/sq. meter)	2989.07
	Pounds/sq. inch	0.433515
Feet/hour	Cm./hr.	30.48
	Cm./min.	0.508
	Cm./sec.	0.0084666
	Feet/min.	0.0166666
	Inches/hr.	12
	Kilometers/hr.	0.0003048
	Kilometers/min.	5.08×10^{-6}
	Knots (Int.)	0.0001645788
	Miles/hr.	0.000189393

To convert from	To	Multiply by
	Miles/min.	3.156565×10^{-6}
	Miles/sec.	5.2609428×10^{-8}
Feet/minute	Cm./sec.	0.508
	Feet/sec.	0.0166666
	Kilometers/hr.	0.018288
	Meters/min.	0.3048
	Meters/sec.	0.00508
	Miles/hr.	0.01136363
Feet/second	Cm./sec.	30.48
	Kilometers/hr.	1.09728
	Kilometers/min.	0.018288
	Meters/min.	18.288
	Miles/hr.	0.68181818
	Miles/min.	0.01136363
Feet/(sec. × sec.)	Kilometers/(hr. × sec.)	1.09728
	Meters/(sec. × sec.)	0.3048
	Miles/(hr. × sec.)	0.68181818
Firkins (Brit.)	Bushels (Brit.)	1.125
	Cu. cm.	40914.79
	Cu. feet	1.444892
	Firkins (U.S.)	1.200949
	Gallons (Brit.)	9
	Liters	40.91364
	Pints (Brit.)	72
Firkins (U.S.)	Barrels (U.S., dry)	0.29464286
	Barrels (U.S., liq.)	0.28571429
	Bushels (U.S.)	0.96678788
	Cu. feet	1.203125
	Firkins (Brit.)	0.8326747
	Liters	34.06775
	Pints (U.S., liq.)	72
Foot-candles	Lumens/sq. ft.	1
	Lumens/sq. meter	10.763910
	Lux	10.763910
	Milliphots	1.0763910
Foot-lamberts	Candles/sq. cm.	0.00034262591
	Candles/sq. ft.	0.31830989
	Millilamberts	1.0763910
	Lamberts	0.0010763910
	Lumens/sq. ft.	1

To convert from	To	Multiply by
Foot-poundals	B.t.u.	3.99678×10^{-5}
	B.t.u. (IST.)	3.99417×10^{-5}
	B.t.u. (mean)	3.99104×10^{-5}
	Cal.	0.0100717
	Cal. (IST.)	0.0100651
	Cal. (mean)	0.0100573
	Cu. cm.-atm.	0.415890
	Cu. ft.-atm.	1.46870×10^{-5}
	Dyne-cm.	4.2140110×10^{5}
	Ergs	4.2140110×10^{5}
	Foot-pounds	0.0310810
	Hp.-hours	1.56974×10^{-8}
	Joules	0.042140110
	Joules (Int.)	0.0421332
	Kg.-meters	0.00429710
	Kw.-hours	1.17056×10^{-8}
	Liter-atm.	0.000415879
	B.t.u.	0.00128593
Foot-pounds	B.t.u. (IST.)	0.00128509
	B.t.u. (mean)	0.00128408
	Cal.	0.324048
	Cal. (IST.)	0.323836
	Cal. (mean)	0.323582
	Cal. (20°C.)	0.324211
	Kilocal.	0.000324048
	Kilocal. (IST.)	0.000323836
	Kilocal. (mean)	0.000323582
	Cu. ft.-atm.	0.000472541
	Dyne-cm.	1.35582×10^{7}
	Ergs	1.35582×10^{7}
	Foot-poundals	32.1740
	Gram-cm.	13,825.5
	Hp.-hours	5.05050×10^{-7}
	Joules	1.35582
	Kg.-meters	0.138255
	Kw.-hours	3.76616×10^{-7}
	Kw.-hours (Int.)	3.76554×10^{-7}
	Liter-atm.	0.0133805
	Newton-meters	1.3558180

To convert from	To	Multiply by
	Lb. H_2O evap. from and at 212°F.	1.3245×10^{-6}
	Watt-hours	0.000376616
Foot-pounds/hr.	B.t.u./min.	2.14321×10^{-5}
	B.t.u. (mean)/min.	2.14013×10^{-5}
	Cal./min.	0.00540080
	Cal. (mean)/min.	0.00539304
	Ergs/min.	2.25970×10^{5}
	Foot-pounds/min.	0.0166666
	Horsepower	5.050505×10^{-7}
	Horsepower (metric)	5.12055×10^{-7}
	Kilowatts	3.76616×10^{-7}
	Watts	0.000376616
	Watts (Int.)	0.000376554
Foot-pounds/min.	B.t.u./sec.	2.14321×10^{-5}
	B.t.u. (mean)/sec.	2.14013×10^{-5}
	Cal./sec.	0.00540080
	Cal. (mean)/sec.	0.00539304
	Ergs/sec.	2.25970×10^{5}
	Foot-pounds/sec.	0.0166666
	Horsepower	3.030303×10^{-5}
	Horsepower (metric)	3.07233×10^{-5}
	Joules/sec.	0.0225970
	Joules (Int.)/sec.	0.0225932
	Kilowatts	2.25970×10^{-5}
	Watts	0.0225970
Foot-pounds/lb.	B.t.u./lb.	0.00128593
	B.t.u. (IST.)/lb.	0.00128509
	B.t.u. (mean)/lb.	0.00128408
	Cal/gm.	0.000714404
	Cal. (IST.)/gram	0.000713937
	Cal. (mean)/gram	0.000713377
	Hp.-hr./lb.	5.05050×10^{-7}
	Joules/gram	0.00298907
	Kg.-meters/gram	0.000304800
	Kw.-hr./gram	8.30296×10^{-10}
Foot-pounds/sec.	B.t.u./min.	0.0771556
	B.t.u. (mean)/min.	0.0770447
	B.t.u./sec.	0.00128593

To convert from	To	Multiply by
	B.t.u. (mean)/sec.	0.00128408
	Cal./sec.	0.324048
	Cal. (mean)/sec.	0.323582
	Ergs/sec.	1.35582×10^7
	Gram-cm./sec.	13,825.5
	Horsepower	0.00181818
	Joules/sec.	1.35582
	Kilowatts	0.00135582
	Watts	1.35582
	Watts (Int.)	1.35559
Furlongs	Centimeters	20,116.8
	Chains (Gunter's)	10
	Chains (Ramden's)	6.6
	Feet	660
	Inches	7920
	Meters	201.168
	Miles (naut., Int.)	0.10862203
	Miles (statute)	0.125
	Rods	40
	Yards	220
Gallons (Brit.)	Barrels (Brit.)	0.027777
	Bushels (Brit.)	0.125
	Cu. centimeters	4546.087
	Cu. feet	0.1605436
	Cu. inches	277.4193
	Drachms (Brit. fluid)	1280
	Firkins (Brit.)	0.111111
	Gallons (U.S., liq.)	1.200949
	Gills (Brit.)	32
	Liters	4.545960
	Minims (Brit.)	76,800
	Ounces (Brit., fluid)	160
	Ounces (U.S., fluid)	153.7215
	Pecks (Brit.)	0.5
	Lb. of H_2O (62°F.)	10
Gallons (U.S., dry)	Barrels (U.S., dry)	0.038095592
	Barrels (U.S., liq.)	0.036941181
	Bushels (U.S.)	0.125

To convert from	To	Multiply by
	Cu. centimeters	4404.8828
	Cu. feet	0.15555700
	Cu. inches	268.8025
	Gallons (U.S., liq.)	1.16364719
	Liters	4.404760
Gallons (U.S., liq.)	Acre-feet	3.0688833×10^{-6}
	Barrels (U.S., liq.)	0.031746032
	Barrels (petroleum, U.S.)	0.023809524
	Bushels (U.S.)	0.10742088
	Cu. centimeters	3785.4118
	Cu. feet	0.133680555
	Cu. inches	231
	Cu. meters	0.0037854118
	Cu. yards	0.0049511317
	Gallons (Brit.)	0.8326747
	Gallons (U.S., dry)	0.85936701
	Gallons (wine)	1
	Gills (U.S.)	32
	Liters	3.785306
	Minims (U.S.)	61,440
	Ounces (U.S., fluid)	128
	Pints (U.S., liq.)	8
	Quarts (U.S., liq.)	4
Gallons (U.S.) of H_2O (4°C.)	Lb. of H_2O	8.34517
Gallons (U.S.) of H_2O (60°F.)	Lb. of H_2O	8.33717
Gallons (U.S.)/day	Cu. ft./hr.	0.0055700231
Gallons (Brit.)/hr.	Cu. meters/min.	7.576812×10^{-5}
Gallons (U.S.)/hr.	Acre-feet/hr.	3.0688833×10^{-6}
	Cu. ft./hr.	0.1336805
	Cu. meters/min.	6.3090197×10^{-5}
	Cu. yd./min.	8.2518861×10^{-5}
	Liters/hr.	3.785306
Gal. (Brit.)/sec.	Cu. cm./sec.	4546.087
Gal. (U.S.)/sec.	Cu. cm./sec.	3785.4118
	Cu. ft./min.	8.020833
	Cu. yd./min.	0.29706790
	Liters/min.	227.1183

To convert from	To	Multiply by
Gammas	Grams	1×10^{-6}
	Micrograms	1
Gausses	E.M. cgs. units of magnetic flux density	1
	E.S. cgs. units	3.335635×10^{-11}
	Gausses (Int.)	0.999670
	Maxwells/sq. cm.	1
	Lines/sq. cm.	1
	Lines/sq. inch	6.4516
Gausses (Int.)	Gausses	1.000330
Gausses/oersted	E.M. cgs. units of permeability	1
	E.S. cgs. units	1.112646×10^{-21}
Geepounds	Slugs	1
	Kilograms	14.5939
Gigameters	Meters	1×10^{9}
Gilberts	Abampere-turns	0.079577472
	Ampere-turns	0.79577472
	E.M. cgs. units of mmf., or magnetic potential	1
	E.S. cgs. units	2.997930×10^{10}
	Gilberts (Int.)	1.000165
Gilberts (Int.)	Gilberts	0.999835
Gilberts/cm.	Ampere-turns/cm.	0.79577472
	Ampere-turns/in.	2.0212678
	Oersteds	1
Gilberts/maxwell	Ampere-turns/weber	7.957747×10^{7}
	E.M. cgs. units of reluctance	1
	E.S. cgs. units	8.987584×10^{20}
Gills (Brit.)	Cu. cm.	142.0652
	Gallons (Brit.)	0.03125
	Gills (U.S.)	1.200949
	Liters	0.1420613
	Ounces (Brit., fluid)	5
	Ounces (U.S., fluid)	4.803764
	Pints (Brit.)	0.25
Gills (U.S.)	Cu. cm.	118.29412
	Cu. inches	7.21875
	Drams (U.S., fluid)	32

To convert from	To	Multiply by
	Gallons (U.S., liq.)	0.03125
	Gills (Brit.)	0.8326747
	Liters	0.1182908
	Minims (U.S.)	1920
	Ounces (U.S., fluid)	4
	Pints (U.S., liq.)	0.25
	Quarts (U.S., liq.)	0.125
Gons (Grades)	Circles	0.0025
	Circumferences	0.0025
	Degrees	0.9
	Minutes	54
	Radians	0.015707963
	Revolutions	0.0025
	Seconds	3240
Grains	Carats (metric)	0.32399455
	Drams (apoth. *or* troy)	0.016666
	Drams (avdp.)	0.036571429
	Dynes	63.5460
	Grams	0.06479891
	Milligrams	64.79891
	Ounces (apoth. *or* troy)	0.0020833
	Ounces (avdp.)	0.0022857143
	Pennyweights	0.041666
	Pounds (apoth. *or* troy)	0.000173611
	Pounds (avdp.)	0.00014285714
	Scruples (apoth.)	0.05
	Tons (metric)	6.479891×10^{-8}
Grains/cu. ft.	Grams/cu. meter	2.2883519
Grains/gal. (U.S.)	Parts/million	17.11854
	Pounds/million gal.	142.8571
Grams-force	Dynes	980.665
	Newtons	9.80665×10^{-3}
Grams	Carats (metric)	5
	Decigrams	10
	Dekagrams	0.1
	Drams (apoth. *or* troy)	0.25720597
	Drams (avdp.)	0.56438339
	Dynes	980.665
	Grains	15.432358
	Kilograms	0.001

To convert from	To	Multiply by
	Micrograms	1×10^6
	Myriagrams	0.0001
	Ounces (apoth. *or* troy)	0.032150737
	Ounces (avdp.)	0.035273962
	Pennyweights	0.64301493
	Poundals	0.0709316
	Pounds (apoth. *or* troy)	0.0026792289
	Pounds (avdp.)	0.0022046226
	Scruples (apoth.)	0.77161792
	Tons (metric)	1×10^{-6}
Grams/cm.	Dynes/cm.	980.665
	Grams/inch	2.54
	Kg./km.	100
	Kg./meter	0.1
	Poundals/inch	0.180166
	Pounds/ft.	0.067196898
	Pounds/inch	0.0055997415
	Tons (metric)/km.	0.1
Grams/(cm. × sec.)	Poises	1
	Lb./(ft. × sec.)	0.06719690
Grams/cu. cm.	Dynes/cu. cm.	980.665
	Grains/milliliter	15.43279
	Grams/milliliter	1.000028
	Poundals/cu. inch	1.16236
	Pounds/circ. mil-ft.	3.4049170×10^{-7}
	Pounds/cu. ft.	62.427961
	Pounds/cu. inch	0.036127292
	Pounds/gal. (Brit.)	10.02241
	Pounds/gal. (U.S., dry)	9.7111064
	Pounds/gal. (U.S., liq.)	8.3454044
Grams/cu. meter	Grains/cu. ft.	0.43699572
Grams/liter	Parts/million	1000
	Lb./1000 cu. ft.	0.06242621
	Lb./gal. (U.S.)	8.345171
Grams/milliliter	Grams/cu. cm.	0.999972
	Pounds/cu. ft.	62.42621
	Pounds/gallon (U.S.)	8.345171
Grams/sq. cm.	Atmospheres	0.000967841
	Bars	0.000980665

To convert from	To	Multiply by
	Cm. of Hg (0°C.)	0.0735559
	Dynes/sq. cm.	980.665
	In. of Hg (32°F.)	0.0289590
	Kg./sq. meter	10
	Mm. of Hg (0°C.)	0.735559
	Poundals/sq. inch	0.457623
	Pounds/sq. inch	0.014223343
Grams/ton (long)	Milligrams/kg.	0.98420653
Grams/ton (short)	Milligrams/kg.	1.1023113
Grams-cm.	B.t.u.	9.30113×10^{-8}
	B.t.u. (IST.)	9.29505×10^{-8}
	B.t.u. (mean)	9.28776×10^{-8}
	Cal.	2.34385×10^{-5}
	Cal. (IST.)	2.34231×10^{-5}
	Cal. (mean)	2.34048×10^{-5}
	Cal. (15°C.)	2.34284×10^{-5}
	Cal. (20°C.)	2.34502×10^{-5}
	Kilocal.	2.34385×10^{-8}
	Kilocal. (IST.)	2.34231×10^{-8}
	Kilocal. (mean)	2.34048×10^{-8}
	Dyne-cm.	980.665
	Ergs	980.665
	Foot-poundals	0.00232715
	Foot-pounds	7.2330138×10^{-5}
	Hp.-hours	3.65303×10^{-11}
	Joules	9.80665×10^{-5}
	Kw.-hours	2.72407×10^{-11}
	Kw.-hours (Int.)	2.72362×10^{-11}
	Newton-meters	9.80665×10^{-5}
	Watt-hours	2.72407×10^{-8}
Gram-cm./sec.	B.t.u./sec.	9.30113×10^{-8}
	Cal./sec.	2.34385×10^{-5}
	Ergs-sec.	980.665
	Foot-pounds/sec.	7.2330138×10^{-5}
	Horsepower	1.31509×10^{-7}
	Joules/sec.	9.80665×10^{-5}
	Kilowatts	9.80665×10^{-8}
	Kilowatts (Int.)	9.80503×10^{-8}
	Watts	9.80665×10^{-5}

To convert from	To	Multiply by
Gram/sq. cm.	Pounds/sq. inch	0.000341717
Gram wt.-sec./sq. cm.	Poises	980.665
Gravitational constants	Cm./(sec. × sec.)	980.621
	Ft./(sec. × sec.)	32.1725
Hands	Centimeters	10.16
	Inches	4
Hectares	Acres	2.4710538
	Ares	100
	Sq. cm.	1×10^8
	Sq. feet	107639.10
	Sq. meters	10,000
	Sq. miles	0.0038610216
	Sq. rods	395.36861
Hectograms	Grams	100
	Poundals	7.09316
	Pounds (apoth *or* troy)	0.26792289
	Pounds (avdp.)	0.22046226
Hectoliters	Bushels (Brit.)	2.749694
	Bushels (U.S.)	2.837839
	Cu. cm.	1.00028×10^5
	Cu. feet	3.531566
	Gallons (U.S., liq.)	26.41794
	Liters	100
	Ounces (U.S.) fluid	3381.497
	Pecks (U.S.)	11.35136
Hectometers	Centimeters	10,000
	Decimeters	1000
	Dekameters	10
	Feet	328.08399
	Meters	100
	Rods	19.883878
	Yards	109.3613
Hectowatts	Watts	100
Hefner units	Candles (English)	0.86
	Candles (German)	0.85
	Candles (Int.)	0.90
	10-cp. pentane candles	0.090

To convert from	To	Multiply by
Henries	Abhenries	1×10^9
	E.M. cgs. units	1×10^9
	E.S. cgs. units	1.112646×10^{-12}
	Henries (Int.)	0.999505
	Millihenries	1000
	Mks. (r *or* nr) units	1
	Stathenries	1.112646×10^{-12}
Henries (Int.)	Henries	1.000495
Henries/meter	Cgs. units of permeability	795,774.72
	E.M. cgs. units	795,774.72
	E.S. cgs. units	8.854156×10^{-16}
	Gausses/oersted	795,774.72
	Mks. (nr) units	0.079577472
	Mks. (r) units	1
Hogsheads	Butts (Brit.)	0.5
	Cu. feet	8.421875
	Cu. inches	14,553
	Cu. meters	0.23848094
	Gallons (Brit.)	52.458505
	Gallons (U.S.)	63
	Gallons (wine)	63
	Liters	238.47427
Horsepower	B.t.u. (mean)/hr.	2542.48
	B.t.u./min.	42.4356
	B.t.u. (mean)/sec.	0.706243
	Cal./hr.	6.41616×10^5
	Cal. (IST.)/hr.	6.41196×10^5
	Cal. (mean)/hr.	6.40693×10^5
	Cal./min.	10,693.6
	Cal. (IST.)/min.	10,686.6
	Cal. (mean)/min.	10.678.2
	Ergs/sec.	7.45700×10^9
	Foot-pounds/hr.	1,980,000
	Foot-pounds/min.	33,000
	Foot-pounds/sec.	550
	Horsepower (boiler)	0.0760181
	Horsepower (electric)	0.999598
	Horsepower (metric)	1.01387
	Joules/sec.	745.700
	Kilowatts	0.745700

To convert from	To	Multiply by
	Kilowatts (Int.)	0.745577
	Tons of refrig. (U.S., comm.)	0.21204
	Watts	745.700
Horsepower (boiler)	B.t.u. (mean)/hr.	33,445.7
	Cal./min.	140,671.6
	Cal. (mean)/min.	140,469.4
	Cal. (15°C.)/min.	140,611.1
	Cal. (20°C.)/min.	140,742.2
	Ergs/sec.	9.80950×10^{10}
	Foot-pounds/min.	434,107
	Horsepower	13.1548
	Horsepower (electric)	13.1495
	Horsepower (metric)	13.3372
	Horsepower (water)	13.1487
	Joules/sec.	9809.50
	Kilowatts	9.809.50
	Lb. H_2O evap. per hr. from and at 212°F	34.5
Horsepower (electric)	B.t.u./hr.	2547.16
	B.t.u. (IST.)/hr.	2545.50
	B.t.u. (mean)/hr.	2543.50
	Cal./sec.	178.298
	Kilocal./hr.	641.874
	Ergs/sec.	7.46×10^{9}
	Foot-pounds/min.	33,013.3
	Foot-pounds/sec.	550.221
	Horsepower	1.00040
	Horsepower (boiler)	0.0760487
	Horsepower (metric)	1.0142777
	Horsepower (water)	0.999942
	Joules/sec.	746
	Kilowatts	0.746
	Watts	746
Horsepower (metric)	B.t.u/hr.	2511.31
	B.t.u. (IST.)/hr.	2509.66
	B.t.u. (mean)/hr.	2507.70
	Cal./hr.	6.32838×10^{5}

To convert from	To	Multiply by
	Cal. (IST.)/hr.	6.32425×10^5
	Cal. (mean)/hr.	6.31929×10^5
	Ergs/sec.	7.35499×10^9
	Foot-pounds/min.	32,548.6
	Foot-pounds/sec.	542.476
	Horsepower	0.986320
	Horsepower (boiler)	0.0749782
	Horsepower (electric)	0.985923
	Horsepower (water)	0.985866
	Kg.-meter/sec.	75
	Kilowatts	0.735499
	Watts	735.499
Horsepower (water)	Foot-pounds/min.	33,015.2
	Horsepower	1.00046
	Horsepower (boiler)	0.0760531
	Horsepower (electric)	1.00006
	Horsepower (metric)	1.01434
	Kilowatts	0.746043
Horsepower-hours	B.t.u.	2546.14
	B.t.u. (IST.)	2544.47
	B.t.u. (mean)	2542.48
	Cal.	641,616
	Cal. (IST.)	641,196
	Cal. (mean)	640,693
	Foot-pounds	1.98×10^6
	Joules	2.68452×10^6
	Kg.-meters	273,745
	Kw.-hours	0.745700
	Watt-hours	745.700
Hp.-hr./lb.	B.t.u./lb.	2546.14
	Cal./gram	1414.52
	Cu. ft.-(lb./sq. in.)/lb.	13,750
	Foot-pounds/lb.	1,980,000
	Joules/gram	5918.35
Hours (mean solar)	Days (mean solar)	0.0416666
	Days (sidereal)	0.041780746
	Hours (sidereal)	1.00273791
	Minutes (mean solar)	60
	Minutes (sidereal)	60.164275

To convert from	To	Multiply by
	Seconds (mean solar)	3600
	Seconds (sidereal)	3609.8565
	Weeks (mean calendar)	0.0059523809
Hours (sidereal)	Days (mean solar)	0.41552899
	Days (sidereal)	0.0416666
	Hours (mean solar)	0.99726957
	Minutes (mean solar)	59.836174
	Minutes (sidereal)	60
Hundredweights (long)	Kilograms	50.802345
	Pounds	112
	Quarters (Brit., long)	4
	Quarters (U.S., long)	0.2
	Tons (long)	0.05
Hundredweights (short)	Kilograms	45.359237
	Pounds (advp.)	100
	Quarters (Brit., short)	4
	Quarters (U.S., short)	0.2
	Tons (long)	0.044642857
	Tons (metric)	0.045359237
	Tons (short)	0.05
Inches	Ångström units	2.54×10^8
	Centimeters	2.54
	Chains (Gunter's)	0.00126262
	Cubits	0.055555
	Fathoms	0.013888
	Feet	0.083333
	Feet (U.S. Survey)	0.083333167
	Links (Gunter's)	0.126262
	Links (Ramden's)	0.083333
	Meters	0.0254
	Mils	1000
	Picas (printer's)	6.0225
	Points (printer's)	72.27000
	Wavelength of orange-red line of krypton 86	41,929.399

To convert from	To	Multiply by
	Wavelength of the red line of cadmium	39,450.369
	Yards	0.027777
Inches of Hg (32°F.)	Atmospheres	0.0334211
	Bars	0.0338639
	Dynes/sq. cm.	33,863.9
	Ft. of air (1 atm., 60°F.)	926.24
	Ft. of H_2O (39.2°F.)	1.132957
	Grams/sq. cm.	34.5316
	Kg./sq. meter	345.316
	Mm. of Hg (60°C.)	25.4
	Ounces/sq. inch	7.85847
	Pascals	3386.39
Inches of Hg (32°F.)	Pounds/sq. ft.	70.7262
Inches of Hg (60°F.)	Atmospheres	0.0333269
	Dynes/sq. cm.	39,768.5
	Grams/sq. cm.	34.4343
	Mm. of Hg (60°F.)	25.4
	Ounces/sq. inch	7.83633
	Pounds/sq. ft.	70.5269
Inches of H_2O (4°C.)	Atmospheres	0.0024582
	Dynes/sq. cm.	2490.82
	In. of Hg (32°F.)	0.0735539
	Kg./sq. meter	25.3993
	Ounces/sq. ft.	83.2350
	Ounces/sq. inch	0.578020
	Pascals	249.089
	Pounds/sq. ft.	5.20218
	Pounds/sq. inch	0.03612628
Inches/hr.	Cm./hr.	2.54
	Feet/hr.	0.0833333
	Miles/hr.	1.578282×10^{-5}
Inches/min.	Cm./hr.	152.4
	Feet/hr.	5
	Miles/hr.	0.000946969
Joules (abs.)	B.t.u.	0.000948451
	B.t.u. (IST.)	0.000947831

To convert from	To	Multiply by
	B.t.u. (mean)	0.000947088
	Cal.	0.239006
	Cal. (IST.)	0.238849
	Cal. (mean)	0.238662
	Cal. (15°C.)	0.238903
	Cal. (20°C.)	0.239126
	Kilocal. (mean)	0.000238662
	Cu. ft.-atm.	0.000348529
	Ergs	1×10^{7}
	Foot-poundals	23.730360
	Foot-pounds	0.737562
	Gram-cm.	10,197.16
	Hp.-hours	3.72506×10^{-7}
	Joules (Int.)	0.999835
	Kg.-meters	0.1019716
	Kw.-hours	2.7777×10^{-7}
	Liter-atm.	0.00986895
	Volt-coulombs (Int.)	0.999835
	Watt-hours (abs.)	0.0002777777
	Watt-hours (Int.)	0.000277732
	Watt-sec.	1
	Watt-sec. (Int.)	0.999835
Joules (Int.)	B.t.u.	0.000948608
	B.t.u. (IST.)	0.000947988
	B.t.u. (mean)	0.000947244
	Cal.	0.239045
	Cal. (IST.)	0.238888
	Cal. (mean)	0.238702
	Cu. cm.-atm.	9.87086
	Cu. ft.-atm.	0.000348586
	Dyne-cm.	1.000165×10^{7}
	Ergs	1.000165×10^{7}
	Foot-poundals	23.73428
	Foot-pounds	0.737684
	Gram-cm.	10,198.8
	Joules (abs.)	1.000165
	Kw.-hours	2.77824×10^{-7}
	Liter-atm.	0.00987058
	Volt-coulombs	1.000165
	Volt-coulombs (Int.)	1

To convert from	To	Multiply by
	Watt-sec.	1.000165
	Watt-sec. (Int.)	1
Joules/(abcoulomb) × °F.)	Joules/(coulomb × °C.)	0.18
Joules/amp.-hr.	Joules/abcoulomb	0.002777
	Joules/statcoulomb	9.265653×10^{-14}
Joules/coulomb	Joules/abcoulomb	10
	Volts	1
Joules/(coulomb × °F.)	Joules/(coulomb × °C.)	1.8
Joules/ °C	B.t.u./ °F	0.000526917
	Cal./ °C.	0.239006
	Cal. (mean)/ °C	0.238662
Joules/electronic charge	Joules/abcoulomb	6.24196×10^{19}
Joules/(electronic charge × °C.)	Joules/(coulomb × °C.)	6.24196×10^{18}
Joules/(gram × °C.)	B.t.u./(lb. × °F.)	0.239006
	Cal./(gram × °C.)	0.239006
Joules (Int.)/(gram × °C.)	B.t.u./(lb. × °F.)	0.239045
	Cal. (mean)/(gram × °C.)	0.238702
Joules/sec. (abs.)	B.t.u./min.	0.0569071
	Cal./min	14.3403
	Kilocal./min	0.0143403
	Kilocal. (mean)/min.	0.0143197
	Dyne-cm./sec.	1×10^{7}
	Ergs/sec.	1×10^{7}
	Foot-pounds/sec.	0.737562
	Gram-cm./sec.	10,197.16
	Horsepower	0.00134102
	Watts	1
	Watts (Int.)	0.999835
Joules (Int.)/sec.	B.t.u./min.	0.0569165
	B.t.u. (mean)/min.	0.0568347
	Cal./min.	14.3427
	Kilocal./min.	0.0143427
	Dyne-cm./sec.	1.000165×10^{7}
	Ergs/sec.	1.000165×10^{7}
	Foot-pounds/min.	44.2610

To convert from	To	Multiply by
	Foot-pounds/sec.	0.737684
	Gram-cm./sec.	10,198.8
	Horsepower	0.00134124
	Watts	1.000165
	Watts (Int.)	1
Kilderkins (Brit.)	Cu. cm.	81,829.57
	Cu. feet	2.889784
	Cu. inches	4993.55
	Cu. meters	0.08182957
	Gallons (Brit.)	18
Kilocalories	B.t.u.	3.9683207
	B.t.u. (IST.)	3.96573
	B.t.u. (mean)	3.96262
	B.t.u. (60°F.)	3.96709
	Kilocal.	1000
	Kilocal. (mean)	0.998563
	Kilocal. (15°C.)	0.999570
	Kilocal. (20°C.)	1.00050
	Cu. cm.-atm.	41,292.86
	Ergs	4.184×10^{10}
	Foot-poundals	99,287.8
	Foot-pounds	3085.96
	Gram-cm.	4.26649×10^{7}
	Hp.-hours	0.00155857
	Joules	4184
	Kw.-hours	0.001162222
	Liter-atm.	41.2917
	Watt-hours	1.162222
Kilocalories (mean)	B.t.u.	3.97403
	B.t.u. (IST.)	3.97144
	B.t.u. (mean)	3.9683207
	B.t.u. (60°F.)	3.97280
	Cal.	1001.44
	Cal. (IST.)	1000.78
	Cal. (mean)	1000
	Cal. (15°C.)	1000.10
	Cal. (20°C.)	1001.94
	Ergs	4.19002×10^{10}
	Foot-poundals	99,430.8
	Foot-pounds	3090.40

To convert from	To	Multiply by
	Gram-cm.	4.27263×10^{7}
	Hp.-hours	0.00156081
	Joules	4190.02
	Kg.-meters	427.263
	Kw.-hours (Int.)	0.00116370
	Liter-atm.	41.3511
	Watt-hours	1.16390
Kilocalories/hr.	Watts	1.162222
Kilocalories/min.	Kg. ice melted/min.	0.012548
	Lb. ice melted/min.	0.027665
	Watts	69.7333
Kilograms-force	Dynes	9.80665×10^{5}
	Newtons	9.80665
	Pounds-force	2.20462
	Poundals	70.9316
Kilograms-force/sq. cm.	Pascals (N/sq. meter)	98,066.5
Kilograms	Drams (apoth. *or* troy)	257.20597
	Drams (avdp.)	564.38339
	Dynes	980,665
	Grains	15,432.358
	Hundredweights (long)	0.019684131
	Hundredweights (short)	0.022046226
	Ounces (apoth. *or* troy)	32.150737
	Ounces (avdp.)	35.273962
	Pennyweights	643.01493
	Poundals	70.931635
	Pounds (apoth. *or* troy)	2.6792289
	Pounds (avdp.)	2.2046226
	Quarters (Brit., long)	0.078736522
	Quarters (U.S., long)	0.0039368261
	Scruples (apoth.)	771.61792
	Slugs	0.06852177
	Tons (long)	0.00098420653
	Tons (metric)	0.001
	Tons (short)	0.0011023113
Kilograms/cu. meter	Grams/cu. cm.	0.001
	Lb./cu. ft.	0.062427961
	Lb./cu. inch	3.6127292×10^{-5}

To convert from	To	Multiply by
Kg. of ice melted/ hr.	Tons of refrig. (U.S., comm.)	0.026336
Kilograms/sq. cm.	Atmospheres	0.967841
	Bars	0.980665
	Cm. of Hg (0°C.)	73.5559
	Dynes/sq. cm.	980,665
	Ft. of H_2O (39.2°F.)	32.8093
	In. of Hg (32°F.)	28.9590
	Pounds/sq. inch	14.223343
Kilograms/sq. meter	Atmospheres	9.67841×10^{-5}
	Bars	9.80665×10^{-5}
	Dynes/sq. cm.	98.0665
	Ft. of H_2O (39.2°F.)	0.00328093
	Grams/sq. cm.	0.1
	In. of Hg (32°F.)	0.00289590
	Mm. of Hg (0°C.)	0.0735559
	Pounds/sq. ft.	0.20481614
	Pounds/sq. in.	0.0014223343
Kilograms/sq. mm.	Pounds/sq. ft.	204,816.14
	Pounds/sq. in.	1422.3343
	Tons (short)/sq. in.	0.71116716
Kilogram sq. cm.	Pounds sq. ft.	0.0023730360
	Pounds sq. in.	0.34171719
Kilogram-meters	B.t.u. (mean)	0.00928776
	Cal. (mean)	2.34048
	Kilocal. (mean)	0.00234048
	Cu. ft.-atm.	0.00341790
	Dynes-cm.	9.80665×10^{7}
	Ergs	9.80665×10^{7}
	Foot-poundals	232.715
	Foot-pounds	7.23301
	Gram-cm.	100,000
	Hp.-hours	3.65304×10^{-6}
	Joules	9.80665
	Joules (Int.)	9.80503
	Kw.-hours	2.72407×10^{-6}
	Liter-atm.	0.0967814
	Newton-meters	9.80665
	Watt-hours	0.00272407
	Watt-hours (Int.)	0.00272362

To convert from	To	Multiply by
Kilogram-meters/ sec.	Watts	9.80665
Kilolines	Maxwells	1000
	Webers	1×10^{-5}
Kiloliters	Cu. centimeters	1.000028×10^{6}
	Cu. feet	35.31566
	Cu. inches	61,025.45
	Cu. meters	1.000028
	Cu. yards	1.307987
	Gallons (Brit.)	219.9755
	Gallons (U.S., dry)	227.0271
	Gallons (U.S., liq.)	264.1794
	Liters	1000
Kilometers	Astronomical units	6.68878×10^{-9}
	Centimeters	100,000
	Feet	3280.8399
	Feet (U.S. Survey)	3280.833
	Light years	1.05702×10^{-13}
	Meters	1000
	Miles (naut., Int.)	0.53995680
	Miles (statute)	0.62137119
	Myriameters	0.1
	Rods	198.83878
	Yards	1093.6133
Kilometers/hr.	Cm./sec.	27.7777
	Feet/hr.	3280.8399
	Feet/min.	54.680665
	Knots (Int.)	0.53995680
	Meters/sec.	0.277777
	Miles (statute)/hr.	0.62137119
Kilometers/(hr. × sec.)	Cm./(sec. × sec.)	27.7777
	Ft./(sec. × sec.)	0.91134442
	Meters/(sec. × sec.)	0.277777
Kilometers/min.	Cm./sec.	1666.666
	Feet/min.	3280.8399
	Kilometers/hr.	60
	Knots (Int.)	32.397408
	Miles/hr.	37.282272
	Miles/min.	0.62137119

To convert from	To	Multiply by
Kilovolts/cm.	Abvolts/cm.	1×10^{11}
	Microvolts/meter	1×10^{11}
	Millivolts/meter	1×10^{8}
	Statvolts/cm.	3.335635
	Volts/inch	2540
Kilowatts	B.t.u./hr.	3414.43
	B.t.u. (IST.)/hr.	3412.19
	B.t.u. (mean)/hr.	3409.52
	B.t.u. (mean)/min.	56.8253
	B.t.u. (mean)/sec.	0.947088
	Cal. (mean)/hr.	859,184
	Cal. (mean)/min.	14,319.7
	Cal. (mean)/sec.	238.662
	Kilocal. (mean)/hr.	859.184
	Kilocal. (mean)/min.	14.3197
	Kilocal. (mean)/sec.	0.238662
	Cu. ft.-atm./hr.	1254.70
	Ergs/sec.	1×10^{10}
	Foot-poundals/min.	1.42382×10^{6}
	Foot-pounds/hr.	2.65522×10^{6}
	Foot-pounds/min.	44,253.7
	Foot-pounds/sec.	737.562
	Gram-cm./sec.	1.019716×10^{7}
	Horsepower	1.34102
	Horsepower (boiler)	0.101942
	Horsepower (electric)	1.34048
	Horsepower (metric)	1.35962
	Joules/hr.	3.6×10^{6}
	Joules (IST.)/hr.	3.59941×10^{6}
	Joules/sec.	1000
	Kg.-meters/hr.	3.67098×10^{5}
	Kilowatts (Int.)	0.999835
	Watts (Int.)	999.835
Kilowatts (Int.)	B.t.u./hr.	3414.99
	B.t.u. (IST.)/hr.	3412.76
	B.t.u. (mean)/hr.	3410.08
	B.t.u. (mean)/min.	56.8347
	B.t.u. (mean)/sec.	0.947244
	Cal. (mean)/hr.	859,326
	Cal. (mean)/min.	14,322.1

To convert from	To	Multiply by
	Kilocal./hr.	860.563
	Kilocal. (IST.)/hr.	860
	Kilocal (mean)/hr.	859.326
	Cu. cm.-atm./hr.	3.55351×10^7
	Cu. ft.-atm./hr.	1254.91
	Ergs/sec.	1.000165×10^{10}
	Foot-poundals/min.	1.42406×10^6
	Foot-pounds/min.	44,261.0
	Foot-pounds/sec.	737.684
	Gram-cm./sec.	1.01988×10^7
	Horsepower	1.34124
	Horsepower (boiler)	0.101959
	Horsepower (electric)	1.34070
	Horsepower (metric)	1.35985
	Joules/hr.	3.60059×10^6
	Joules (Int.)/hr.	3.6×10^6
	Kg.-meters/hr.	367,158
	Kilowatts	1.000165
Kilowatt-hours	B.t.u. (mean)	3409.52
	Cal. (mean)	859,184
	Foot-pounds	2.65522×10^6
	Hp.-hours	1.34102
	Joules	3.6×10^6
	Kg.-meters	367,098
	Lb. H_2O evap. from and at 212°F	3.5168
	Watt-hours	1000
	Watt-hours (Int.)	999.835
Kilowatt-hours (Int.)	B.t.u. (mean)	3410.08
	Cal. (IST.)	860,000
	Cal. (mean)	859,326
	Cu., cm.-atm.	3.55351×10^7
	Cu. ft.-atm.	1254.91
	Foot-pounds	2.65566×10^6
	Hp.-hours	1.34124
	Joules	3.60059×10^6
	Joules (Int.)	3.6×10^6
	Kg.-meters	367,158

To convert from	To	Multiply by
Kw.-hr./gram	B.t.u./lb.	1.54876×10^6
	B.t.u. (IST.)/lb.	1.54774×10^6
	B.t.u. (mean)/lb.	1.54653×10^6
	Cal./gram	860,421
	Cal. (mean)/gram	859,184
	Cu. cm.-atm./gram	3.55292×10^7
	Cu. ft.-atm./lb.	569,124
	Hp.-hr./lb.	608.277
	Joules/gram	3.6×10^6
Knots (Int.)	Cm./sec.	51.4444
	Feet/hr.	6076.1155
	Feet/min.	101.26859
	Feet/sec.	1.6878099
	Kilometers/hr.	1.852
	Meters/min.	30.8666
	Meters/sec.	0.514444
	Miles (naut., Int.)/hr.	1
	Miles (statute)/hr.	1.1507794
Lamberts	Candles/sq. cm.	0.31830989
	Candles/sq. ft.	295.71956
	Candles/sq. inch	2.0536081
	Foot-lamberts	929.0304
	Lumens/sq. cm.	1
Lasts (Brit.)	Liters	2909.414
Leagues (naut., Brit.)	Feet	18,240
	Kilometers	5.559552
	Leagues (naut., Int.)	1.0006393
	Leagues (statute)	1.151515
	Miles (statute)	3.454545
Leagues (naut., Int.)	Fathoms	3038.0577
	Feet	18,228.346
	Kilometers	5.556
	Leagues (statute)	1.1507794
	Miles (statute)	3.4523383
Leagues (statute)	Fathoms	2640
	Feet	15,840
	Kilometers	4.828032
	Leagues (naut., Int.)	0.86897625

To convert from	To	Multiply by
	Miles (naut., Int.)	2.6069287
	Miles (statute)	3
Light years	Astronomical units	63,279.5
	Kilometers	9.46055×10^{12}
	Miles (statute)	5.87851×10^{12}
Lines	Maxwells	1
Lines (Brit.)	Centimeters	0.211666
	Inches	0.083333
Lines/sq. cm.	Gausses	1
Lines/sq. inch	Gausses	0.15500031
	Webers/sq. inch	1×10^{-8}
Links (Gunter's)	Chains (Gunter's)	0.01
	Feet	0.66
	Feet (U.S. Survey)	0.65999868
	Inches	7.92
	Meters	0.201168
	Miles (statute)	0.000125
	Rods	0.04
Links (Ramden's)	Centimeters	30.48
	Chains (Ramden's)	0.01
	Feet	1
	Inches	12
Liters	Bushels (Brit.)	0.2749694
	Bushels (U.S.)	0.02837839
	Cu. centimeters	1000.028
	Cu. feet	0.03531566
	Cu. inches	61.02545
	Cu. meters	0.001000028
	Cu. yards	0.001307987
	Drams (U.S., fluid)	270.5198
	Gallons (Brit.)	0.2199755
	Gallons (U.S., dry)	0.2270271
	Gallons (U.S., liq.)	0.2641794
	Gills (Brit.)	7.039217
	Gills (U.S.)	8.453742
	Hogsheads	0.004193325
	Minims (U.S.)	16,231.19
	Ounces (Brit., fluid)	35.19609
	Ounces (U.S., fluid)	33.81497
	Pecks (Brit.)	0.1099878

To convert from	To	Multiply by
	Pecks (U.S.)	0.1135136
	Pints (Brit.)	1.759804
	Pints (U.S., dry)	1.816217
	Pints (U.S., liq.)	2.113436
	Quarts (Brit.)	0.8799021
	Quarts (U.S., dry)	0.9081084
	Quarts (U.S., liq.)	1.056718
Liters/min.	Cu. ft./min.	0.03531566
	Cu. ft./sec.	0.0005885943
	Gal. (U.S., liq.)/min.	0.2641794
Liters/sec.	Cu. ft./min.	2.118939
	Cu. ft./sec.	0.03531566
	Cu. yards/min.	0.07847923
	Gal. (U.S., liq.)/min.	15.85077
	Gal. (U.S., liq.)/sec.	0.2641794
Liter-atm.	B.t.u.	0.0961045
	B.t.u. (IST.)	0.0960417
	B.t.u. (mean)	0.0959664
	Cal.	24.2179
	Cal.(IST.)	24.2021
	Cal. (mean)	24.1831
	Cu. ft.-atm.	0.0353157
	Foot-poundals	2404.55
	Foot-pounds	74.7356
	Hp.-hours	3.77452×10^{-5}
	Joules	101.328
	Joules (Int.)	101.311
	Kg.-meters	10.3326
	Kw.-hours	2.81466×10^{-5}
Liter-atm. (lat. 45°)	Joules	101.323
Lumens	Candle power (spher.)	0.079577472
Lumens (at 5550 Å)	Watts	0.0014705882
Lumens/sq. cm.	Lamberts	1
	Phots	1
Lumens/(sq. cm. × steradian)	Lamberts	3.1415927
Lumens/sq. ft.	Foot-candles	1
	Foot-lamberts	1
	Lumens/sq. meter	10.763910

To convert from	To	Multiply by
Lumens/(sq. ft. × steradian)	Millilamberts	3.3815822
Lumens/sq. meter	Foot-candles	0.09290304
	Lumens/sq. ft.	0.09290304
	Phots	0.0001
Lux	Foot-candles	0.09290304
	Lumens/sq. meter	1
	Phots	0.0001
Maxwells	E.M. cgs. units of induction flux	1
	E.S. cgs. units	3.335635×10^{-11}
	Gauss-sq. cm.	1
	Lines	1
	Maxwells (Int.)	0.999670
	Volt-seconds	1×10^{-8}
	Webers	1×10^{-8}
Maxwells (Int.)	Maxwells	1.000330
Maxwells/sq. cm.	Maxwells/sq. in.	6.4516
	Maxwells (Int.)/sq. cm.	0.999670
Maxwells (Int.)/sq. cm.	Maxwells/sq. cm.	1.000330
Maxwells/sq. inch	Maxwells/sq. cm.	0.15500031
Megalines	Maxwells	1×10^{6}
MegaPascals	Bars	10
	Newtons/sq. mm.	1
	Pascals	1×10^{6}
Megmhos/cm.	Abmhos/cm.	0.001
	Megmhos/inch cube	2.54
	(Microhm-cm.)$^{-1}$	1
Megmhos/inch	Megmhos/cm.	0.39370079
	(Microhm-inches)$^{-1}$	1
Megohms	Microhms	1×10^{12}
	Ohms	1×10^{6}
	Statohms	1.112646×10^{-6}
Megohms^{-1}	Micromhos	1
Meters	Ångström units	1×10^{10}
	Centimeters	100
	Chains (Gunter's)	0.049709695

To convert from	To	Multiply by
	Chains (Ramden's)	0.032808399
	Fathoms	0.54680665
	Feet	3.2808399
	Feet (U.S. Survey)	3.280833
	Furlongs	0.0049709695
	Inches	39.370079
	Kilometers	0.001
	Links (Gunter's)	4.9709695
	Links (Ramden's)	3.2808399
	Megameters	1×10^{-6}
	Miles (naut., Brit.)	0.00053961182
	Miles (naut., Int.)	0.00053995680
	Miles (statute)	0.00062137119
	Millimeters	1000
	Millimicrons	1×10^{9}
	Mils	39,370.079
	Rods	0.19883878
	Yards	1.0936133
Meters of Hg (0°C.)	Atmospheres	1.3157895
	Ft. of H_2O (60°F.)	44.6474
	In. of Hg (32°F.)	39.370079
	Kg./sq. cm.	1.35951
	Pounds/sq. inch	19.3368
Meters/hr.	Feet/hr.	3.2808399
	Feet/min.	0.054680665
	Knots (Int.)	0.00053995680
	Miles (statute)/hr.	0.00062137119
Meters/min.	Cm./sec.	1.666666
	Feet/min.	3.2808399
	Feet/sec.	0.054680665
	Kilometers/hr.	0.06
	Knots (Int.)	0.032397408
	Miles (statute)/hr.	0.037282272
Meters/sec.	Feet/min.	196.85039
	Feet/sec.	3.2808399
	Kilometers/hr.	3.6
	Kilometers/min.	0.06
	Miles (statute)/hr.	2.2369363
Meters/(sec. × sec.)	Kilometers/(hr. × sec.)	3.6
	Miles/(hr. × sec.)	2.2369363

To convert from	To	Multiply by
Meter-candles	Lumens/sq. meter	1
Mhos	Abmhos	1×10^{-9}
	Cgs. units of conductance	1
	E.M. cgs. units	1×10^{-9}
	E.S. cgs. units	8.987584×10^{11}
	Mhos (Int.)	1.000495
	Mks. (r *or* nr) units	1
	$Ohms^{-1}$	1
	Siemen's units	1
	Statmhos	8.987584×10^{11}
Mhos (Int.)	Abmhos	9.99505×10^{-10}
	Mhos	0.999505
Mhos/meter	Abmhos/cm.	1×10^{-11}
	Mhos (Int.)/meter	1.000495
Mho-ft./circ. mil	Mhos/cm.	6.0153049×10^{6}
Microfarads	Abfarads	1×10^{-15}
	Farads	1×10^{-6}
	Statfarads	8.987584×10^{5}
Micrograms	Grams	1×10^{-6}
	Milligrams	0.001
Microhenries	Henries	1×10^{-6}
	Stathenries	1.112646×10^{-18}
Microhms	Abohms	1000
	Megohms	1×10^{-12}
	Ohms	1×10^{-6}
	Statohms	1.112646×10^{-18}
Microhm-cm.	Abohm-cm.	1000
	Circ. mil-ohms/ft.	6.0153049
	Microhm-inches	0.39370079
	Ohm-cm.	1×10^{-6}
Microhm-inches	Circ. mil-ohms/ft.	15.278875
	Michrom-cm.	2.54
Micromicrofarads	Farads	1×10^{-12}
Micromicrons	Ångström units	0.01
	Centimeters	1×10^{-10}
	Inches	$3.9370079 \times 10^{-11}$
	Meters	1×10^{-12}
	Microns	1×10^{-6}
Microns	Ångström units	10,000

To convert from	To	Multiply by
Microns	Centimeters	0.0001
	Feet	3.2808399×10^{-5}
	Inches	3.9370079×10^{-5}
	Meters	1×10^{-6}
	Millimeters	0.001
	Millimicrons	1000
Miles (naut., Brit.)	Cable lengths (Brit.)	8.4444
	Fathoms	1013.333
	Feet	6080
	Meters	1853.184
	Miles (Adm., Brit.)	1
	Miles (naut., Int.)	1.0006393
	Miles (statute)	1.151515
Miles (naut., Int.)	Cable lengths	8.4390493
	Fathoms	1.012.6859
	Feet	6076.1155
	Feet (U.S. Survey)	6076.1033
	Kilometers	1.852
	Leagues (naut., Int.)	0.333333
	Meters	1852
	Miles (geographical)	1
	Miles (naut., Brit.)	0.99936110
	Miles (statute)	1.1507794
Miles (statute)	Centimeters	160,934.4
	Chains (Gunter's)	80
	Chains (Ramden's)	52.8
	Feet	5280
	Feet (U.S. Survey)	5279.9894
	Furlongs	8
	Inches	63,360
	Kilometers	1.609344
	Light years	1.70111×10^{-12}
	Links (Gunter's)	8000
	Meters	1609.344
	Miles (naut., Brit.)	086842105
	Miles (naut., Int.)	0.86897624
	Myriameters	0.1609344
	Rods	320
	Yards	1760

To convert from	To	Multiply by
Miles/hr.	Cm./sec.	44.704
	Feet/hr.	5280
	Feet/min.	88
	Feet/sec.	1.466666
	Kilometers/hr.	1.609344
	Knots (Int.)	0.86897624
	Meters/min.	26.8224
	Miles/min.	0.0166666
Miles/(hr. × min.)	Cm./(sec. × sec.)	0.7450666
Miles/(hr. × sec.)	Cm./(sec. × sec.)	44.704
	Ft./(sec. × sec.)	1.466666
	Kilometers/(hr. × sec.)	1.609344
	Meters/(sec. × sec.)	0.44704
Miles/min.	Cm./sec.	2682.24
	Feet/hr.	316,800
	Feet/sec.	88
	Kilometers/min.	1.609344
	Knots (Int.)	52.138574
	Meters/min.	1609.344
	Miles/hr.	60
Millibars	Atmospheres	0.000986923
	Bars	0.001
	Baryes	1000
	Dynes/sq. cm.	1000
	Grams/sq. cm.	1.019716
	In. of Hg (32°F.)	0.0295300
	Pascals	100
	Pounds/sq. ft.	2.088543
	Pounds/sq. inch	0.0145038
Milligrams	Carats (1877)	0.004871
	Carats (metric)	0.005
	Drams (apoth. *or* troy)	0.00025720597
	Drams (advp.)	0.00056438339
	Grains	0.015432358
	Grams	0.001
	Ounces (apoth. *or* troy)	3.2150737×10^{-5}
	Ounces (avdp.)	3.5273962×10^{-5}
	Pennyweights	0.00064301493
	Pounds (apoth. *or* troy)	2.6792289×10^{-6}

To convert from	To	Multiply by
	Pounds (avdp.)	2.2046226×10^{-6}
	Scruples (apoth.)	0.00077161792
Milligrams/assay ton	Milligrams/kg.	34.285714
	Ounces (troy)/ton (avdp.)	1
Milligrams/gm.	Dynes/cm.	0.980665
	Pounds/inch	5.5997415×10^{-6}
Milligrams/gram	Carats (parts gold per 24 of mixture)	0.024
	Grams/ton (short)	907.18474
	Milligrams/assay ton	29.166666
	Ounces (avdp.)/ton (long)	35.84
	Ounces (avdp.)/ton (short)	32
	Ounces (troy)/ton (long)	32.6666
	Ounces (troy)/ton (short)	29.1666
Milligrams/inch	Dynes/cm.	0.386089
	Dynes/inch	0.980665
	Grams/cm.	0.00039370079
	Grams/inch	0.0001
Milligrams/kg.	Pounds (avdp.)/ton (short)	0.002
Milligrams/liter	Grains/gal. (U.S.)	0.05841620
	Grams/liter	0.001
	Parts/million	1
	Lb./cu. ft.	6.242621×10^{-5}
Milligrams/mm.	Dynes/cm.	9.80665
Millihenries	Abhenries	1×10^{6}
	Henries	0.001
	Stathenries	1.112646×10^{-15}
Millilamberts	Candles/sq. cm.	0.00031830989
	Candles/sq. inch	0.0020536081
	Foot-lamberts	0.9290304
	Lamberts	0.001
	Lumens/sq. cm.	0.001
	Lumens/sq. ft.	0.9290304
Milliliters	Cu. cm.	1.000028
	Cu. inches	0.06102545
	Drams (U.S., fluid)	0.2705198
	Gills (U.S.)	0.008453742
	Liters	0.001
	Minims (U.S.)	16.23119
	Ounces (Brit., fluid)	0.03519609

To convert from	To	Multiply by
	Ounces (U.S., fluid)	0.03381497
	Pints (Brit.)	0.001759804
	Pints (U.S., liq.)	0.002113436
Millimeters	Ångström units	1×10^7
	Centimeters	0.1
	Decimeters	0.01
	Dekameters	0.0001
	Feet	0.0032808399
	Inches	0.039370079
	Meters	0.001
	Microns	1000
	Mils	39.370079
	Wavelength of orange-red line of krypton 86	1650.76373
	Wavelength of red line of cadmium	1553.16413
Millimeters of Hg (0°C)	Atmospheres	0.0013157895
	Bars	0.00133322
	Dynes/sq. cm.	1333.224
	Grams/sq. cm.	1.35951
	Kg./sq. meter	13.5951
	Pascals	133.3224
	Pounds/sq. ft.	2.78450
	Pounds/sq. inch	0.0193368
	Torrs	1
Millimeters of H_2O (4°C)	Pascals	9.80665
Millimicrons	Ångström units	10
	Centimeters	1×10^{-7}
	Inches	3.9370079×10^{-8}
	Microns	0.001
	Millimeters	1×10^{-6}
	Nanometers	1
Milliphots	Foot-candles	0.9290304
	Lumens/sq. ft.	0.9290304
	Lumens/sq. meter	10
	Lux	10
	Phots	0.001

To convert from	To	Multiply by
Millivolts	Statvolts	3.335635×10^{-6}
	Volts	0.001
Minims (Brit.)	Cu. cm.	0.05919385
	Cu. inches	0.003612230
	Milliliters	0.5919219
	Ounces (Brit., fluid)	0.0020833333
	Scruples (Brit., fluid)	0.05
Minims (U.S.)	Cu. cm.	0.061611520
	Cu. inches	0.0037597656
	Drams (U.S., fluid)	0.0166666
	Gallons (U.S., liq.)	1.6276042×10^{-5}
	Gills (U.S.)	0.0005208333
	Liters	6.160979×10^{-5}
	Milliliters	0.06160979
	Ounces (U.S., fluid)	0.002083333
	Pints (U.S., liq.)	0.0001302083
Minutes (angular)	Degrees	0.0166666
	Quadrants	0.000185185
	Radians	0.00029088821
	Seconds (angular)	60
Minutes (mean solar)	Days (mean solar)	0.00069444444
	Days (sidereal)	0.00069634577
	Hours (mean solar)	0.0166666
	Hours (sidereal)	0.016712298
	Minutes (sidereal)	1.00273791
Minutes (sidereal)	Days (mean solar)	0.00069254831
	Minutes (mean solar)	0.99726957
	Months (mean calendar)	2.2768712×10^{-5}
	Seconds (sidereal)	60
Minutes/cm.	Radians/cm.	0.00029088821
Months (lunar)	Days (mean solar)	29.530588
	Hours (mean solar)	708.73411
	Minutes (mean solar)	42524.047
	Seconds (mean solar)	2.5514428×10^{6}
	Weeks (mean calendar)	4.2186554
Months (mean calendar)	Days (mean solar)	30.416666
	Hours (mean solar)	730
	Months (lunar)	1.0300055

To convert from	To	Multiply by
	Weeks (mean calendar)	4.3452381
	Years (calendar)	0.08333333
	Years (sidereal)	0.083274845
	Years (tropical)	0.083278075
Myriagrams	Grams	10,000
	Kilograms	10
	Pounds (avdp.)	22.046226
Nanometers	Ångström	10
	Micrometer	0.001
	Mil	3.937008×10^{-5}
	Millimicron	1
Newtons	Dynes	1×10^{5}
	Kilograms-force	0.1019716
	Poundals	7.23301
	Pounds-force	0.224809
Newton-meters	Dyne-cm.	1×10^{7}
	Gram-cm.	10,197.162
	Kg.-meters	0.10197162
	Pound-feet	0.73756215
Newtons/sq. meter	Pascals	1
Newtons/sq. mm	MegaPascals	1
Noggins (Brit.)	Cu. cm.	142.0652
	Gallons (Brit.)	0.03125
	Gills (Brit.)	1
Oersteds	Ampere-turns/inch	2.0212678
	Ampere-turns/meter	79.577472
	E.M. cgs. units of magnetic field intensity	1
	E.S. cgs. units	2.997930×10^{10}
	Gilberts/cm.	1
	Oersteds (Int.)	1.000165
Oersteds (Int.)	Oersteds	0.999835
Ohms	Abohms	1×10^{9}
	Cgs. units of resistance	1
	Megohms	1×10^{-6}
	Microhms	1×10^{6}
	Ohms (Int.)	0.999505
	Statohms	1.112646×10^{-12}

To convert from	To	Multiply by
Ohms (Int.)	Ohms	1.000495
Ohms (mil, foot)	Circ. mil-ohms/ft.	1
	Ohm-cm.	1.6624261×10^{-7}
Ohm-cm.	Circ. mil-ohms/ft.	6.0153049×10^{6}
	Microhm-cm.	1×10^{6}
	Ohm-inches	0.39370079
Ohm-inches	Ohm-cm.	2.54
Ohm-meters	Abohm-cm.	1×10^{11}
	E.M. cgs. units	1×10^{11}
	E.S. cgs. units	1.112646×10^{-10}
	Mks. units	1
	Statohm-cm.	1.112646×10^{-10}
Ounces (apoth. *or* troy)	Dekagrams	1.7554286
	Drams (apoth. *or* troy)	8
Drams	(avdp.)	17.554286
	Grains	480
	Grams	31.103486
	Milligrams	31,103.486
	Ounces (avdp.)	1.0971429
	Pennyweights	20
	Pounds (apoth. *or* troy)	0.0833333
	Pounds (avdp.)	0.068571429
	Scruples (apoth.)	24
	Tons (short)	3.4285714×10^{-5}
Ounces (advp.)	Drams (apoth. *or* troy)	7.201666
	Drams (avdp.)	16
	Grains	437.5
	Grams	28.349523
	Hundredweights (long)	0.00055803571
	Hundredweights (short)	0.000625
	Ounces (apoth. *or* troy)	0.9114583
	Pennyweights	18.229166
	Pounds (apoth. *or* troy)	0.075954861
	Pounds (avdp.)	0.0625
	Scruples (apoth.)	21.875
	Tons (long)	2.7901786×10^{-5}
	Tons (metric)	2.8349527×10^{-5}
	Tons (short)	3.125×10^{-5}

To convert from	To	Multiply by
Ounces (Brit., fluid)	Cu. cm.	28.41305
	Cu. inches	1.733870
	Drachms (Brit., fluid)	8
	Drams (U.S., fluid)	7.686075
	Gallons (Brit.)	0.00625
	Milliliters	28.41225
	Minims (Brit.)	480
	Ounces (U.S., fluid)	0.9607594
Ounces (U.S., fluid)	Cu. cm.	29.573730
	Cu. inches	1.8046875
	Cu. meters	2.9573730×10^{-5}
	Drams (U.S., fluid)	8
	Gallons (U.S., dry)	0.0067138047
	Gallons (U.S., liq.)	0.0078125
	Gills (U.S.)	0.25
	Liters	0.029572702
	Minims (U.S.)	480
	Ounces (Brit., fluid)	1.040843
	Pints (U.S., liq.)	0.625
	Quarts (U.S., liq.)	0.03125
Ounces/sq. inch	Dynes/sq. cm.	4309.22
	Grams/sq. cm.	4.3941849
	In. of H_2O (39.2°F.)	1.73004
	In. of H_2O (60°F.)	1.73166
	Pounds/sq. ft.	9
	Pounds/sq. inch.	0.0625
Ounces (avdp.)/ton (long)	Milligrams/kg.	27.901786
Ounces (avdp.)/ton (short)	Milligrams/kg.	31.25
Paces	Centimeters	76.2
	Chains (Gunter's)	0.0378788
	Chains (Ramden's)	0.025
	Feet	2.5
	Hands	7.5
	Inches	30
	Ropes (Brit.)	0.125

To convert from	To	Multiply by
Palms	Centimeters	7.62
	Chains (Ramden's)	0.0025
	Cubits	0.1666666
	Feet	0.25
	Hands	0.75
	Inches	3
Parsecs	Kilometers	3.08374×10^{12}
	Light years	3.26164
	Miles (statute)	1.91615×10^{12}
Parts/million	Grains/gal. (Brit.)	0.07015488
	Grains/gal. (U.S.)	0.05841620
	Grams/liter	0.001
	Milligrams/liter	1
Pascals	Atmospheres	9.869233×10^{-6}
	Bars	1×10^{-5}
	Dyne/sq. cm.	10
	Feet of H_2O (conv.)	3.34552×10^{-4}
	Inches of Hg (conv.)	2.95300×10^{-4}
	Inches of H_2O (conv.)	4.01463×10^{-3}
	Kilograms-force/sq. cm.	0.01972×10^{-5}
	MegaPascals	1×10^{-6}
	Millibars	0.01
	Mm. of Hg (conv.)	7.50062×10^{-3}
	Mm. of H_2O (conv.)	0.101972
	Newtons/sq.-meter	1
	Newtons/sq. mm.	1×10^{-6}
	Poundals/sq. ft.	0.671969
	Pounds-force/sq. ft.	0.0208854
	Pounds-force/sq. inch	1.45038×10^{-4}
	Tons	7.50062×10^{-3}
Pecks (Brit.)	Bushels (Brit.)	0.25
	Coombs (Brit.)	0.0625
	Cu. cm.	9092.175
	Cu. inches	554.8385
	Gallons (Brit.)	2
	Gills (Brit.)	64
	Hogsheads	0.03812537
	Kilderkins (Brit.)	0.111111
	Liters	9.091920
	Pints (Brit.)	16

To convert from	To	Multiply by
	Quarterns (Brit., dry)	4
	Quarters (Brit., dry)	0.03125
	Quarts (Brit.)	8
	Quarts (U.S., dry)	8.256449
Pecks (U.S.)	Barrels (U.S., dry)	0.076191185
	Bushels (U.S.)	0.25
	Cu. cm.	8809.7675
	Cu. feet	0.311114005
	Cu. inches	537.605
	Gallons (U.S., dry)	2
	Gallons (U.S., liq.)	2.3272944
	Liters	8.809521
	Pints (U.S., dry)	16
	Quarts (U.S., dry)	8
Pennyweights	Drams (apoth. *or* troy)	0.4
	Drams (avdp.)	0.87771429
	Grains	24
	Grams	1.55517384
	Ounces (apoth. *or* troy)	0.05
	Ounces (avdp.)	0.054857143
	Pounds (apoth. *or* troy)	0.0041666
	Pounds (avdp.)	0.0034285714
Perches (masonry)	Cu. feet	24.75
Phots	Foot-candles	929.0304
	Lumens/sq. cm.	1
	Lumens/sq. meter	10,000
	Lux	10,000
Picas (printer's)	Centimeters	0.42175170
	Inches	0.166044
Pints (Brit.)	Cu. cm.	568.26092
	Gallons (Brit.)	0.125
	Gills (Brit.)	4
	Gills (U.S.)	4.803797
	Liters	0.5682450
	Minims (Brit.)	9600
	Ounces (Brit., fluid)	20
	Pints (U.S., dry)	1.032056
	Pints (U.S., liq.)	1.200949
	Quarts (Brit.)	0.5
	Scruples (Brit., fluid)	480

To convert from	To	Multiply by
Pints (U.S., dry)	Bushels (U.S.)	0.015625
	Cu. cm.	550.61047
	Cu. inches	33.6003125
	Gallons (U.S., dry)	0.125
	Gallons (U.S., liq.)	0.14545590
	Liters	0.5505951
	Pecks (U.S.)	0.0625
	Quarts (U.S., dry)	0.5
Pints (U.S., liq.)	Cu. cm.	473.17647
	Cu. feet	0.016710069
	Cu. inches	28.875
	Cu. yards	0.00061889146
	Drams (U.S., fluid)	128
	Gallons (U.S., liq.)	0.125
	Gills (U.S.)	4
	Liters	0.4731632
	Milliliters	473.1632
	Minims (U.S.)	7680
	Ounces (U.S., fluid)	16
	Pints (Brit.)	0.8326747
	Quarts (U.S., liq.)	0.5
Planck's constant	Erg-seconds	6.6255×10^{-27}
	Joule-seconds	6.6255×10^{-34}
	Joule-sec./Avog. No. (chem.)	3.9905×10^{-10}
Points (printer's)	Centimeters	0.03514598
	Inches	0.013837
	Picas	0.0833333
Poises	Cgs. units of absolute viscosity	1
	Grams/(cm. × sec.)	1
Poise-cu. cm./gram	Sq. cm./sec	1
Poise-cu. ft./lb.	Sq. cm./sec.	62.427960
Poise-cu. in./gram	Sq. cm./sec	16.387064
Poles/sq. cm.	E.M. cgs. units of magnetization	1
Pottles (Brit.)	Gallons (Brit.)	0.5
	Liters	2.272980

To convert from	To	Multiply by
Poundals	Grams-force	14.0981
	Newtons	0.1382550
	Pounds-force	0.0310810
Poundals/sq. ft.	Pascals	1.488164
Pounds (apoth. *or* troy)	Drams (apoth. *or* troy)	96
	Drams (avdp.)	210.65143
	Grains	5760
	Grams	373.24172
	Kilograms	0.37324172
	Ounces (apoth. *or* troy)	12
	Ounces (avdp.)	13.165714
	Pennyweights	240
	Pounds (avdp.)	0.8228571
	Scruples (apoth.)	288
	Tons (long)	0.00036734694
	Tons (metric)	0.00037324172
	Tons (short)	0.00041142857
Pounds (avdp.)	Drams (apoth. *or* troy)	116.6666
	Drams (avdp.)	256
	Grains	7000
	Grams	453.59237
	Hundredweights (long)	0.00892857
	Hundredweights (short)	0.01
	Kilograms	0.45359237
	Ounces (apoth. *or* troy)	14.583333
	Ounces (avdp.)	16
	Pound-force	1
	Pennyweights	291.6666
	Poundals	32.1740
	Pounds (apoth. *or* troy)	1.215277
	Scruples (apoth.)	350
	Slugs	0.0310810
	Tons (long)	0.00044642857
	Tons (metric)	0.00045359237
	Tons (short)	0.0005
Pounds-force	Kilograms-force	0.453592
	Newtons	4.44822

To convert from	To	Multiply by
	Poundals	32.1740
	Pounds (avdp.)	1
Pounds-force/sq. ft.	Pascals	47.8803
Pounds-force/sq. inch	Pascals	6894.76
	MegaPascals	0.00689476
Pounds of H_2O evap. from and at 212°F.	B.t.u.	970.9
	B.t.u. (IST.)	970.2
	B.t.u. (mean)	969.4
	Joules	1.0237×10^6
	Joules (Int.)	1.0234×10^6
Pounds/cu. ft.	Grams/cu. cm.	0.016018463
	Kg./cu. meter	16.018463
Pounds/cu. inch	Grams/cu. cm.	27.679905
	Grams/liter	27.68068
	Kg./cu. meter	27679.905
Pounds/gal. (Brit.)	Pounds/cu. ft.	6.228839
Pounds/gal. (U.S., liq.)	Grams/cu. cm.	0.11982643
	Pounds/cu. ft.	7.4805195
Pounds/inch	Grams/cm.	178.57967
	Grams/ft.	5443.1084
	Grams/inch	453.59237
	Ounces/cm.	6.2992
	Ounces/inch	16
	Pounds/meter	39.370079
Pounds/minute	Kilograms/hr.	27.2155422
	Kilograms/min.	0.45359237
Pounds of H_2O (39.2°F.)/minute	Cu. ft./min.	0.01601891
	Gal. (U.S.)/min.	0.1198298
	Liters/min.	0.45359237
Pounds/sq. ft.	Atmospheres	0.000472541
	Bars	0.000478803
	Cm. of Hg (0°C.)	0.0359131
	Dynes/sq. cm.	478.803
	Ft. of air (1 atm., 60°F.)	13.096
	Grams/sq. cm.	0.48824276

To convert from	To	Multiply by
	In. of Hg (32°F.)	0.0141390
	In. of H_2O (39.2°F.)	0.192227
	Kg./sq. meter	4.8824276
	Mm. of Hg (0°C.)	0.359131
Pounds/sq. inch	Atmospheres	0.0680460
	Bars	0.0689476
	Cm. of Hg (0°C.)	5.17149
	Cm. of H_2O (4°C.)	70.3089
	Dynes/sq. cm.	68,947.6
	Grams/sq. cm.	70.306958
	In. of Hg (32°F.)	2.03602
	In. of H_2O (39.2°F.)	27.6807
	Kg./sq. cm.	0.070306958
	Mm. of Hg (0°C.)	51.7149
Pounds-force-sec/sq. ft.	Poises	478.803
Pounds-force-sec/sq. in.	Poises	68,947.6
Puncheons (Brit.)	Cu. meters	0.31797510
	Gallons (Brit.)	69.94467
	Gallons (U.S.)	84
Quadrants	Minutes	5400
	Radians	1.5707963
Quarterns (Brit., dry)	Buckets (Brit.)	0.125
	Bushels (Brit.)	0.0625
	Cu. cm.	2273.044
	Gallons (Brit.)	0.5
	Liters	2.272980
	Pecks (Brit.)	0.25
Quartens (Brit., liq.)	Cu. cm.	142.0652
	Gallons (Brit.)	0.03125
	Liters	0.1420613
Quarters (U.S., long)	Kilograms	254.0117272
	Pounds (avdp.)	560
Quarters (U.S., short)	Kilograms	226.796185
	Pounds	500

To convert from	To	Multiply by
Quarts (Brit.)	Cu. cm.	1136.522
	Cu. inches	69.35482
	Gallons (Brit.)	0.25
	Gallons (U.S., liq.)	0.3002373
	Liters	1.136490
	Quarts (U.S., dry)	1.032056
	Quarts (U.S., liq.)	1.200949
Quarts (U.S., dry)	Bushels (U.S.)	0.03125
	Cu. cm.	1101.2209
	Cu. feet	0.038889251
	Cu. inches	67.200625
	Gallons (U.S., dry)	0.25
	Gallons (U.S., liq.)	0.29091180
	Liters	1.1011901
	Pecks (U.S.)	0.125
	Pints (U.S., dry)	2
Quarts (U.S., liq.)	Cu. cm.	946.35295
	Cu. feet	0.033420136
	Cu. inches	57.75
	Drams (U.S., fluid)	256
	Gallons (U.S., dry)	0.21484175
	Gallons (U.S., liq.)	0.25
	Gills (U.S.)	8
	Liters	0.9463264
	Ounces (U.S., fluid)	32
	Pints (U.S., liq.)	2
	Quarts (Brit.)	0.8326747
	Quarts (U.S., dry)	0.8593670
Quintals (metric)	Grams	100,000
	Hundredweights (long)	1.9684131
	Kilograms	100
	Pounds (avdp.)	220.46226
Radians	Circumferences	0.15915494
	Degrees	57.295779
	Minutes	3437.7468
	Quadrants	0.63661977
	Revolutions	0.15915494
	Seconds	206,264.81

To convert from	To	Multiply by
Radians/cm.	Degrees/cm.	57.295779
	Degrees/ft.	1746.3754
	Degrees/inch	145.53128
	Minutes/cm.	3437.7468
Radians/sec.	Degrees/sec.	57.295779
	Revolutions/min.	9.5492966
	Revolutions/sec.	0.15915494
Radians/(sec. × sec.)	Revolutions/(min. × min.)	572.95779
	Revolutions/(min. × sec.)	9.5492966
	Revolutions/(sec. × sec.)	0.15915494
Register tons	Cu. feet	100
	Cu. meters	2.8316847
Revolutions	Degrees	360
	Grades	400
	Quadrants	4
	Radians	6.2831853
Reyns	Centipoises	6.89476×10^6
Rhes	Poises^{-1}	1
Rods	Centimeters	502.92
	Chains (Gunter's)	0.25
	Chains (Ramden's)	0.165
	Feet	16.5
	Feet (U.S. Survey)	16.499967
	Furlongs	0.025
	Inches	198
	Links (Gunter's)	25
	Links (Ramden's)	16.5
	Meters	5.0292
	Miles (statute)	0.003125
	Perches	1
	Yards	5.5
Rods (Brit., volume)	Cu. feet	1000
	Cu. meters	28.316847
Roentgen	Coulombs/kilogram	2.58×10^{-4}
Roods (Brit.)	Acres	0.25
	Ares	10.117141
	Sq. perches	40
	Sq. yards	1210

To convert from	To	Multiply by
Ropes (Brit.)	Feet	20
	Meters	6.096
	Yards	6.6666666
Scruples (apoth.)	Drams (apoth. *or* troy)	0.333333
	Drams (avdp.)	0.73142857
	Grains	20
	Grams	1.2959782
	Ounces (apoth. *or* troy)	0.041666
	Ounces (avdp.)	0.045714286
	Pennyweights	0.833333
	Pounds (apoth. *or* troy)	0.003472222
	Pounds (avdp.)	0.0028571429
Scruples (Brit., fluid)	Minims (Brit.)	20
Seams (Brit.)	Bushels (Brit.)	8
	Cu. feet	10.27479
	Liters	290.9414
Seconds (angular)	Degrees	0.000277777
	Minutes	0.0166666
	Radians	4.8481368×10^{-6}
Seconds (mean solar)	Days (mean solar)	1.1574074×10^{-5}
	Days (sidereal)	1.1605763×10^{-5}
	Hours (mean solar)	0.0002777777
	Hours (sidereal)	0.00027853831
	Minutes (mean solar)	0.0166666
	Minutes (sidereal)	0.016712298
	Seconds (sidereal)	1.00273791
Seconds (sidereal)	Days (mean solar)	1.1542472×10^{-5}
	Days (sidereal)	1.1574074×10^{-5}
	Hours (mean solar)	0.00027701932
	Hours (sidereal)	0.000277777
	Minutes (mean solar)	0.016621159
	Minutes (sidereal)	0.0166666
	Seconds (mean solar)	0.99726957
Shakes	Seconds	1×10^{-8}
Siemen's units	(Same as mhos)	

To convert from	To	Multiply by
Skeins	Feet	360
	Meters	109.728
Slugs	Geepounds	1
	Kilograms	14.5939
	Pounds (avdp.)	32.1740
Slugs/cu. ft.	Grams/cu. cm.	0.515379
Space (entire)	Hemispheres	2
	Steradians	12.566371
Spans	Centimeters	22.86
	Fathoms	0.125
	Feet	0.75
	Inches	9
	Quarters (Brit. linear)	1
Spherical right angles	Hemispheres	0.25
	Spheres	0.125
	Steradians	1.5707963
Sq. centimeters	Ares	1×10^{-6}
	Circ. mm.	127.32395
	Circ. mils	197,352.52
	Sq. chains (Gunter's)	2.4710538×10^{-7}
	Sq. chains (Ramden's)	1.0763910×10^{-7}
	Sq. decimeters	0.01
	Sq. feet	0.0010763910
	Sq. ft. (U.S. Survey)	0.0010763867
	Sq. inches	0.15500031
	Sq. meters	0.0001
	Sq. mm.	100
	Sq. mils	155,000.31
	Sq. rods	3.9536861×10^{-6}
	Sq. yards	0.00011959900
Sq. chains (Gunter's)	Acres	0.1
	Sq. feet	4356
	Sq. ft. (U.S. Survey)	4355.9826
	Sq. inches	627,264
	Sq. links (Gunter's)	10,000
	Sq. meters	404.68564
	Sq. miles	0.00015625

To convert from	To	Multiply by
	Sq. rods	16
	Sq. yards	484
Sq. chains (Ramden's)	Acres	0.22956841
	Sq. feet	10,000
	Sq. ft. (U.S. Survey)	9999.9600
	Sq. inches	1.44×10^6
	Sq. links (Ramden's)	10,000
	Sq. meters	929.0304
	Sq. miles	0.00035870064
	Sq. rods	36.730946
	Sq. yards	1111.111
Sq. decimeters	Sq. cm.	100
	Sq. inches	15.500031
Square degrees	Steradians	0.00030461742
Sq. dekameters	Acres	0.024710538
	Ares	1
	Sq. meters	100
	Sq. yards	119.59900
Sq. feet	Acres	2.295684×10^{-5}
	Ares	0.0009290304
	Sq. cm.	929.0304
	Sq. chains (Gunter's)	0.00022956841
	Sq. ft. (U.S. Survey)	0.99999600
	Sq. inches	144
	Sq. links (Gunter's)	2.2956841
	Sq. meters	0.09290304
	Sq. miles	3.5870064×10^{-8}
	Sq. rods	0.0036730946
	Sq. yards	0.111111
Sq. feet (U.S. Survey)	Acres	$2.29569330 \times 10^{-5}$
	Sq. centimeters	929.03412
	Sq. chains (Ramden's)	0.00010000040
Sq. hectometers	Sq. meters	10,000
Sq. inches	Circ. mils	1,273,239.5
	Sq. cm.	6.4516
	Sq. chains (Gunter's)	1.5942251×10^{-6}
	Sq. decimeters	0.064516
	Sq. feet	0.0069444

To convert from	To	Multiply by
	Sq. ft. (U.S. Survey)	0.0069444167
	Sq. links (Gunter's)	0.01594225
	Sq. meters	0.00064516
	Sq. miles	$2.4909767 \times 10^{-10}$
	Sq. mm.	645.16
	Sq. mils	1×10^{6}
Sq. inches/sec.	Sq. cm./hr.	23,225.76
	Sq. cm./sec.	6.4516
	Sq. ft./min.	0.416666
Sq. kilometers	Acres	247.10538
	Sq. feet	1.0763910×10^{7}
	Sq. ft. (U.S. Survey)	1.0763867×10^{7}
	Sq. inches	1.5500031×10^{9}
	Sq. meters	1×10^{6}
	Sq. miles	0.38610216
	Sq. yards	1.1959900×10^{6}
Sq. links (Gunter's)	Acres	1×10^{-5}
	Sq. cm.	404.68564
	Sq. chains (Gunter's)	0.0001
	Sq. feet	0.4356
	Sq. ft. (U.S. Survey)	0.43559826
	Sq. inches	62.7264
Sq. links (Ramden's)	Acres	2.2956841×10^{-5}
	Sq. feet	1
Sq. meters	Acres	0.00024710538
	Ares	0.01
	Hectares	0.0001
	Sq. cm.	10,000
	Sq. feet	10.763910
	Sq. inches	1550.0031
	Sq. kilometers	1×10^{-6}
	Sq. links (Gunter's)	24.710538
	Sq. links (Ramden's)	10.763910
	Sq. miles	3.8610216×10^{-7}
	Sq. mm.	1×10^{6}
	Sq. rods	0.039536861
	Sq. yards	1.1959900
Sq. miles	Acres	640
	Hectares	258.99881
	Sq. chains (Gunter's)	6400

To convert from	To	Multiply by
	Sq. feet	2.7878288×10^{7}
	Sq. ft. (U.S. Survey)	2.78288×10^{7}
	Sq. kilometers	2.5899881
	Sq. meters	2,589,988.1
	Sq. rods	102,400
	Sq. yards	3.0976×10^{6}
Sq. millimeters	Circ. mm.	1.2732395
	Circ. mils	1973.5252
	Sq. cm.	0.01
	Sq. inches	0.0015500031
	Sq. meters	1×10^{-6}
Sq. mils	Circ. mils	1.2732395
	Sq. cm.	6.4516×10^{-6}
	Sq. inches	1×10^{-6}
	Sq. mm.	0.00064516
Sq. rods	Acres	0.00625
	Ares	0.2529285264
	Hectares	0.002529285264
	Sq. cm.	252,928.5264
	Sq. feet	272.25
	Sq. ft. (U.S. Survey)	272.24891
	Sq. inches	39,204
	Sq. links (Gunter's)	625
	Sq. links (Ramden's)	272.25
	Sq. meters	25.29285264
	Sq. miles	9.765625×10^{-6}
	Sq. yards	30.25
Sq. yards	Acres	0.00020661157
	Ares	0.0083612736
	Hectares	8.3612736×10^{-5}
	Sq. cm.	8361.2736
	Sq. chains (Gunter's)	0.0020661157
	Sq. chains (Ramden's)	0.0009
	Sq. feet	9
	Sq. ft. (U.S. Survey)	8.9999640
	Sq. inches	1296
	Sq. links (Gunter's)	20.661157
	Sq. links (Ramden's)	9
	Sq. meters	0.83612736
	Sq. miles	$3.228305785 \times 10^{-7}$

To convert from	To	Multiply by
	Sq. perches (Brit.)	0.033057851
	Sq. rods	0.033057851
Statamperes	Abamperes	3.335635×10^{-11}
	Amperes	3.335635×10^{-10}
	E.M. cgs. units of current	3.335635×10^{-11}
	E.S. cgs. units	1
Statcoulombs	Ampere-hours	9.265653×10^{-14}
	Coulombs	3.335635×10^{-10}
	Electronic charges	2.082093×10^{9}
	E. M. cgs. units of electric charge	3.335635×10^{-11}
Statfarads	E.M. cgs. units of capacitance	1.112646×10^{-21}
	E.S. cgs. units	1
	Farads	1.112646×10^{-12}
	Microfarads	1.112646×10^{-6}
Stathenries	Abhenries	8.987584×10^{20}
	E.M. cgs. units of inductance	8.987584×10^{20}
	E.S. cgs. units	1
	Henries	8.987584×10^{11}
	Millihenries	8.987584×10^{14}
Statohms	Abohms	8.987584×10^{20}
	E.S. cgs. units	1
	Ohms	8.987584×10^{11}
Statvolts	Abvolts	2.997930×10^{10}
	Volts	299.7930
Statvolts/cm.	Volts/cm.	299.7930
	Volts/inch	761.4742
Statvolts/inch	Volts/cm.	118.0287
Steradians	Hemispheres	0.15915494
	Solid angles	0.079577472
	Spheres	0.079577472
	Spher. right angles	0.63661977
	Square degrees	3282.8063
Steres	Cubic meters	1
	Decisteres	10
	Dekasteres	0.1
	Liters	999.972

To convert from	To	Multiply by
Stilbs	Candles/sq. cm.	1
	Candles/sq. inch	6.4516
	Lamberts	3.1415927
Stokes	Cgs. units of kinematic viscosity	1
	Sq. cm./sec.	1
	Sq. inches/sec.	0.15500031
	Poise cu. cm./gram	1
Stones (Brit., legal)	Centals (Brit.)	0.14
Tons (long)	Dynes	9.96402×10^8
	Hundredweights (long)	20
	Hundredweights (short)	22.4
	Kilograms	1016.0469
	Ounces (avdp.)	35,840
	Pounds (apoth. *or* troy)	2722.22
	Pounds (avdp.)	2240
	Tons (metric)	1.1060469
	Tons (short)	1.12
Tons (metric)	Dynes	9.80665×10^8
	Grams	1×10^6
	Hundredweights (short)	22.046226
	Kilograms	1000
	Ounces (avdp.)	35,273.962
	Pounds (apoth. *or* troy)	2679.2289
	Pounds (avdp.)	2204.6226
	Tons (long)	0.98420653
	Tons (short)	1.1023113
Tons (short)	Dynes	8.89644×10^8
	Hundredweights (short)	20
	Kilograms	907.18474
	Ounces (avdp.)	32,000
	Pounds (apoth. *or* troy)	2430.555
	Pounds (avdp.)	2000
	Tons (long)	0.89285714
	Tons (metric)	0.90718474
Tons of refrig. (U.S., comm.)	B.t.u. (IST.)/hr.	12,000
	B.t.u. (IST.)/min.	200
	Kilocal. (IST.)/hr.	3023.949

To convert from	To	Multiply by
	Horsepower	4.71611
	Kg. of ice melted/hr.	37.971
	Lb. of ice melted/hr.	83.711
Tons of refrig. (U.S., std.)	B.t.u. (IST.)	288,000
	B.t.u. (mean)	287,774
	Kilocal. (IST.)	72,574.8
	Kilocal. (mean)	72,517.9
	Lb. of ice melted	2009.1
Tons (long)/sq. ft.	Atmospheres	1.05849
	Dynes/sq. cm.	1.07252×10^6
	Grams/sq. cm.	1093.6638
	Pounds/sq. ft.	2240
Tons (short)/sq. ft.	Atmospheres	0.945082
	Dynes/sq. cm.	957.605
	Grams/sq. cm.	976.486
	Pounds/sq. inch	13.8888
Tons (long)/sq. in.	Atmospheres	152.423
	Dynes/sq. cm.	1.54443×10^8
	Grams/sq. cm.	157,487.59
Tons (short)/sq. in.	Dynes/sq. cm.	1.37895×10^8
	Kg./sq. mm.	1406.139
	Pounds/sq. inch	2000
Torrs (*or* Tors)	Millimeters of Hg (0°C.)	1
	Pascals	133.3224
Townships (U.S.)	Acres	23,040
	Sections	36
	Sq. miles	36
Tuns	Gallons (U.S.)	252
	Hogsheads	4
Volts	Abvolts	1×10^8
	Mks. (r *or* nr) units	1
	Statvolts	0.003335635
	Volts (Int.)	0.999670
Volts (Int.)	Volts	1.000330
Volts/°C.	Joules/(coulomb × °C.)	1
Volt-coulombs	Joules(Int.)	0.999835
Volt-coulombs (Int.)	Joules	1.000165

To convert from	To	Multiply by
Volt-electronic charge-seconds	Planck's constant	2.41814×10^{14}
Volt-faraday (chem.)-seconds	Planck's constant	1.45650×10^{38}
Volt-faraday (phys.)-seconds	Planck's constant	1.45690×10^{38}
Volt-seconds	Maxwells	1×10^{8}
Watts	B.t.u./hr.	3.41443
	B.t.u. (mean)/hr.	3.40952
	B.t.u. (mean)/min.	0.0568253
	B.t.u./sec.	0.000948451
	B.t.u. (mean)/sec.	0.000947088
	Cal./hr.	860.421
	Cal. (mean)/hr.	859.184
	Cal. (20°C.)/hr.	860.853
	Cal./min.	14.3403
	Cal. (IST.)/min.	14.3310
	Cal. (mean)/min.	14.3197
	Cal., *kh.*/min.	0.0143403
	Kilocal. (IST.)/min.	0.0143310
	Kilocal. (mean)/min.	0.0143197
	Ergs/sec.	1×10^{7}
	Foot-pounds/min.	44.2537
	Horsepower	0.00134102
	Horsepower (boiler)	0.000101942
	Horsepower (elec.)	0.00134048
	Horsepower (metric)	0.00135962
	Joules/sec.	1
	Kilowatts	0.001
	Liter-atm./hr.	35.5282
Watts (Int.)	B.t.u./hr.	3.41499
	B.t.u. (mean)/hr.	3.41008
	B.t.u./min.	0.569165
	B.t.u. (mean)/min.	0.0568347
	Cal./hr.	860.563
	Cal. (mean)/hr.	859.326
	Kilocal./min.	0.0143427
	Kilocal. (IST.)/min.	0.0143333

To convert from	To	Multiply by
	Kilocal. (mean)/min.	0.0143221
	Ergs/sec.	1.000165×10^7
	Joules (Int.)/sec.	1
	Watts	1.000165
Watts/sq. cm.	B.t.u./(hr. × sq. ft.)	3172.10
	Cal./(hr. × sq. cm.)	860.421
	Ft.-lb./(min. × sq. ft.)	41,113.1
Watts/sq. in.	B.t.u./(hr. × sq. ft.)	491.677
	Cal./(hr. × sq. cm.)	133.365
	Ft.-lb./(min. × sq. ft.)	6372.54
Watt-hours	B.t.u.	3.41443
	B.t.u. (mean)	3.40952
	Cal.	860.421
Watt-hours	Kilocal. (mean)	0.859184
	Cal. (mean)	859.184
	Foot-pounds	2655.22
	Hp.-hours	0.00134102
	Joules	3600
	Joules (Int.)	3599.41
	Kg.-meters	367.098
	Kw.-hours	0.001
	Watt-hours (Int.)	0.999835
Watt-sec.	Foot-pounds	0.737562
	Gram-cm.	10,197.16
	Joules	1
	Liter-atm.	0.00986895
	Volt-coulombs	1
Wavelength of orange-red line of krypton 86	Ångström units	6057.80211
	Millimeters	0.000605780211
Wavelength of red line of cadmium	Ångström units	6438.4696
	Millimeters	0.00064384696
Webers	Cgs. units of induction flux	1×10^8
	E.M. cgs. units of induction flux	1×10^8
	Lines	1×10^8
	Maxwells	1×10^8
	Mks. units of induction flux	1

To convert from	To	Multiply by
	Mks. units of magnetic charge	0.079577472
	Mks. units of magnetic charge	1
	Volt-seconds	1
Webers/sq. cm.	Gausses	1×10^8
	Lines	1×10^8
	Lines/sq. inch	6.4516×10^8
Webers/sq. in.	Gausses	1.5500031×10^7
Weeks (mean calendar)	Days (mean solar)	7
	Days (sidereal)	7.0191654
	Hours (mean solar)	168
	Hours (sidereal)	168.45997
	Minutes (mean solar)	10,080
	Minutes (sidereal)	10,107.598
	Months (lunar)	0.23704235
	Months (mean calendar)	0.23013699
	Years (calendar)	0.019178082
	Years (sidereal)	0.019164622
	Years (tropical)	0.019165365
Weys (Brit., mass.)	Pounds (avdp.)	252
X units	Meters	1.00202×10^{-13}
Yards	Centimeters	91.44
	Chains (Gunter's)	0.4545454
	Chains (Ramden's)	0.03
	Cubits	2
	Fathoms	0.5
	Feet	3
	Feet (U.S. Survey)	2.9999940
	Furlongs	0.00454545
	Inches	36
	Meters	0.9144
	Poles (Brit.)	0.181818
	Quarters (Brit., linear)	4
	Rods	0.181818
	Spans	4

To convert from	To	Multiply by
Years (calendar)	Days (mean solar)	365
	Hours (mean solar)	8760
	Minutes (mean solar)	525,600
	Months (lunar)	12.360065
	Months (mean calendar)	12
	Seconds (mean solar)	3.1536×10^7
	Weeks (mean calendar)	52.142857
	Years (sidereal)	0.99929814
	Years (tropical)	0.99933690
Years (leap)	Days (mean solar)	366
Years (sidereal)	Days (mean solar)	365.25636
	Days (sidereal)	366.25640
	Years (calendar)	1.0007024
	Years (tropical)	1.0000388
Years (tropical)	Days (mean solar)	365.24219
	Days (sidereal)	366.24219
	Hours (mean solar)	8765.8126
	Hours (sidereal)	8789.8126
	Months (mean calendar)	12.007963
	Seconds (mean solar)	3.1556926×10^7
	Seconds (sidereal)	3.1643326×10^7
	Weeks (mean calendar)	52.177456
	Years (calendar)	1.0006635
	Years (sidereal)	0.99996121

Bibliography

Abdullah, A.J. 1955. "Some Aspects of the Dynamics of Tornadoes." *Monthly Weather Review* 83/4:83–94.

American Architectural Manufacturers Association. 1985. *Design Windloads for Buildings and Boundary Layer Wind Tunnel Testing*, AAMA Aluminum Curtain Wall Series No. 11.

American Standards Association. Sectional Committee on Building Code Requirements for Minimum Design Loads in Buildings. 1945. (A58.1-1945).

American Standards Association. Sectional Committee on Building Code Requirements for Minimum Design Loads in Buildings. 1955. "American Standard Building Code Requirements for Minimum Design Loads in Buildings and Other Structures" (A58.1-1955).

American Standards Association. Sectional Committee on Building Code Requirements for Minimum Design Loads in Buildings. 1972. "American National Standards Building Code Requirements for Minimum Design Loads in Buildings and Other Structures" (A58.1-1972).

American Society of Civil Engineers. 1940. "Wind Bracing in Steel Buildings" (Final Report, ASCE Subcommittee No. 31, Committee on Steel), *Trans. ASCE*, 105:1713.

American Society of Civil Engineers. 1961. "Wind Forces on Structures." Final Report of the Task Committee on Wind Forces of the Committee on Loads and Stresses of the Structural Division. *Trans. ASCE* 125/11:1124–98.

American Society of Civil Engineers. 1987. *Wind Loading and Wind-Induced Structural Response*.

American Society of Civil Engineers. 1990. *Minimum Design Loads for Buildings and Other Structures*. Revision and Redesignation of ANSI A58.1-1982. Approved December 1988, Published July 1990 (ANSI/ASCE 7-88).

American Society of Testing Materials. 1992. *Annual Book of ASTM Standards*. Philadelphia: ASTM.

Associate Committee on the National Building Code, National Research Council of Canada. 1990. *Supplement to the National Building Code of Canada.* Ottawa: National Research Council of Canada.

Asp, M.O. 1950. "Tornadoes in Oklahoma, 1875–1949." *Monthly Weather Review* 23–26. February.

Avila, Lixion A., and Pasch Richard J. 1992. "Atlantic Hurricanes: Bob Hogs the Headlines." *Weatherwise* February/March.

Aynsley, R.M., Melbourne, W., and Vickery, B.J. 1977. *Architectural Aerodynamics.* London: Applied Science Publishers Ltd.

Baier, J. 1896. "Low Pressure in the St. Louis Tornado." *Monthly Weather Review* 332. September.

Banks, D.C. 1963. "A Study of Bearing Capacity in Sands Under Dynamic Loadings." MS thesis, Georgia Institute of Technology, Atlanta.

Battan, L.J. 1959. "Duration of tornadoes." *Bull. Amer. Meteor. Soc.* 40:340–42.

Betchel Power Corporation. 1974. *Tornado and Extreme Wind Design Criteria for Nuclear Power Plants.* Topical Report BC-TOP-3A, rev. 3, San Francisco, California, August.

Bigler, S.G. 1975. *Tornado Damage Surveys.* A&M Project 156, Scientific Report No. 1. Department of Oceanography and Meteorology. College Station: Texas A&M University.

Booker, C.A. 1954. "On Transmission Towers Destroyed by the Worcester, Mass. Tornado of June 9, 1953." *Bulletin of the American Meteorological Society* 225–29, May.

Boscovich. 1749. *Sopra il turbine che la notte tra gli XI, e XII Giugno de MDCCXLIX danneggio una gran parte di Roma* [*About the Tornado that on the Night of June 11–12, 1749, Damaged a Large Part of Rome*]. Rome.

British Standards Institution. 1972. *Basic Data for the Design of Buildings* (Chapter 5: "Loading," Part 2: "Wind Loads").

Brooks, C.F. 1949. *Why the Weather?* New York: Harcourt, Brace & Co.

Brooks, E.M. 1951. Tornadoes and Related Phenomena. In *Compendium of Meteorology.* Boston: American Meteorological Society.

California Institute of Public Affairs. 1981. *The Nuclear Power Issue: A Guide to Who Is Doing What in the US and Abroad.* Claremont: California Institute of Public Affairs.

Carr, J.A. 1952. 1952. Preliminary Reports on the Tornadoes of March 21–22, 1952. *Monthly Weather Review* 50–58.

Casagrande, A., and Shannon, W.L. 1949. "Strength of Soils Under Dynamic Loads." *Trans. Amer. Soc. Civ. Eng.* 114:755–72.

Case, Bob. 1990. "Hurricanes: Strong Storms Out of Africa." *Weatherwise* February.

Chang, C.C. 1969. "First Real Man-Made Tornado Is Generated in Laboratory 'Cage' by Space Scientists at the Catholic University of America." News item released by the Office of Public Information of the Catholic University of America.

Cheung, J.C.K., and Melbourne, W.H. 1984. *Cladding Pressures on Rectangular High Rise Buildings.* First Workshop on Wind Engineering and Industrial Aerodynamics, CSIRO, 3–5 July 1984, Highett, Victoria.

Clarke, R.H. 1962. "Severe Local Wind Storms in Australia." *Division of Meteorological Physics.* Technical Paper No. 13. CSIRO. Melbourne, Australia.

Cook, N.J. 1985. *The Designer's Guide to Wind Loading of Building Structure.* Part 1: *Background, Damage Survey, Wind Data, and Structural Classification.* Part 2: *Static Structures.* London: Butterworth's.

Cook, R.A. 1993. *Hurricane Andrew: Recommendations for Building Codes and Building Code Enforcement.* College of Engineering, University of Florida, January 31.

Crowther, Hugh G. 1993. "Tornadoes Hit New Heights." *Weatherwise* February/March.

Cunningham Jr., T.P. March 1993. *Roof Sheathing Fastening Schedules For Wind Uplift.* American Plywood Association Report T92-28. Tacoma, WA: The American Plywood Association.

Dade County Grand Jury. 1992. *In the Circuit Court of the Eleventh Judicial Circuit of Florida in and for the County of Dade.* Filed December 14.

Dames and Moore. 1972. "Tornado Wind Damage Probability and Recurrence for West Valley, New York." U.S. Nuclear Regulatory Commission, docket no. 50-201.

Davenport, A.G. 1961. "The Application of Statistical Concepts to the Wind Loading of Structures." *Proc. Inst. Civil Engr.* 19:449–71.

Davenport, A.G. 1967. *Gust Loading Factors. J. Struc. Div., ASCE* 93/ST3:11–34.

Davenport, A.G. 1971. "On the Statistical Prediction of Structural Performance in the Wind Environment." Meeting Preprint 1420, ASCE National Structural Engineering Meeting, 19–23 April 1971, Baltimore.

Davenport, A.G. 1977. "The Prediction of Risk Under Wind Loading." Pp. 511–38 in *Proceedings of the Second International Conference on Structural Safety and Reliability.* September 1977, Munich.

Davenport, A.G., and Hill-Carroll, P.J. 1986. "Damping in Tall Buildings: Its Variability and Treatment in Design Proceedings." In *Proceedings of ASCE Seminar on Motion in Wind.* Seattle.

Davenport, A.G. Surry, D., and Stathopoulos, T. 1977. *Wind Loads on Low-Rise Buildings.* Ontario: University of Western Ontario.

Dergarabedian, P., and Fendell, F.E. 1971. "A Method for Rapid Estimation of Maximum Wind Speeds in Tornadoes and Hurricanes." *Monthly Weather Review* 99:143–45.

Durst, C.S. 1960. "Wind Speeds Over Short Periods of Time." *Met. Mag.* 89:181.

Evesson, D.T. 1969. "Tornado Occurrences in New South Wales." *Australian Meteorological Magazine* 17/3:143–65.

Evesson, D.T. 1970. *Tornado Occurrences in New South Wales.* Bureau of Meteorology Working Paper No. 130. Sydney, Australia. August 18.

Farmer, P. 1986. *Wind Energy Nineteen Seventy-Five to Nineteen Eighty-Five: A Bibliography.* New York: Springer-Verlag.

FEMA [Federal Emergency Management Agency]. 1992a. *Interagency Hazard Mitigation Team Report.* Report No. 955-DR-FL. Atlanta, GA: FEMA.

FEMA [Federal Emergency Management Agency]. 1992b. *Preliminary Report in Response to Hurricane Andrew, Dade County, Florida.* September 30.

Ferguson, Edward, and Ostby, Frederick. 1991. "Tornadoes of 1990." *Weatherwise* April.

Ferguson, Edward, Schaefer, T.J., Weiss, S.J., and Wilson, L.F. 1983. "The Year of the Tornado." *Weatherwise* February.

Ferrell, W. 1889. *A Popular Treatise on the Winds*. New York: John Wiley & Sons.

Finley, J.P. 1884. *Report on the Character of Six Hundred Tornadoes*. Professional Papers of the Signal Service No. VII. Available at the NOAA Library, Silver Spring, Maryland.

Flora, S.D. 1953. *Tornadoes of the United States*. Norman: University of Oklahoma Press.

Frankenfield, H.C. 1896. "The Tornado of May 27 at St. Louis, Missouri." *Monthly Weather Review* 77–81. March.

Fujita, T.T. 1960. *A Detailed Analysis of the Forgo Tornadoes of June 20, 1957*. U.S. Weather Bureau Research Paper No. 42. Washington, DC: U.S. Weather Bureau.

Fujita, T.T. 1971. *Proposed Characterization of Tornadoes and Hurricanes by Area and Intensity*. SMRP Research Paper No. 91. Chicago: University of Chicago.

Fujita, T.T. 1976. *Assessment of Knowledge and Implications for man*. In *Proceedings of the Symposium on Tornadoes*, 22–24, June Texas Tech University).

Fujita, T.T., Bradbury, D.L., and Black, P.G. 1967. *Estimation of Tornado Wind Speed from Characteristic Ground Marks*. Satellite and Mesometeorology Research Project Paper 69, Department of Geophysical Sciences, University of Chicago. Paper presented at the Fifth International Conference on Severe Local Storms, St. Louis, Missouri.

Fujita, T.T., and Pearson, A.D. 1973. "Results of FPP classification of 1971 and 1972 tornadoes." Pp. 142–45 in *Preprints of the Eighth International Conference on Severe Local Storms*. October 15–17, Denver, Colorado. Boston: American Meteorological Society.

Funk and Wagnall. 1993. *The World Almanac and Book of Facts*. Mahwah, NJ: Funk and Wagnall.

Gates, D.M. 1972. *Man and His Environment: Climate*. New York: Harper & Row.

Gilbert, E.G., and Walker, J.M. 1966. *Tornado at the Royal Horticultural Society's Garden*. London: Wiley.

Glaser, A.H. 1959. "An Observational Deduction of the Structure of a Tornado Vortex: Cumulus Dynamics." In *Proceedings of the First Conference on Cumulus Convection*, May 19–22, Portsmouth, New Hampshire. New York: Pergamon Press.

Greiner, Leo, A., Halmstad, John T., and Leftwich Jr., Preston, W. 1990. *Severe Local Storm Warning Verification: 1989*. National Severe Storms Forecast Center, Kansas City, Missouri 64106-2897, U.S. Department of Commerce, National Oceanic and Atmospheric Administration, National Weather Service, May.

Gunther, Emil B. 1985. "Cyclones in the Eastern Pacific: A 'Normal Season.'" *Weatherwise* February.

Hardy, R.M. 1957. "Engineering Problems Involving Preconsolidated Clay Shells." *Trans. Eng. Inst. Can.* 1:5–14.

Harkness, G.S. 1894. "The Tornado at Little Rock, Arkansas, October 2, 1894." Editorial. *Monthly Weather Review* 413–14. October.

Hawkins, H.F., and Imbembo, S.M. 1976. "Structure of a Small Intense Hurricane." *Monthly Weather Review* 104/4:418–42.

Hazen, H.A. 1890. *The Tornado*. New York: N.D.C. Hodges.

Hebert, Jarrell, and Mayfield. 1993. "The Deadliest, Costliest, and Most Intense U.S. Hurricanes of This Century." NOAA Technical Memo NHC-31, February.

Hoecker Jr., W.H. 1960. "Wind Speed and Airflow Patterns in the Dallas Tornado of April 2, 1957." *Monthly Weather Review* 167–80. May.

Hoecker Jr., W.H. 1961. Three-Dimensional Pressure Pattern of the Dallas Tornado and Some Resultant Implications. *Monthly Weather Review* 533–42.

Hoecker Jr., W.H. 1967. Personal communication with the author.

Holmes, J.D. 1988. "Distribution of Peak Wind Loads on a Low-Rise Building." In *Proceedings of the Seventh International Conference on Wind Engineering, Part 2*. 6–10 July 1988, Aachen, Germany.

Holmes, J.D., Jackson, P.S., and Melbourne, W.H., eds. 1984. *Wind Engineering 1983: Proceedings of the 6th International Conference on Wind Engineering, Gold Coast, Australia, March 21–25, and Auckland, New Zealand, April 6–7, 1983*, 3 vols. New York: Elsevier Science Publishing Company.

Holmes, J.D., Melbourne, W.H., and Walker, G.R. 1990. *A Commentary on the Australian Standard for Wind Loads*. Melbourne: Australian Wind Engineering Society.

Houghton, E.L., and Carruthers, N.B. 1976. *Wind Forces on Buildings and Structures: An Introduction*. New York: John Wiley and Sons.

Howe, G.M. 1975. "Tornado Path Sizes." *J. Appl. Meteorol.* 13:343–47.

Hudak, Mary. 1992. *Interagency Hazard Mitigation Team Report*. August.

Humphreys, W.J. 1940. *Physics of the Air*. New York: McGraw-Hill.

International Conference of Building Officials. 1991. *Uniform Building Code*. Whittier, CA: ICBO.

Janeselli, R. 1970. "Il tornado che colpi la laguna di Venezia I–11 Settembre 1970." *Annali Di Geofisica*. Revista dell'Instituto Nazionale Di Geofisica, Roma.

Jarvinen, B.J., and Lawrence, M.B. 1985. "An Evaluation of the SLOSH Storm Surge Model." *Bulletin of the American Meteorological Society* 66:1408–1411.

Jeary, A.P., and Ellis, B.R. 1983. "On Predicting the Response of Tall Buildings to Wind Excitation." *Journal of Wind Engineering and Industrial Aerodynamics* 13:173–82.

Jorgensen, David P. 1982. *The Structure of Typhoon Irma (1974) As Revealed by 700-mb Aircraft Data*. NOAA Technical Report ERL:418, AOML:32.

Keith, Edward L., and Rose, John D. 1992. *Hurricane Andrew: Structural Performance of Buildings in Southern Florida (August 24, 1992)*. APA Report T92-21. Tacoma, WA: The American Plywood Association.

Keller, D., and Vonnegut, B. 1976. "Wind Speed Estimation Based on the Penetration of Straws and Splinters Into Wood." *Weatherwise* 29/5.

Kert, T.H., and Dean, T.J. 1963. *Report on a Tornado at Malta*.

Kessler, Edwin. 1967. Personal communication with Peter A. Morris, Director of the Division of Reactor Licensing of the U.S. Atomic Energy Commission, February 21.

Kessler, Edwin. 1970. "Tornadoes." *Bull. Amer. Meteor. Soc.* 926–36.

Kessler, Edwin. 1977. *National Severe Storms Laboratory, History and FY 1976 Program.* Norman, OK: National Severe Storms Laboratory.

Kessler, Edwin, and Lee, T.J. 1976. *Normalized Indices of Destruction and Deaths by Tornadoes.* NOAA Technical Memo ERL NSSL-77. Norman, OK: National Severe Storms Laboratory.

Kramer, C., and Gerhardt, H.J., eds. 1989. *Advances in Wind Engineering: Proceedings of the 7th International Congress on Wind Engineering, Aachen, FRG, July 6–10, 1987*, 3 vols. New York: Elsevier Science Publishing Company.

Krayer, William R., and Marshall, Richard D. 1992. "Gust Factors Applied to Hurricane Winds." *Bull. Amer. Meteor. Soc.* 73/5, May.

Lacy, R.E. 1968. Tornadoes in Britain During 1963–66.

Lamb, H.H. 1957. *Tornadoes in England, May 21, 1950.* Geophysical Memoirs No. 99. London: Meteorological Office, Air Ministry.

Lamb, H.H. 1965. *Tornadoes Destruction of Trees at Shalford Park, Guildford.*

Lamy, François. 1689. *Conjectures physiques sur les collonnes de nue* [*Theories on Tornadoes*]. Paris.

Lawrence, Miles. 1985. "The Atlantic Hurricane Season." *Weatherwise* February.

Lawrence, Miles. 1987. "The Atlantic Hurricane Season." *Weatherwise* February.

Lawrence, Miles. 1990. "Eastern Pacific Tropical Cyclones." *Weatherwise* February.

Lawrence, Miles, and Rappaport, Edward N. 1993. "Eastern Pacific Hurricanes." *Weatherwise* February/March.

Lawson, T.V. 1980. *Wind Effects on Buildings*, Vol. 1: *Design Applications*. London: Applied Science Publishers.

Lawson, T.V. *Wind Effects on Buildings*, Vol. 2: *Statistics and Meteorology*. London: Applied Science Publishers.

Leftwich, Preston W., Wilson, Larry F., and Crowther, Hugh G. 1992. "Tornadoes: Record Pace Continues." *Weatherwise* February/March.

Lewis, D. Philip. 1993. *Final Report, Governor's Disaster Planning and Response Review Committee.* January 15. Tallahassee: The State of Florida.

Logie, J. 1919. "Note on Tornadoes." *Quart. J. R. Meteor. Soc.* 45:317.

Long, R.R. 1960. *Tornadoes and Dust Whirls*. Johns Hopkins University, Department of Mechanical Engineering Technical Report No. 10 (ONR Series) or No. 13 (CWB Series).

MacDonald, Angus J. 1975. *Wind Loading on Buildings*. New York: John Wiley and Sons.

Martin, J.R. 1940. *Tornadoes in the United States, 1916–1937, Inclusive*. Mimeographed summary in NOAA Library, Washington, DC.

Mayfield, Max. 1992. "Eastern Pacific Hurricanes: Long Season, but Normal Numbers." *Weatherwise* February/March.

Mayfield, Max, and Avila, Lixion A. 1993. "Atlantic Hurricanes." *Weatherwise* February/March.

McDonald, J.R., Mehta, K.C., and Minor, J.C. 1975. *Development of Wind-Speed Risk Models for the Savannah River Plant Site*. Report for E.I. du Pont de Nemours & Company under purchase order no. AXC 672-W.

Meaden, G.T. 1976. "Tornadoes in Britain: Their Intensities and Distribution in Space and Time." *J. Meteorol.* 1:242–51.

Melaragno, Michele G. 1968a. *Outdoor Tornado Shelters for Residential Areas*. Manhattan: Kansas State Printing Service, Kansas State University.

Melaragno, Michele G. 1968b. *Tornado Forces and Their Effects on Buildings*. Manhattan: Kansas State Printing Service, Kansas State University.

Melaragno, Michele G. 1970. *Tornado Forces and Their Effects on Buildings*, 2nd ed. Washington, DC: Office of Civil Defense, January.

Melaragno, Michele G. 1972. "Dwelling Structures for the Great Plains, 1st ed.," *Building Science*, 7:87–92.

Melaragno, Michele G. 1973a. "Better Buildings in Tornadoes and Hurricane Zones." *The Florida Architect* 23/2:19–21, March–April.

Melaragno, Michele G. 1973b. "Some Aspects of Tornado Protection in the Design of Structures." *Ingegneria Civile*, No. 43.

Melaragno, Michele G. 1973c. "New Buildings Can Survive Tornadoes." *Texas Architect* 23/3, May–June.

Melaragno, Michele G. 1976. "Structural Design and Australian Tornadoes." *Architectural Science Review* 19/4:78–82, December.

Melaragno, Michele G. 1982. *Wind in Architectural and Environmental Design*. New York: Van Nostrand Reinhold.

Melaragno, Michele G. 1983. "Design for Tornadoes." Engineering Paper No. 765. *Journal of the Institute of Structural Engineers (London)*, September.

Miami Herald. 1992. *What Went Wrong*, 20 December.

Montanari, Geminiano. 1694. *Le forze di Eolo, Dialogo fisico-matematico sopra gli effetti del vortice, o sia Turbine … che il gierno 29 iuglio 1686 ha scorso e flagellato molte ville, e luoghi del territorio di Mantova, Padova, Verona, etc.* [*The forces of Aeolus: Physico-Mathematical Dialogue About the Effects of the Vortex, i.e., Tornado, that on July 29, 1686, Struck and Plagued Many Villas and Locations in the Territories of Mantua, Padua, Verona, etc.*]. Venice.

National Oceanic and Atmospheric Administration. 1966–80. *Climatological Data: National Summary*. Washington, DC: NOAA. (monthly publication)

National Oceanic and Atmospheric Administration. 1966–80. *Storm Data*. Washington, DC: NOAA. (monthly publication)

National Oceanic and Atmospheric Administration. 1993. *Hurricane Research Division, Fiscal Year 1992 Programs, Fiscal Year 1993 Projections*. Atlantic Oceanographic and Meteorological Laboratory. Miami, Florida. January.

National Oceanic and Atmospheric Administration, National Severe Storms Laboratory. 1990. *Annual Report, FY 1990*. Norman, OK: National Severe Storms Laboratory.

National Oceanic and Atmospheric Administration, National Severe Storm Laboratories. 1992. *Overview of National Severe Storm Laboratory: Science and Organization, FY 1991–1992*. Norman, OK: National Severe Storms Laboratory.

Outram, T.S. 1904. Storm of August 20, 1904, in Minnesota. *Monthly Weather Review* 365–66. August.

Pearson, A.D. 1971. "Statistics on Tornadoes that Caused Fatalities, 1960–1970." Pp. 194–197 in *Preprints of the Seventh International Conference on Severe Local Storms*. 5–7 October, Kansas City, Missouri. Boston: American Meteorological Society.

Peltier. 1840. *Meteorologie. Observations et recherches experimentales sur les causes qui concourent à la formation des trombes.* [*Observations and Experimental Research on the Causes Which Contribute to the Formation of Tornadoes*]. Paris.

Precast-Prestressed Concrete Institute (PPCI). 1985. *PCI Design Handbook.* Winter Park, FL: PPCI.

Quigley, R.M., and Vogan, R.W. 1970. "Black Shale Heaving at Ottawa, Canada." *Canadian Geotechnical Journal* 7/2:106–12.

Reinhold, T.A., Vickery, P.J., and Powell, M.D. 1992. "Wind Speeds in Hurricane Andrew: Myths and Reality." Hot Topics Session, ACI Annual Convention, San Juan, Puerto Rico. October 26.

Reynolds, G.W. 1958. "Venting and Other Building Practices as Practical Means of Reducing Damage from Tornado Low Pressure." *Bull. Amer. Meteorol. Soc.* 14–20. January.

Rose, John D. 1993. *Hurricane Iniki: Structural Performance of Buildings on Kauai, Hawaii (September 11, 1992).* APA Report T93-1. Tacoma, WA: The American Plywood Association.

Sachs, Peter. 1978. *Wind Forces in Engineering*. Elmsford, NY: Pergamon Press.

Schafer, J.T., Kelly, D.L., and Abbey Jr., R.F. 1979. "Tornado Track Characteristics and Hazard Probabilities." In *Preprints of the Fifth International Conference on Wind Engineering*. 8–14 July, Fort Collins, Colorado. Boston: American Meteorological Society.

Seed, H.B., and Lundgren, R. 1954. "Investigation of the Effect of Transient Loading on the Strength and Deformation Characteristics of Saturated Sands." *Proc. Amer. Soc. Testing Materials* 54:1288–1306.

Seelye, C.J. 1945. "Tornadoes in New Zealand." *New Zealand Journal of Science and Technology* 27:166–74.

Segman, R. 1971. Wet Tornadoes. NOAA, pp. 45–47.

Segner Jr., E.P. 1960. *Estimates of Minimum Wind Forces Causing Structural Damage*. U.S. Weather Bureau Research Paper No. 41. Washington, DC: U.S. Weather Bureau.

Shea, D.J., and Gray, W.M. 1973. "The Hurricane's Inner Core Region, I: Symmetric and Asymmetric Structures." *Journal of the Atmospheric Sciences* 30/8:1544–64.

Sheets, Robert C. 1990. *The National Hurricane Center: Past, Present, and Future*. Boston: American Meteorological Society.

Sheets, Robert C. 1992. Personal communication. October 23.

Simiu, Emil, and Scanlan, Robert H. 1986. *Wind Effects on Structures: An Introduction To Wind Engineering*. New York: John Wiley and Sons.

Skaggs, R.H. 1970. "On tornado probabilities." *Proc. Assn. Amer. Geog.* 2:123–26.

Standards Association of Australia. "Interim 350." Homebush, New South Wales.

Standards Association of Australia. "Australian Standard CA34 Part II — 1971. Homebush, New South Wales.

Standards Association of Australia. "Australian Standard 1170: Part 2, the 1973 Edition." Homebush, New South Wales.

Standards Association of Australia. 1989. "Minimum Design Load on Structures: SAA Loading Code." Homebush, New South Wales.

Standards Association of Australia. 1990. "Australian Standard 1170: Minimum Design Loads on Structures, SAA Loading Code Part 2, Wind Loads" (AS 1170 Part 2). Homebush, New South Wales: Standards Association of Australia.

Standerwick, D.C. 1993. Personal communication with the author.

Swanson, A.E., Stippich, R.E., and Bates, F.C. 1967. Paper Presented at the Fifth International Conference on Severe Local Storms, St. Louis, Missouri.

Sykes, W. 1967. Letter to the Editor. *Weatherwise* 271. December.

Teesdale, L.V. 1928. "Tornado-Resistant Construction of Buildings Possible by Venting." *Southern Lumberman* 40–42. July 7.

Tepper, Morris. 1958. "Tornadoes." *Scientific American* May.

The State of Florida. 1993. *The Final Report of the Governor's Disaster Planning and Response Review Committee.* Tallahassee, January 15.

Thom, H.C.S. 1963. "Tornado probabilities." *Monthly Weather Review* 730–36. October–December.

Tissell, John R. 1993. *Structural Panel Shear Walls.* APA Research Report 154. Tacoma, WA: The American Plywood Association.

Tissell, John R., and Elliott, James R. 1993. *Plywood Diaphragms.* APA Report 138. Tacoma, WA: The American Plywood Association.

Tissell, John R., and Rose, John D. 1988. *Plywood End Shear Walls in Mobile Homes.* APA Report 151. Tacoma, WA: The American Plywood Association.

U.S. Atomic Energy Commission. 1974. *Design Basis: Tornadoes for Nuclear Power Plants.* Regulatory Guide 1.76. Washington, DC: U.S. Government Printing Office.

U.S. Department of Commerce Code Committee. 1924. "Minimum Live Loads Allowable for Use in Design of Buildings." Washington, DC: The National Bureau of Standards.

U.S. Department of Health, Education, and Welfare. 1979. *Morbidity and Mortality Report* 21/17.

U.S. Department of Housing and Urban Development. 1966. "HUD Recommends Ways to Beat Storm Damage." News item released October 7.

U.S. Weather Bureau. 1965. *Tornado Facts.* Washington, DC: U.S. Weather Bureau.

Van Tassel, E.L. 1955. "The North Platte Valley Tornado Outbreak of June 27, 1955." *Monthly Weather Review* 255–64. November.

Vonnegut, B. 1960. "Electrical Theory of Tornadoes." *J. Geophys. Res.* 65:203–12.

Vonnegut, B., and Weyer, J.R. 1966. "Luminous Phenomena in Nocturnal Tornadoes." *Science* 1213–20.

Walkimoto, R., and Black, P.G. 1993. *Damage Survey of Hurricane Andrew and Its Relationship to the Radar-Detected Eyewall.* May 12.

Wegener, A. 1917. *Wind und Wasserhosen im Europa.* Braunschweig: F. Vieweg & Sohn.

Whitman, R.V. 1957. "The Behavior of Soils Under Transient Loading." Pp. 207–12 in *Proceedings of the Fourth International Conference on Soil Mechanics and Foundation Engineering (London)*. Vol. 1.

Whitman, R.V., and Healy, K.A. 1962. "Shear Strength of Sands During Rapid Loadings." *Proc. Amer. Soc. Civ. Eng.* 88(SM2):99–133.

Wilson, J.W., and Morgan Jr., G.M. 1971. "Long-Track Tornadoes and Their Significance." Pp. 183–86 in *Preprints of the Seventh Conference on Severe Local Storms*, October 5–7, Kansas City, Missouri. Boston: American Meteorological Society.

Wolford, L.V. 1960. *Tornado Occurrences in the United States*. Technical Paper No. 20, revised. Washington, DC: U.S. Weather Bureau.

Woodard, J.M. 1964. "An Investigation of the Dynamic Bearing Capacity of Footings on Sand." MS thesis, Georgia Institute of Technology, Atlanta.

Zollo, Ronald F. 1993. *Hurricane Andrew, August 24, 1992: Structural Performance of Buildings in Dade Country, Florida.* Technical Report No. CEN 93-1. Coral Gables, FL: University of Miami.

INDEX

D

E

F

M

N

O

P

R

S

T

U